微软 Azure 实战参考

李 竞 陈勇华 编著

北京航空航天大学出版社

内 容 简 介

本书介绍了微软 Azure 云平台的 IAAS 和 PAAS 中的具体服务,主要包括:云存储(Azure Storage)、App Service、云数据库(Azure SQL Database)、分布式缓存、分布式消息队列服务(Service Bus)、分布式基础架构服务(Service Fabric)、大数据处理平台 HDInsight、数据分析服务 Power BI、人工智能/机器学习(Machine Learning)、多媒体服务等。读者可以从中学到 Azure 云平台的基本概念、基本操作方法,学会如何利用云平台的资源,为进一步开发复杂的应用打下基础。

本书适用于云技术行业使用微软 Azure 云平台开发应用程序的技术人员。

图书在版编目(CIP)数据

微软 Azure 实战参考 / 李竞,陈勇华编著. -- 北京：北京航空航天大学出版社,2019.5

ISBN 978-7-5124-2990-1

Ⅰ. ①微… Ⅱ. ①李… ②陈… Ⅲ. ①机器学习 Ⅳ. ①TP181

中国版本图书馆 CIP 数据核字(2019)第 069432 号

版权所有,侵权必究。

微软 Azure 实战参考

李 竞　陈勇华　编著

责任编辑　刘晓明

*

北京航空航天大学出版社出版发行

北京市海淀区学院路 37 号(邮编 100191)　http://www.buaapress.com.cn
发行部电话:(010)82317024　传真:(010)82328026
读者信箱: emsbook@buaacm.com.cn　邮购电话:(010)82316936
三河市华骏印务包装有限公司印装　各地书店经销

*

开本:710×1 000　1/16　印张:38.75　字数:826 千字
2019 年 6 月第 1 版　2019 年 6 月第 1 次印刷　印数:3 000 册
ISBN 978-7-5124-2990-1　定价:99.00 元

若本书有倒页、脱页、缺页等装质量问题,请与本社发行部联系调换。联系电话:(010)82317024

前　言

时至今日，云计算已经在全球深入普及，各云计算的厂商在云计算的市场上已经竞争了好几年。目前市场上的云计算商家包括谷歌、微软、亚马逊、阿里、腾讯等。这些公有云平台主要提供了 IAAS、PAAS 服务，还有另外一些厂商提供了 SAAS 服务，如 Salesforce。各个云计算商家在激烈的竞争中不断地改进他们的服务，对其服务也推陈出新，将一些旧的服务慢慢地废弃，然后推出一些更强大、更先进、更方便的服务以飨用户。

本书介绍微软为用户提供的最新的 Azure 云服务的主要方面，包括 IAAS、PAAS。IAAS（基础设施即服务），指的是厂商将硬件以可计价的方式提供给用户租用，如 CPU、内存、硬盘空间等。在 Azure 上计算机都是虚拟的，不管是普通 PC，还是高配置的服务器，都是采用虚拟机技术虚拟出来的硬件，我们看到的都是虚拟硬件。Azure 按规格型号来计价，某些型号的虚拟机器的价格贵一些；有的配置低，价格就便宜一些。其计费方式一般是按每月多少美金的形式计价。除了虚拟 PC、虚拟服务器外，IAAS 还包括虚拟的网络设施，如防火墙、公网 IP、域名、交换机、路由器等。用户可以通过付费的方式使用这些资源，而不必关心这些资源具体在哪里，也不必担心数据丢失的问题。因为微软的基础设施保证了这些硬件资源的多用户共享，同时又互不干扰。此外，备份了用户的数据，所以不会因为硬件资源的损坏而导致数据的丢失。PAAS（平台即服务）指的是在 IAAS 之上，厂商还提供了其他一些服务，如便宜的存储、消息总线、可伸缩的网站服务、安全登录授权、分布式的数据库等，这些服务就构成了 PAAS。这些平台即服务通常是有别于其他云计算厂商的，使用 Azure 的平台即服务就得使用 Azure 的 SDK 来进行编程，比如存取存储器里的资源，就得通过 Azure SDK 来进行。

全书基于微软的云计算平台介绍其 IAAS、PAAS 两大部分的内容，一一介绍 Azure 平台的服务。介绍的方式是先介绍基础概念，然后介绍一些深入的知识，再介绍一些实例代码演示如何使用该服务来完成我们的工作。

第 1 章概述，主要讲述一些 Azure 全平台共通的基础概念。

第 2 章 Azure 云存储，讲述云存储的特性、使用方法。

第 3 章 App Service，讲述具有可水平和垂直扩展的 App Service，以及如何开发和配置一个 App Service。

第 4 章 Azure SQL Database，讲述 Azure 平台上 SQL Database 与本地 SQL Database 的异同，以及一些常用操作等。

第 5 章 Azure Active Directory，讲述 Azure 平台的 ID 账号管理、安全、授权、登录机制。

第 6 章 Service Bus，讲述 Azure 平台的消息总线服务，以及如何存储消息和传递消息的机制。

第 7 章 Service Fabric，讲述利用 Azure 平台的一些基础服务构成一些定制化的服务。

第 8 章 Media Service，讲述利用媒体服务来构建一些多媒体应用。

第 9 章 HDInsight，讲述利用 HDInsight 来进行大数据处理。

第 10 章 Azure Automation，讲述利用 Automation 对 Azure 进行自动化管理。

第 11 章 Power BI，讲述利用 Power BI 来做商业智能的数据挖掘、数据分析。

第 12 章 部署 Linux 服务器到 Azure，讲述如何部署 Linux 服务器到 Azure 平台。

第 13 章 部署 MongoDB 到 Azure，讲述如何部署分布式 MongoDB 到 Azure 平台。

第 14 章 Azure 人工智能，讲述 Azure 为人工智能提供的解决方案，以及如何利用 Azure 人工智能解决我们的实际问题。

各章独立性比较强，如果读者对某一章内容比较熟悉，则可跳过该章，阅读其他章。每一章都用实战的方式来讲解，会启动相关的工具，如 Visual Studio 2017，来演示一步步的操作，读者可从中获得实际操作经验；同时，有详细的讲解，读者可以知其所以然，明白为什么这么做等。

本书采用 Visual Studio 2017 来举例操作。其他版本的 Visual Studio 与之类似。

本书适用于云技术行业使用微软 Azure 云平台开发应用程序的技术人员。

艾锐势科技深圳有限公司 WA 组的同事们为本书的写作提供了咨询和支持；微软公司开发经理杨海峰先生审阅了本书，提出了很好的意见，在此对他们表示衷心的感谢！

作　者

2019 年 3 月

目 录

第 1 章 概 述 1
- 1.1 订阅 Subscription 4
- 1.2 支付模式 5
- 1.3 资源类型 5
- 1.4 虚拟机 8
- 1.5 虚拟防火墙、虚拟路由器和虚拟负载均衡 8
- 1.6 公网 IP 9
- 1.7 认识 Azure 界面 9
- 1.8 在 Visual Studio 中操作 Azure 资源 19
- 1.9 在 SQL Server Management Studio 中操作 Azure 资源 21

第 2 章 Azure 云存储 23
- 2.1 Azure Table Storage 23
- 2.2 Azure Blob Storage 50
- 2.3 Azure Queue Storage 58
- 2.4 Azure File Storage 62
- 2.5 Azure Storage CDN 65

第 3 章 App Service 69
- 3.1 App Service 简介 69
- 3.2 用 ASP.NET Core 开发一个 App Service 69
- 3.3 调试 71
- 3.4 部署 74
- 3.5 在大流量下的垂直扩展和水平扩展 90
- 3.6 创建 Redis 缓存 92
- 3.7 用 Asp.net core 开发 Azure App Service 架构需考虑的一些要点 102

3.8 用 Java 开发一个 Azure App Service ·················· 104

3.9 用 Python 开发一个 Azure App Service ················ 112

第 4 章 Azure SQL Database ·················· 121

4.1 Azure SQL Database 对比 SQL Server ················ 121

4.2 创建一个 SQL Azure 服务器 ·················· 127

4.3 创建一个 SQL Azure 数据库 ·················· 131

4.4 用 SQL Server Management Studio 链接 Azure SQL Database ······· 133

4.5 在 Azure SQL Database 中执行的 TSQL ··············· 136

4.6 用 Visual Studio 开发应用链接 Azure SQL Database ··········· 145

4.7 把本地的 SQL Server 数据库迁移到 Azure SQL 云数据库上 ········ 156

4.8 Azure SQL Database Data Sync 数据同步功能 ············ 159

第 5 章 Azure Active Directory ·················· 170

5.1 简　介 ·················· 170

5.2 开始使用 Azure AD ·················· 172

5.3 Azure AD 流程 ·················· 183

5.4 JWT Token ·················· 190

5.5 使用 Azure AD 样例 ·················· 193

第 6 章 Service Bus ·················· 201

6.1 基　础 ·················· 201

6.2 队列（Queue）入门 ·················· 206

6.3 用 Visual Studio 开发 Service Bus Queue ·············· 220

6.4 Service Bus Queue 和 Storage Queue 的区别 ············· 234

6.5 用 Visual Studio 开发 Service Bus Topic ·············· 239

6.6 用 Visual Studio 开发 Service Bus Relay On ············· 243

第 7 章 Service Fabric ·················· 248

7.1 基　础 ·················· 248

　7.1.1 开发环境准备 ·················· 248

　7.1.2 基本概念和俯瞰图 ·················· 251

7.2 Stateless Service 服务的创建 ·················· 252

7.3 Stateful 服务的创建 ·················· 283

7.4 调试 Service Fabric 应用 ·················· 288

第 8 章 Media Service ·················· 290

8.1 基　础 ·················· 290

8.1.1	什么是 Media Service	290
8.1.2	用 Media Service 能做什么	290
8.1.3	Media Service 概览	290
8.2	视频点播 Video-On-Demand	297
8.3	视频直播 Live Streaming	314

第 9 章 HDInsight 326

9.1	简 介	326
9.2	用 HDInsight 创建 Hadoop	326
9.3	Apache Hive 简介	346
9.4	使用 Hadoop 的 Pig 和 MapReduce	360
9.5	用 HDInsight 创建 Spark	369

第 10 章 Azure Automation 379

10.1	简 介	379
10.2	创建 Azure Automation 账号	382
10.3	创建 Azure Automation Runbook	385
10.4	Azure Automation Runbook 的类型	388
10.5	创建图形化 Runbook	391
10.6	创建 PowerShell Runbook	400
10.7	创建 PowerShell 工作流 Runbook	407

第 11 章 Power BI 413

11.1	简 介	413
11.2	Power BI Desktop	413
11.2.1	安装 Power BI Desktop	413
11.2.2	Power BI Desktop 界面	415
11.2.3	Query Editor	417
11.2.4	创建 Report 和数据可视化	427
11.2.5	发布报表到 Power BI Service	434
11.2.6	Power BI M Language	435
11.3	Power BI Service	441
11.3.1	导入 Dataset	442
11.3.2	创建 Report	443
11.3.3	创建 Dashboard	445
11.3.4	创建 Workbook	447
11.3.5	Data Gateway	448
11.4	Azure 中的 Power BI Embedded	449
11.4.1	在 Azure Portal 中创建 Power BI Embedded 容量	450

11.4.2 将一个 Power BI 报告、仪表板或瓦片嵌入到客户的应用程序中 …… 452

第 12 章 部署 Linux 服务器到 Azure … 469

12.1 虚拟机简介 …… 469
12.2 用 Azure Portal 创建虚拟机 …… 471
12.3 用 Azure CLI 创建 Linux 虚拟机 …… 474
12.4 Automation Configuration …… 477
12.5 用 Azure CLI 2.0 创建 Azure VM 的自定义虚拟机镜像 …… 483
12.6 用 Azure CLI 2.0 创建和部署高可用虚拟机 …… 486
12.7 用 Azure CLI 2.0 创建虚拟机自动伸缩集并在 Linux 上部署高度可用的应用程序 …… 488
12.8 用 Azure CLI 2.0 在 Azure 中负载平衡虚拟机,创建一个高度可用的应用程序 …… 493

第 13 章 部署 MongoDB 到 Azure … 501

13.1 MongoDB 简介 …… 501
13.2 单个 MongoDB 部署 …… 503
13.3 用 gz 发行包的方式安装 MongoDB …… 505
13.4 MongoDB 的管理 …… 507
13.5 MongoDB 集群的部署 …… 525
13.6 Azure Cosmos DB 提供的 MongoDB API …… 527
　13.6.1 Azure Cosmos DB:移植一个已有的 Node.js MongoDB Web 应用 …… 528
　13.6.2 将数据迁移到 Azure Cosmos DB MangoDB API 账户 …… 537

第 14 章 Azure 人工智能 … 538

14.1 人工智能基础 …… 538
14.2 Azure 人工智能简介 …… 539
14.3 在 Azure 门户中创建认知服务账户 …… 540
14.4 第一个 Azure 人工智能程序——在计算机视觉中使用 REST API 和 Python 分析远程图像 …… 543
14.5 利用认知服务语音 SDK 在浏览器中识别 JavaScript 中的语音 …… 548
14.6 利用 LUIS 根据文字内容理解用户所需 …… 553
14.7 用 Python 调用文本分析认知服务 …… 558
14.8 用 Python 调用 Bing Web 搜索 API …… 566
14.9 用 Bing Web 搜索 API 创建单页应用程序 …… 567
14.10 利用 Azure 门户开始 Azure 机器学习 …… 578
14.11 利用 Azure 机器学习服务训练图像分类模型 …… 585
14.12 在 Azure 容器实例(ACI)中训练图像分类模型 …… 594
14.13 在 Azure 容器实例(ACI)中部署图像分类模型 …… 602

第 1 章 概 述

Azure平台提供的服务包括IAAS和PAAS。IAAS(基础设施即服务),指的是厂商将硬件以可计价的方式提供出来供用户租用,如CPU、内存、硬盘空间等。在Azure上计算机都是虚拟出来的,不管是普通PC,还是高配置的服务器,都是采用虚拟机技术虚拟出来的硬件,我们看到的都是虚拟硬件。Azure按规格型号来计价,某些型号的虚拟机器的价格贵一些;有的配置低,价格就便宜一些。其计费方式一般是按每月多少美金的形式计价。除了虚拟PC、虚拟服务器外,IAAS还包括虚拟的网络设施,如防火墙、公网IP、域名、交换机、路由器等都是虚拟的。用户可以通过付费的方式使用这些资源,而不必关心这些资源具体在哪里,也不必担心数据丢失的问题。因为微软的基础设施保证了这些硬件资源的多用户共享,同时又互不干扰;并且备份了用户的数据,所以不会因为硬件资源的损坏而导致数据丢失。PAAS(平台即服务)指的是在IAAS之上,厂商还提供了其他一些服务,如便宜的存储、消息总线、可伸缩的网站服务、安全登录授权、分布式的数据库等服务,这些服务就构成了PAAS。这些平台即服务通常是有别于其他云计算厂商的,使用Azure的平台即服务就得使用Azure的SDK来进行编程,比如存取存放在存储器里的资源,就得通过Azure SDK来进行。

微软在全球部署了4个Azure的环境。

(1) Azure Global

这个是面向全球用户的环境。这个环境在全球有60多个数据中心,用强大的网络将所有数据中心连接起来。用户使用这个环境的资源,可以选择就近的一个数据中心,部署用户的应用和数据。

(2) Azure Germany

这个是面对德国用户的环境。它基于微软的可信云(trusted cloud)理念,保护用户的安全和隐私,其合规、透明,在德国境内提供数据保存、传输等功能。Azure Germany是物理上与Azure全球隔离开的实例。所有Azure Germany的数据中心都是由可信任的德国公司来运行,并遵循德国的法律,按欧盟的要求设计。Azure Germany适用于那些在EU/EFTA和英国(UK)做生意的人或者组织。

(3) Azure China

这个是由21世纪互联与微软联合运营的面对中国用户的环境。这个环境和

Azure Global 是物理隔离的。21世纪互联在国内是一个比较大的互联网提供商，Azure China 由 21 世纪互联负责运营和销售。微软为 Azure China 带来了可靠、灵活、安全、一致，以及效费比高的云计算服务，同时还不断推进技术进步。微软是第一家全球性的云计算服务提供商在中国开设云计算服务。

(4) Azure USGovernment

这个环境是专门为美国政府所打造的。它的用户即美国政府。这也是一个与 Azure Global 物理隔离的环境。其中存储的是美国政府的数据和应用。

在安装了 Azure SDK 之后，就可以在 PowerShell 中运行以下命令来查看有多少个 Azure 环境。

```
Windows PowerShell
Copyright (C) Microsoft Corporation. All rights reserved.

PS C:\Users\Jing>Get-AzureEnvironment


Name                                    : AzureChinaCloud
EnableAdfsAuthentication                : False
ActiveDirectoryServiceEndpointResourceId : https://management.core.chinacloudapi.cn/
AdTenant                                : Common
GalleryUrl                              : https://gallery.azure.com/
ManagementPortalUrl                     : http://go.microsoft.com/fwlink/? LinkId=301902
ServiceManagementUrl                    : https://management.core.chinacloudapi.cn/
PublishSettingsFileUrl                  : http://go.microsoft.com/fwlink/? LinkID=301776
ResourceManagerUrl                      : https://management.chinacloudapi.cn/
SqlDatabaseDnsSuffix                    : .database.chinacloudapi.cn
StorageEndpointSuffix                   : core.chinacloudapi.cn
ActiveDirectoryAuthority                : https://login.chinacloudapi.cn/
GraphUrl                                : https://graph.chinacloudapi.cn/
TrafficManagerDnsSuffix                 : trafficmanager.cn
AzureKeyVaultDnsSuffix                  : vault.azure.cn
AzureKeyVaultServiceEndpointResourceId  : https://vault.azure.cn

Name                                    : AzureCloud
EnableAdfsAuthentication                : False
ActiveDirectoryServiceEndpointResourceId : https://management.core.windows.net/
AdTenant                                : Common
GalleryUrl                              : https://gallery.azure.com/
ManagementPortalUrl                     : http://go.microsoft.com/fwlink/? LinkId=254433
ServiceManagementUrl                    : https://management.core.windows.net/
PublishSettingsFileUrl                  : http://go.microsoft.com/fwlink/? LinkID=301775
ResourceManagerUrl                      : https://management.azure.com/
```

```
SqlDatabaseDnsSuffix                    : .database.windows.net
StorageEndpointSuffix                   : core.windows.net
ActiveDirectoryAuthority                : https://login.microsoftonline.com/
GraphUrl                                : https://graph.windows.net/
TrafficManagerDnsSuffix                 : trafficmanager.net
AzureKeyVaultDnsSuffix                  : vault.azure.net
AzureKeyVaultServiceEndpointResourceId  : https://vault.azure.net


Name                                    : AzureGermanCloud
EnableAdfsAuthentication                : False
ActiveDirectoryServiceEndpointResourceId: https://management.core.cloudapi.de/
AdTenant                                : Common
GalleryUrl                              : https://gallery.azure.com/
ManagementPortalUrl                     : http://portal.microsoftazure.de/
ServiceManagementUrl                    : https://management.core.cloudapi.de/
PublishSettingsFileUrl                  : https://manage.microsoftazure.de/publishsettings/index
ResourceManagerUrl                      : https://management.microsoftazure.de/
SqlDatabaseDnsSuffix                    : .database.cloudapi.de
StorageEndpointSuffix                   : core.cloudapi.de
ActiveDirectoryAuthority                : https://login.microsoftonline.de/
GraphUrl                                : https://graph.cloudapi.de/
TrafficManagerDnsSuffix                 : azuretrafficmanager.de
AzureKeyVaultDnsSuffix                  : vault.microsoftazure.de
AzureKeyVaultServiceEndpointResourceId  : https://vault.microsoftazure.de


Name                                    : AzureUSGovernment
EnableAdfsAuthentication                : False
ActiveDirectoryServiceEndpointResourceId: https://management.core.usgovcloudapi.net/
AdTenant                                : Common
GalleryUrl                              : https://gallery.azure.com/
ManagementPortalUrl                     : https://manage.windowsazure.us
ServiceManagementUrl                    : https://management.core.usgovcloudapi.net/
PublishSettingsFileUrl                  : https://manage.windowsazure.us/publishsettings/index
ResourceManagerUrl                      : https://management.usgovcloudapi.net/
SqlDatabaseDnsSuffix                    : .database.usgovcloudapi.net
StorageEndpointSuffix                   : core.usgovcloudapi.net
ActiveDirectoryAuthority                : https://login.microsoftonline.us/
GraphUrl                                : https://graph.windows.net/
TrafficManagerDnsSuffix                 : usgovtrafficmanager.net
AzureKeyVaultDnsSuffix                  : vault.usgovcloudapi.net
AzureKeyVaultServiceEndpointResourceId  : https://vault.usgovcloudapi.net
```

可以看到,目前 Azure 有 4 个环境,而且每个环境的参数差别都比较大。第一个是 Azure China 的环境。其 Portal URL 是 https://portal.azure.cn/,注意它是.cn,不是.com。它和 Azure Global 还是不同的。第二个就是 Azure Global 的环境,其 Portal URL 是 https://portal.azure.com/,这是全球用户都可以用的一个公共云环境。第三个是 Azure Germany 的环境。其 Portal URL 是 http://portal.microsoft-azure.de/,后缀名.de 说明一切。最后一个是 Azure USGovernment 的云环境,其 Portal URL 是 https://manage.windowsazure.us。这几个环境都是互不干扰的、物理隔离的环境,都是遵照所在国政府和法律运作的环境。

1.1 订阅 Subscription

要使用 Azure 的服务,就得有相应的订阅。访问 https://portal.azure.com/,就可以访问 Azure Portal 的门户。登录此 Azure Portal 需要一个有订阅 Subscription 的账号。要拥有一个订阅 Subscription,可以有如下几种方式:

① 向 Azure Portal 注册一个账号,绑定一个信用卡。Azure 中,有的是免费使用的资源,有的是付费使用的资源,需按月支付使用费。

② 拥有 MSDN 企业版、专业版订阅 Subscription,这样的用户可以使用 Azure 的服务,而且微软给每个 MSDN 企业版、专业版订阅 Subscription 用户每月 150 美元的信用额度,可以使用这 150 美元来调用 Azure 中的一些付费资源。当用完此 150 美元后,用户只能暂停使用付费资源。等到下一个月,用户又会拥有另一个 150 美元的额度,又可以使用 150 美元的资源。只要不超过 150 美元,用户都不会受限制。

③ 拥有 Office 365 的订阅。这样也可以登录 Azure Portal。

订阅界面如图 1-1 所示。

图 1-1 在 Azure Portal 显示订阅

此界面会显示当前账号拥有的全部 Subscription 权限。一个账号可以有多个 Subscription。资源的使用都是跟着某个 Subscription 的。每个 Subscription 的收费是分开算的。

1.2 支付模式

Azure 使用的是多种支付模式,如按月支付模式、按年支付模式、按使用量支付(Pay as you go)模式。例如:创建虚拟机,选择 Size,如图 1-2 所示。

图 1-2 创建虚拟机时选虚拟机的大小

可以看到它是按月收费模式。

虚拟机或者服务器,可以使用按年支付模式,比如保留三年,三年的价格略微便宜点。

1.3 资源类型

资源可以按如下分类。

1. IAAS 类的资源

(1) 服务器

微软为 Azure 提供了很多物理服务器。这些物理服务器都可以供 Azure 用户使用。Azure 用户不必关心自己的应用究竟部署在哪些物理服务器上。这些对 Azure 用户来说都是透明的。只有微软的工程师才能在系统中看到具体的物理服务器。

(2) 虚拟机

虚拟技术是 IAAS 中的重要一环。虚拟技术可以将物理服务器变成多个逻辑服务器。用户看到的是其在使用的逻辑服务器。对其背后的物理服务器没有任何感知。微软主要使用自家的虚拟技术——微软 Virtual PC 技术。这个 Virtual PC 技术和 Oracle VirtualBox、VMWare 等技术类似。

（3）存 储

其主要是磁盘，还有 NAS 之类的存储资源。这些资源被微软整合到平台中，以各种形式提供出来，如 App Service 中的存储、Azure Storage 中的存储、虚拟机的存储空间、消息队列的存储等。

（4）网 络

微软的 Azure 提供了一些基础架构，有物理的防火墙、路由器、交换机等网络设备，还有公网 IP 等资源。这些资源都可以在用户间共享。用户可以在 Azure Portal 中申请这些网络资源，然后对这些网络资源进行配置。这些配置不是像配置 cisco 路由器那样发命令，而是在 Azure Portal 界面里配置一些参数。这样，这些虚拟的网络设施就都可以为用户所用了。

2. PAAS 类的资源

（1）Service Bus

它提供消息队列、事务机制，以系列的 API 供用户使用消息订阅、接收；支持事务操作，可以建立多个消息队列，不同的客户端可以侦听不同的消息队列。不同的 Pricing Tier 支持的服务级别不同。消息异地分布式存储、集群，支持较大量的消息流量，支持的消息操作比较多。客户端可以是任何编程语言的客户端，只要支持 SDK 或者 RESTFul API 就可调用。

（2）Service Fabric

它提供了一种虚拟服务器扩展的机制，提供了便利的手段帮助用户建立强大的服务器集群，或者提供其他形式的服务器集群，如游戏服务器集群、计算集群等。

（3）App Service

它帮助用户实现 Web 应用、Web API 的部署、自动化可伸缩可扩展、安全管理、域名绑定、公网 IP，可以用多种编程语言来编写 App Service 的应用，包括 C♯、Java、Python、PHP 等。

（4）Azure SQL Database

它类似 MS SQL Server 的实现。其大部分语法和用处都和 MS SQL Server 相同。只有一些仅适合 Azure 云计算环境的地方有所不同。Azure SQL Database 有多种 Pricingtier，不同的服务级别不一样。一般 Premium 以上的级别可以保证异地灾备，异地热备份。Azure SQL Database 是一个微软管理的集群，可以保证数据的完整性，数据丢失的概率很小，可用性很高。

（5）Azure Redis Cache

Azure 版的 Redis Cache，有集群设置，一般 Premium 级别以上的可以有地理位置上的不同分布，出现服务中断的概率很小。Redis Cache 可存储的数据量是很大的。Redis 是典型的 key-value 存储，又加上有集群设置，这些 key-value 可以在主从集群上分布开。Redis 本身负责将 key-value 键值进行复制和分发，在响应客户端请求时，根据主从的策略查找用户所需的 key-value 对。

(6) Azure Active Directory

微软提供的类似 MS Windows Active Directory 云计算环境下的 Active Directory。Azure Active Directory 有更多的协议,如 OAuth 2.0、OpenID Connect、WS-Federation、SAML 2.0。MS Windows Active Directory 能支持 SAML 2.0、WS-Federation 协议。Azure Active Directory 能支持将 MS Windows Active Directory 的用户数据导入。Azure Active Directory 支持不同的编程语言编写访问 Azure AD 的程序。

(7) Application Insights

这是微软提供的日志,Exception 记录的服务;支持多种客户端,提供了开放的 API,可以用多种编程语言来编写客户端程序,访问 Application Insights 的 API 记录应用程序中发生的事情,如日志记录、抛出的例外(Exception)等,Application Insights 有一些 Pricing tier,不同的 Pricing tier 获得的空间、服务级别、灾备、数据备份等都不同。

(8) Azure Storage

Azure Storage 是微软提供的存储服务,主要包括 Table Storage、Blob Storage、File Share Storage、Queue Storage。Table Storage 提供的是类似关系型数据库的表存储,还可以用类似 MS SQL Server 的表查询语句来查询数据。Table Storage 比 Azure SQL Database 便宜,同时也支持非常大量的数据存储。Table Storage 有自定义的 Partition Key 和 Row Key 来作为主键。

Blob Storage 是二进制文件的存储,可以用来存储数据文件。Blob Storage 提供了 API 供客户端访问。Blob Storage 可以有目录树结构,供用户来存储大量的数据文件。不同的 Pricing Tier 服务级别提供了不同的数据备份、灾难恢复,保证数据的完整性。

File Share Storage 是文件共享服务。用户可以在此服务中建立文件共享,这个文件共享有点类似于 Windows 文件共享。用户可以看到别的用户共享的文件。可以用 SDK 来访问 Azure File Share Storage 的共享文件。

Queue Storage 是类似于 Service Bus 的消息队列机制的服务,但是 Queue Storage 不支持事务操作,而 Service Bus 支持事务操作。Queue Storage 可以建立若干个 Queue,客户端可以针对特定的 Queue 发消息、收消息。这些消息都存储在所申请的 Storage 中。

(9) Media Service

Azure Media Service 可以帮助用户流媒体化视频,建立大型的可伸缩的视频服务。可以将视频缓存在服务器上,利用网络协议广播到用户;也可以实时拍摄、转码、缓冲,发送给用户。用户可以是移动设备上的 App,也可以是 PC 上的浏览器,支持多种客户端同时访问,给开发者提供了 SDK 来开发应用程序;同时,可以处理版权的授权、获取等服务。用户可以购买一些视频,此购买过的视频就在用户账号里。

（10）Azure Machine Learning

Azure 机器学习发展比较快。微软基于一些框架建立了机器学习集群，可以动态地调整大小，根据训练机器学习集群的大数据，可以训练出一些模型，这些模型存储在机器学习的集群中。有些场景，如人工视力，微软已经提供了一些训练好的模型，对一些图片能进行识别，如从图片中识别人物，其表情是微笑的，还是痛苦的；是否为色情图片；是否含有不好的倾向，如暴力、血腥；还有语音识别，官方也提供了训练好的模型，对世界上不同的语言进行了一些识别，所以比较标准的语言发音一般都可以识别。这些训练好的模型，只有在用户使用一段时间以后，才可以了解到其功能，还有局限性边界，才可以根据其现有的特点来进行一些必要的改造。用户需要自己准备一些训练用数据，比如希望此机器学习能识别某些方言，就需要准备比较全面的此方言的语音数据，然后对机器学习集群进行训练。此训练数据越全面，越能得到满意的训练结果。

1.4 虚拟机

虚拟机是 IAAS 的一部分。微软提供了不同的操作系统选择，如 Windows 系列的 Windows Server 的不同版本、Windows 7 的不同版本、Linux 中的 Redhat OS，或者 Ubuntu、Centos 版本。用户可以用一些著名厂商已经建立好的镜像来建立虚拟机。

在安装了操作系统之后，用户可以再去安装别的软件，比如 mysql、hadoop 之类的软件，并且做一些相应的配置，以实现集群的功能。

1.5 虚拟防火墙、虚拟路由器和虚拟负载均衡

Azure 平台并没有直接对应这三项的服务，而是由 Application Gateway 和 Security Center 两项服务完成了防火墙、路由器、负载均衡的功能。这是之前提到网络云资源的一部分。在 Azure 里找到 Application Gateway，Application Gateway 扮演的是虚拟防火墙、虚拟路由器和虚拟负载均衡设备的角色。它是在 7 层网络协议的层面进行负载均衡，在这个 Gateway 之后，微软有物理的防火墙、物理的路由器、物理的负载均衡器在起作用。这些物理的设备都是在数据中心中起作用的，供 Azure 用户共享使用。如果用户需要建立混合的网络，即一部分是 Azure 的网络，一部分是用户公司的网络，而这两个网络本来是在互联网上分割开的，那么就需要建立起一个 VPN（即虚拟专用网络）。在用户的公司网络外围得有一个 VPN Gateway，在微软的 Azure 网络外围，有一个 VPN Gateway，通过在 Azure Portal 中设置 VPN，可以将 VPN 建立好。只有建立好的 VPN 内部才可以使用内部网络 IP 地址来访问 Azure 中的服务器。Azure 的 VPN Gateway 只接受用户公司的 VPN Gateway 的数据包。其他不被允许的 Gateway 的数据包就会被丢弃。同时 Azure 是多用户的,每个用户间的数据是分开的。Azure 不会允许不同用户的服务器之间进行网络访问,其网络是互相隔开的。

Security Center 是用户可以安排其 Azure 资源安全计划的地方,包括防火墙规则,哪些外部 IP 允许访问 App Service,哪些外部 IP 能访问数据库,还有虚拟机的端点的保护。这些端点都是虚拟机对外的接口,虚拟机在建立以后都有一个子域名,此子域名可以在互联网上访问到。因此,需要对虚拟机对外的端点进行保护,只允许特定的人或者 IP 来进行访问。在没有使用 Azure 云平台的时候,可以使用类似于这些功能的软件,如 Keepalived、HAproxy、Nginx 等,实现负载均衡;同时,还需要其他一些软件或者硬件来实现防火墙、路由器的功能。另外,还需要进行单独的防火墙配置、路由器的配置、负载均衡器的配置,如配置防火墙设备或者防火墙服务器。输入命令配置防火墙规则,需要配置路由服务器的路由规则,还要配置 Keepalived、HAProxy 等软件,以实现负载均衡,以及发生单点失败时的工作流量重定向。这些所有的配置,都是由用户手动进行管理,比较繁杂。但是在 Azure 平台上,这些都不需要手动去设置,而是在 Azure Portal 中进行设置,Azure 系统就会自动完成相关软件的配置(如 Application Gateway 可以帮助我们实现负载均衡,Security Center 中的安全策略可以帮助我们实现防火墙等功能),且根本不用了解具体的细节。

1.6 公网 IP

当用户的服务器需要被互联网用户使用时,就需要给此服务器分配公网 IP。因为需要负载均衡,通常需要将此公网 IP 分配给 Application Gateway,然后再给 Application Gateway 进行配置。根据 URL 来配置访问服务器的 Application Gateway,可以制定某一个特定的子 URL 总是访问某一些服务器。比如,访问请求 http://公司名.com/desk1,就总是访问某一个 App Service;而访问请求 http://公司名.com/desk2,就总是访问另外一个 App Service,这样的模式可以将请求进行分担,可以让那些 App Service 不至于压力过大。

如果应用的规模不大,则可以直接将公网 IP 分配给 App Service,这样就省去了 Application Gateway 的功能。App Service 本身就可以进行伸缩,即可以由多台服务器来进行自动伸缩,只是其自动伸缩也有一定的限制,最多 10 台服务器。

1.7 认识 Azure 界面

操作 Azure 的资源有三种方式:在 Azure Portal 中操作,在 Visual Studio 图形化界面里操作,或者是命令行模式在 Power Shell 中进行命令行形式的操作。这里会讲解如何在 Azure Portal 中进行资源的操作,以 Azure Global 为例。读者可能是用 Azure China 环境,但大体上是类似的,只需要相应地进行一些调整即可。

打开浏览器访问 Azure 的 Portal URL:https://portal.azure.com/,然后要求登录,如前文所说,必须得有一个订阅的账号,用此账号登录即可。如果没有账号,可以先注册一个账号。登录 Azure Portal 后,其界面如图 1-3 所示。

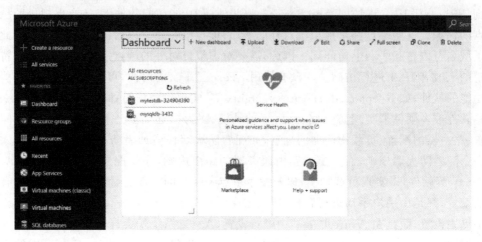

图 1-3　Azure Portal 的界面

大家可以看到 Azure Portal 的界面如下:

左边是主菜单,用来创建资源、管理资源。默认进来以后显示 Dashboard 页面。Dashboard 页面会将用户创建的资源进行列表,方便访问。这些资源不是全部的资源列表,只是最近常访问的资源列表。要查看全部的资源,应选择 All resources 菜单或者 Resource groups 菜单。

Resource groups 菜单的界面如图 1-4 所示。

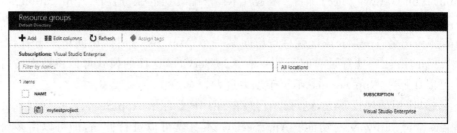

图 1-4　资源组 Resource groups 菜单的界面

所有用户创建的资源组都给列出来了。如前所述,Azure 资源有很多类型,创建这些资源后需要分组进行管理。用户可以按照用途分组,也可以按照类型分组,或者按照环境名分组,一般按照环境名分组的比较多。一套环境就一个组,所有这一套环境的资源都在这个资源组里管理,相对比较好管理,也便于查找、修改及删除。如果一些大的厂商,需要建立比较多的资源和环境,那么这个界面就会有比较多的列表项。

这个界面提供了查找功能,可以输入组名来查找某个资源组;也可以按资源的部署 Region 来查找。

界面上端有 Add 按钮，可用来新建资源组，如图 1－5 所示。

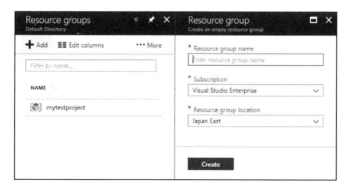

图 1－5　单击 Add 按钮来创建资源组（Resource groups）

注意到这三个输入都是必需的（都有一个红色的＊）。一般要求资源组最好跟资源在同一个 location，如 App Service 在 Japan East，那么其所在的资源组最好也在 Japan East，这样便于管理和操作。因为跨 location 建立资源组的话，有可能某一部分会出现问题。比如资源组在 West US，而组内资源在 Japan East，当没有出问题的时候还好。当 Japan East 出现问题，暂时不能访问的时候，就会出现这个资源似乎是可以访问的，但是试图打开此资源组中的资源的时候，这些资源却不能访问，因为 Japan East 暂时出现了问题。或者是另外一种情况，West US 出现了某些问题，导致资源组没法访问，但是如果单独打开组内的资源，却发现这些资源可以访问。最好是资源组跟其资源在同一个 location。

输入好上述信息就可以单击 Create 按钮来创建 Resource groups 了。

可以单击 Resource groups 的名字，进入此 Resource groups，看到其内所管理的资源，如图 1－6 所示，可以看到其内的资源可以是任何资源类型，如图 1－7 所示。

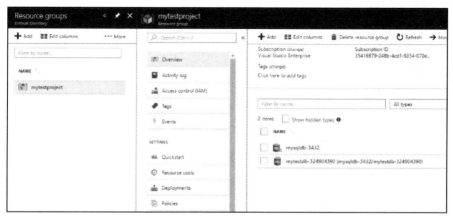

图 1－6　查看资源组中管理的资源

图1-7　显示资源组中不同类型的资源

可以单击 Add 按钮,将一些资源加入此 Resource groups。

可以新建一个资源。微软为用户准备了很多类型的资源,例如:

① Compute(计算)类的资源如图1-8所示。

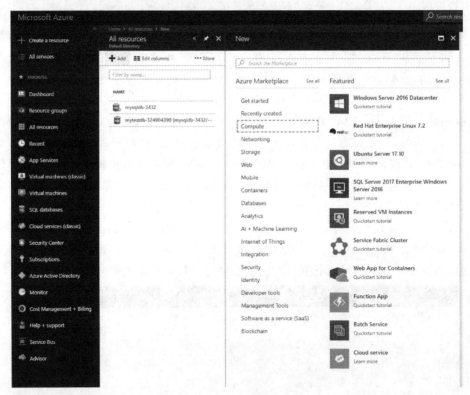

图1-8　创建计算类的资源

② Networking 类的资源如图1-9所示。

③ Storage 类的资源如图1-10所示。

④ Web 类的资源如图1-11所示。

⑤ Containers 类的资源如图1-12所示。

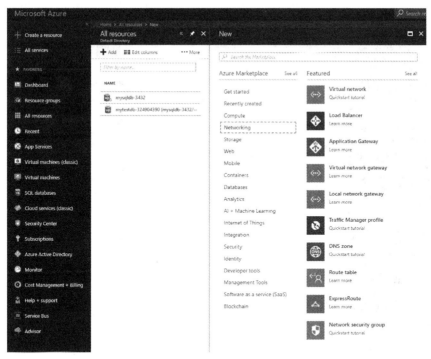

图1-9 创建网络类的资源

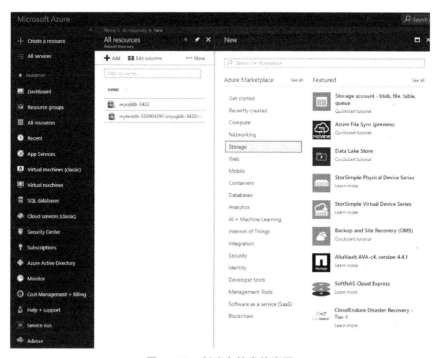

图1-10 创建存储类的资源

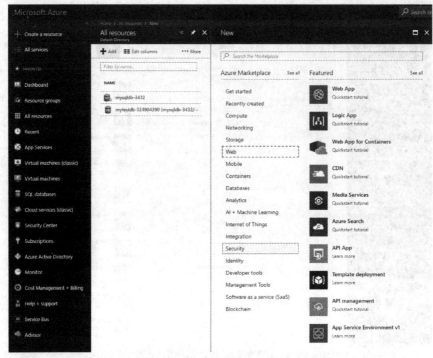

图 1-11 创建 Web 类的资源

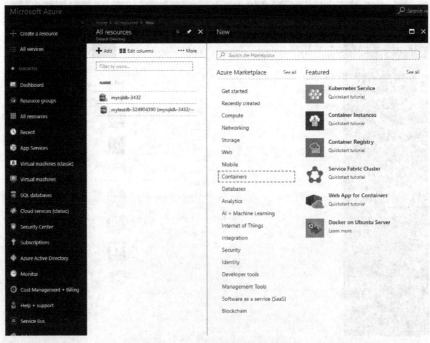

图 1-12 创建容器类的资源

这里有 Kubernetes Service 的容器，还有 Service Fabric Cluster 的容器。

⑥ Databases 类的资源如图 1-13 所示。

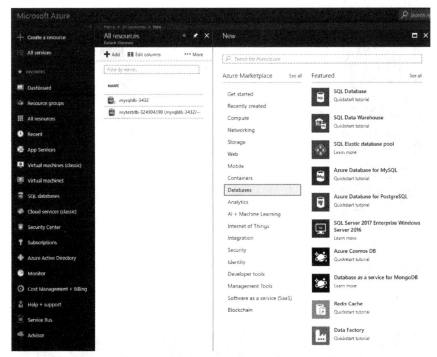

图 1-13 创建数据库类的资源

这里很多服务都是常用的，如 SQL Database、Redis Cache。

⑦ Analytics 类的资源如图 1-14 所示。这里面主要是用于数据分析的资源，如 HDInsight 相当于 hadoop。

⑧ AI＋Machine Learning 类的资源如图 1-15 所示。

这里的资源全都是人工智能和机器学习类的，如 Computer Vision（机器视觉）、Face Recongnition（人脸识别）、Language Understanding（语言理解）等，都是很高大上的题材。

⑨ 区块链类的资源如图 1-16 所示。

1. All Resources 菜单

这里显示所有已创建的资源。会将所有资源列表显示，如图 1-17 所示。

2. App Services 菜单

它是用来创建 Web App 和 Mobile App 的。目前比较多的 Web 应用和移动应用都在 App Services 中创建。

App Services 可以帮助用户快速地建立可伸缩的、安全的、模块化的、分布式的应用。应用可以是 Web Application＋Web API 的形式，也可以是纯 Web API 的形式，也可以是 Mobile Application＋Web API 的形式。

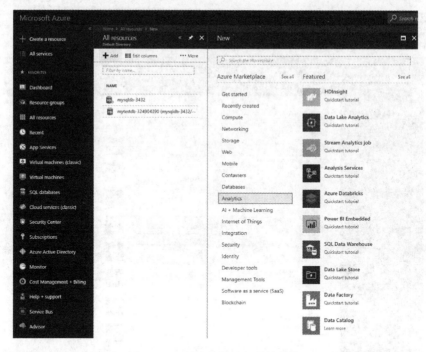

图 1-14 创建数据分析类的资源

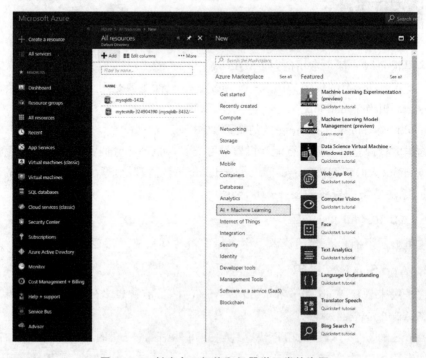

图 1-15 创建人工智能和机器学习类的资源

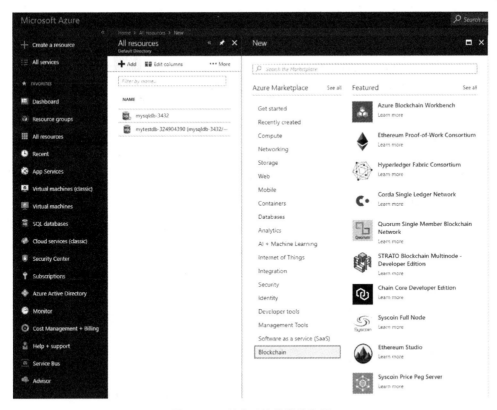

图 1-16　创建区块链类的资源

图 1-17　All Resources 菜单

3. Virtual Machine(Classic)菜单

这是为了让之前在较早 Azure 环境中建立的虚拟机能继续使用而保留下来的，以适应之前较早的 Azure 虚拟机的设置。老式的虚拟机在 Azure 刚开始建立起来的时候有一个 Service Management API(SMAPI)。这个 SMAPI 在当时还是不错的，但还是有局限性，在 Azure 发展起来后，这些基础架构都重新设计了。但是有的虚拟机用户还在引用 SMAPI 等过时的 API，就必须让这部分用户继续使用，所以微软还是保留了这个 Classic 虚拟机。

4. Virtual Machine 菜单

这是最新的虚拟机配置。可以选用较多的操作系统和版本,如 Windows XP、Windows 7、Windows Server 等版本,Linux 可以安装各个厂家的发行版本。

这些新的虚拟机都遵循新的基础架构设计,所以我们后面建立新的虚拟机,最好都用此菜单的功能。

5. SQL Databases 菜单

这是用来建立、管理 Azure SQL Database 的菜单,可以定制 Azure SQL Database 的参数、Pricing Tier、容量、防火墙规则、复制策略、安全策略等。

6. Cloud Service(Classic)菜单

这是为了让一些旧的 Azure 平台云服务能继续运行而保留下来的服务。有一些旧的应用需要依赖旧的 Cloud Service(Classic)上的特性,所以微软还是保留了这个渐渐过时的服务。建议新开发的应用采用新的服务,如 App Service(Web App、Mobile App)。

Cloud Service(Classic)的部署图如图 1-18 所示。

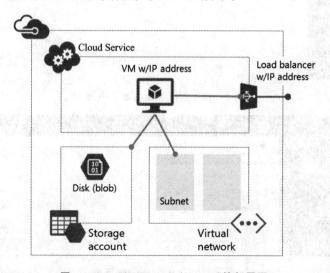

图 1-18 Cloud Service(Classic)的部署图

另外,它还需要指定一些具体的资源,如 Load balancer、应用程序池 App Pool 的虚拟机(VM)、虚拟网络 Virtual network 等。

在使用新一代的 App Service(Web App、Mobile App)时,Load balancer 的资源会被其他更好的服务替代,虚拟网络不需要去进行设计;同时,也不需要去指定具体的虚拟机,Azure 会自动从一个虚拟机池中分配虚拟机给用户使用。

7. Security Center 菜单

这里可以设置安全策略,以保护 Azure 中数据和应用程序的安全,比如防火墙规则、检测入侵的策略、抵御攻击的策略等。

8. Subscriptions 菜单

这个菜单管理当前账号拥有的订阅，也可以新建一个订阅。有些资源可以在不同的订阅下进行管理。比如个人有一个 MSDN 订阅，同时还有一个或者两个公司的订阅，那么使用的时候可以在右上角选择其中之一（如 Switch Directory，如图 1-19 所示）。这样不同的订阅就能互不干扰。

图 1-19　显示账号拥有的订阅列表

9. Azure Active Directory 菜单

它可以用来管理一个或者多个类似于 Windows Active Directory 的目录。可以从已有的 Windows Active Directory 导入用户账号，也可以用手工一个个地创建账号和组。这里有默认的免费版的服务，也有一些需要收费的服务。

10. Monitor 菜单

它用来监测用户的应用和基础设置的健康状态，比如虚拟机的健康状态、App Service 的在线状态。如果有虚拟机宕机，则会进行通知，Azure 会按指定的方式来通知管理员。

11. Azure 的免费服务

微软将 Azure 中的一些服务标记成永久免费。这个新注册账号，可以使用这些免费的服务。当然，这些免费服务都是非常基础的，只有一些基本功能，不能有数据备份，不能保证数据的完整性，没有异地灾难备份，出了问题，客服的响应时间也不一样，毕竟是免费提供的服务。除了这些免费的服务外，还有一些限时免费服务，用户可以在最初的 12 个月免费使用这些服务。

详细的免费资源列表见链接：https://azure.microsoft.com/en-us/free/。

1.8　在 Visual Studio 中操作 Azure 资源

除了在 Azure Portal 中可以操作资源，也可以在 Virtual Studio 中操作 Azure 的资源。首先用账号连接上 Azure，然后打开 Server Explorer 窗口，如图 1-20 所示。

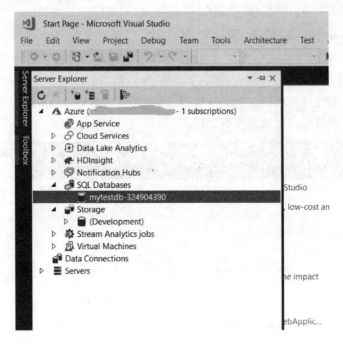

图 1-20　在 Visual Studio 中查看 Azure 中的资源

可以看到已经创建的资源，如此图中的 Azure SQL Databases，有 Connect to Database 按钮，还有另外一个按钮 Connect to Server，如图 1-21 所示。

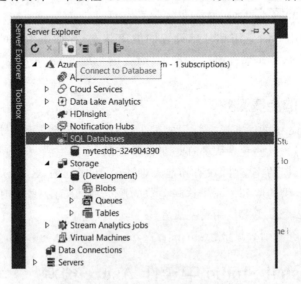

图 1-21　Connect to Database 按钮

按钮 Add SQL Server 用来创建一个新的 SQL Server，这里创建 SQL Server，默认创建的是 Standard 的服务器，如图 1-22 所示。

图 1-22　Add SQL Server 按钮

1.9　在 SQL Server Management Studio 中操作 Azure 资源

首先建立链接,在图 1-23 中。输入我们建立的 Azure SQL Server 的名字、用户名、密码,即可建立链接。

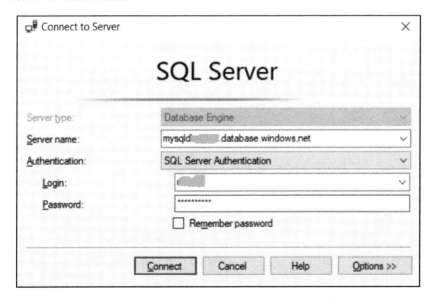

图 1-23　建立到 Azure SQL Database 的链接

如果是第一次建立链接,则需要加上防火墙规则,允许从当前所在的 IP 能链接这个 SQL Server 服务器。要加上防火墙规则,则需要登录 Azure 账号。在弹出的对话框中输入 Azure 账号和密码,下一步就是选择将当前 IP 加入防火墙规则,或者是将当前 IP 所在的地址段(255 个 IP)加入防火墙规则。这样就可以从此 IP 连上 Azure SQL Database 了。此处说的 IP 是指外网 IP,通过路由器、防火墙等,待外网上的出口 IP 连上后,就可以做一些操作了。

比如创建新的数据库,右键单击 Databases,然后选择 New Database,如图 1-24 所示。

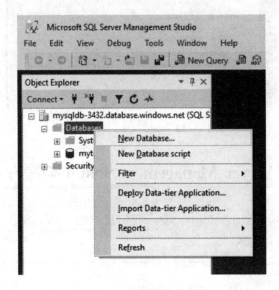

图 1-24　创建新的数据库

这些操作在第 4 章都有详细讲解,可到第 4 章看详细的操作。

第 2 章

Azure 云存储

Azure 云存储主要包括 blob storage、table storage、file storage、queue storage。Azure 云存储是一个比较基础的服务，其价格也相对低；同时，对数据安全性也有比较大的保障，其数据有备份，万一发生物理硬盘的损坏事故，从备份数据中可以很快恢复其数据。

2.1 Azure Table Storage

Azure Table Storage 是一个类似于 nosql 的数据存储，其底层实现原理比较类似 MongoDB。其存取速度快，效费比高，成本显著低于传统的 SQL。

1. Storage account

Azure Table Storage 需要一个 Storage account 来进行访问。可以在 Azure 界面中创建这个账号。

先创建好一个 Resource group，然后在此 Resource group 中创建 Storage account，此 Storage account 下可以存放 table、file、queue、blob。需要先创建 Storage account，在 Azure Portal 中，点击 Create resource 的按钮，Azure 会问需要创建何种资源，输入 Storage account，会出来一个 Storage account(blob、file、table、queue)，选择此项，然后进入如图 2-1 所示的界面。

注意此界面里带"＊"号的都必填或必选，比如输入 Name，一般选择 Resource manager，不选择 Classic。这是因为目前一般需要的是新的 Storage；如果选择 Classic，就会是旧的部署模式。对于新的应用来说，最好都选择 Resource manager。除非有一些旧的应用，才选择 Classic。关于 Account kind，默认是 Storage(general purpose v1)，还有 Storage(general purpose v2) 和 Blobstorage account 的选项。Storage(general purpose v2) 包括另外两项的所有功能，价格也更高。Storage(general purpose v1) 没有 Storage(general purpose v2) 的功能，但是它更便宜，特别是读的比例更高的应用更适合 Storage(general purpose v1)。Storage(general purpose v2) 在 Storage(general purpose v1) 之外还提供了一些最新的功能和 hot/cool 的 access tier，但是其价格也相对高一些。所以可以根据应用的情况来合理选择。

下一项是 Location，可以根据应用的地域来选择，这个可以比较宽松。

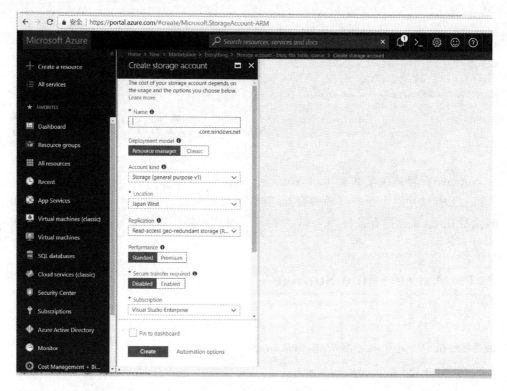

图 2-1 创建 Storage account

下一项是 Replication，有 Locally redundant storage、Geo-redundant storage、Read-access geo-redundant storage 三个选项，Locally redundant storage 是在本地进行复制备份，这样，所有备份都在同一个数据中心里，恢复速度、传输成本都相对较低。但是缺点是如果此数据中心在有灾难的情况下，有可能失去全部的数据备份。因为所有数据备份都在同一个数据中心。Geo-redundant storage 是地理上复制备份，数据备份是在一些地理上不在同一处的数据中心存储，所以数据完全丢失的概率相对小得多，因为地理上的差别，其传输速度没有本地复制得快，所以需要更高的成本来传输数据的备份。Read-access geo-redundant storage 是只读的地理复制备份，主实例是可写的，副实例都是只读的，这样可以让副实例分担一点读的压力。

下一项是 Performance，有两个选择：Standard、Premium。Standard 是标准，价格相对低一些；Premium 是更高级的意思，其价格更高。

下一项是 Secure transfer required，有 Disabled 和 Enabled 两个选择。Disabled 表示不加密，Enabled 表示加密。加密，速度相对慢一些，成本更高一些；不加密，数据有一定可能在传输的时候被泄露。默认是 Disabled。

下一项是 Subscription，这个就是订阅了。商业应用，都会有一个订阅。这个订阅里有一定金额可以用于消费。创建存储也需要用订阅里的金额。如果有多个订阅，则可以从中选择一个。

下一项就是是否创建新 Resource group，这可以根据实际情况来决定，如果还没有 Resource group，则可以创建一个；如果已经有了，可以选择已有的，也可以再新建一个。

下一项是 Virtual network，即虚拟网络，即 VPN，它可以同企业网络连起来，如同在一个网络里一样，这样传输数据不容易被外部攻击者获得。将此 Storage account 与企业的网络建立其 VPN，企业内部的应用访问此 Storage account，如同访问内部网络资源一样。可以用内部网络的 IP 来访问。数据在 Azure 和企业的网络间传输是加密的，因为底层的 VPN 对网络数据都进行了加密。这样保密性更高，但是也需要为此付出更高的成本，比如需要建立 VPN 网关。

这些选项都选择好之后，单击 Create 按钮，即开始配置，过一段时间，即可开始使用。

创建好之后，就会有一个 Storage account name 和 Access keys，这是用来访问 Azure storage 的，如图 2-2 所示。

图 2-2　Storage Account 和相应的访问 Key

Azure Table Storage 是类似 MongoDB 的。对于每一个表，都有 Partition key、Row key。它的层级概念如图 2-3 所示。

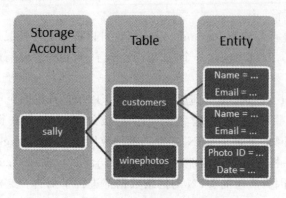

图 2-3 Azure 存储的层级结构

图 2-3 中:

Storage Account 是最上一级概念。Storage Account 是订阅中专用于存储的账号。一个订阅可以有多个 Storage Account,多个 Storage Account 分别按服务级别设置来收费。

Table 是一系列 Entity 的集合,其中包含多个 Entity。

Entity 是一系列 Property 的集合,其中包含多个 Property。一个 Entity 最多可以有 252 个 Property,每一个 Entity 都有三个系统的 Property,即 Partition key、Row key 和 Timestamp。拥有同一个 Partition key 的 Entity 可以查询得更快,同时支持插入、删除等原子操作。Entity 的 Row key 用来在此 Partition 中唯一标识 Entity。

Property 是一个名字和值的对,用来组成 Entity。这点比较像 Redis 等服务器,但是微软的 Table Storage 可以把 Property 组合成 Entity,然后再把 Entity 组合成 Table,这样方便我们进行类似于 SQL 的查询。它的价格也比传统关系型数据库要便宜。

2. 安装开发环境

在使用 Azure Table Storage 之前,需要在 Visual Studio 中安装相关的支持包,新建一个工程或者打开需要访问 Azure Table Storage 的工程。然后打开 Package Manager Console,输入命令:Install-Package WindowsAzure.Storage,如图 2-4 所示。

安装好之后会有 Console 消息说安装成功。然后在工程的 Reference 中会有这项,如图 2-5 所示。

在 packages.config 文件里会有类似这样的一行,如图 2-6 所示。

这样就表示已经成功地加入到工程里来了。下面就是配置 Table Storage 的链接串:

```
<configuration>
    <startup>
```

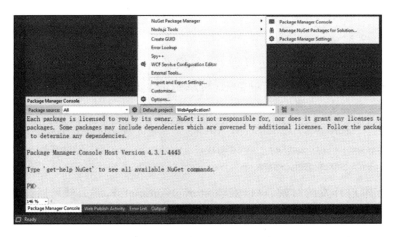

图 2-4 用命令 Install-Package WindowsAzure.Storage 安装存储相关的包

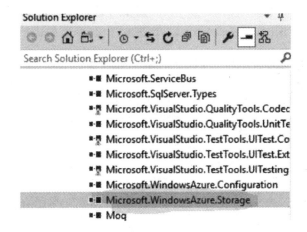

图 2-5 工程中有对应的引用

```
39    ··<package·id="System.Net.Requests"·version="4.3.0"·targetFramework="net452"·/>
40    ··<package·id="System.Spatial"·version="5.8.2"·targetFramework="net452"·/>
41    ··<package·id="WindowsAzure.ServiceBus"·version="4.1.2"·targetFramework="net452"·/>
42    ··<package·id="WindowsAzure.Storage"·version="8.1.3"·targetFramework="net452"·/>
43    ··<package·id="xunit.abstractions"·version="2.0.1"·targetFramework="net452"·/>
44    ··<package·id="xunit.assert"·version="2.2.0"·targetFramework="net452"·/>
45    ··<package·id="xunit.core"·version="2.2.0"·targetFramework="net452"·/>
```

图 2-6 packages.config 文件中关于存储的引用

```
        <supportedRuntime version = "v4.0"
sku = ".NETFramework,Version = v4.5.2" />
    </startup>
    <appSettings>
        <add key = "StorageConnectionString"
value = "DefaultEndpointsProtocol = https;AccountName = account - name;AccountKey = account -
key" />
```

```
</appSettings>
</configuration>
```

注：.net framework 版本可以是 4.5.x 或者 4.6.x。

用实际的 storage account name 替换这里的 account-name，用实际的 key 来替换这里的 account-key，例如：

```
<addkey = "StorageConnectionString"value = "DefaultEndpointsProtocol = https;Account-Name = storagesample;AccountKey = GMuzNHjlB3S9itqZJHHCnRkrokLkcSyW7yK9BRbGp0-ENePunLPwBg-pxV1Z/pVo9zpem/2xSHXkMqTHHLcx8XRA == " />
```

如果在本地开发的时候，可以安装 storage emulator，那么就可以在本地建立存储了。其链接串是这样的：

```
<addkey = "StorageConnectionString"value = "UseDevelopmentStorage = true;" />
```

当部署到实际的 Azure 云环境的时候，需要将这些配置信息改成 Azure 云环境的链接信息。启动 Storage Emulator 后，出现如图 2-7 所示信息。

图 2-7 Storage Emulator 启动后的界面

至此可以用上述链接串来链接本地的存储。用 Storage Explorer 来访问这些本地的 Storage 或者 Azure 中的云存储。先介绍一下用 Storage Explorer 链接本地的存储。

Storage Explorer 启动后的界面如图 2-8 所示。

它会显示出 Development 的存储，即本地的开发用存储。一开始 Blob、Queues、Tables 都是空的，因为没有内容。开发用存储和 Azure 平台的存储是很相似的，可以用 SDK 来访问，唯一的区别是一个在本地 PC，一个在 Azure 平台。右键单击 Tables，选 Create Table，如图 2-9 所示。

输入此 Table 的名字，回车，即创建了此 Table，如图 2-10 所示。

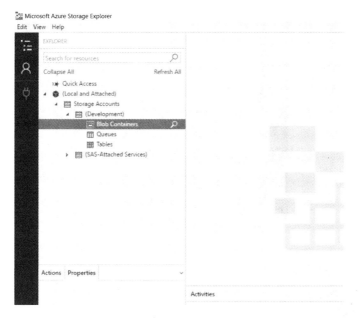

图 2-8　Storage Explorer 的界面

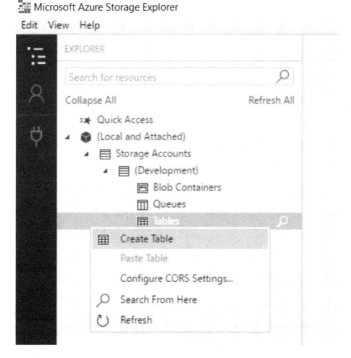

图 2-9　选 Create Table 菜单来创建 Azure Storage Table

图 2-10　创建好的 Azure Storage Table

此时 Table 中还没有任何一个 Entity。这个 Table 与 SQL Server 的表是不同的。SQL Server 的表创建后就必须定义字段,所有字段都定义名字、长度、类型等,而 Table 为空的时候可以不用定义字段,只有加入 Entity 之后其字段才会定义。这些 Entity 的字段就是上述的 Property。目前只能单击 Add Property 按钮,如图 2-11 所示。

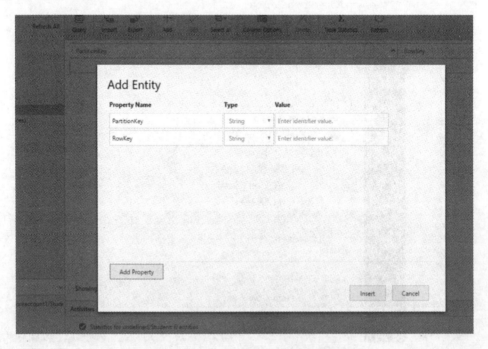

图 2-11　往 Azure Storage Table 中加入 Entity

如上所述,每个 Entity 都有 PartitionKey 和 RowKey,这两个是不能改的。我们可以加入其他的 Property，单击 Add Property 按钮，输入 Property 名字，选择类型，其类型只有图 2-12 中的这几个。虽有点少，但常用的基本都有了。Value 是输入该 Property 的值。这就像新增一条记录似的，需要输入值，如图 2-12 所示。

Add Entity

Property Name	Type	Value
PartitionKey	String	Enter identifier value.
RowKey	String	Enter identifier value.
name	String ▼ (String / Boolean / Binary / DateTime / Double / Guid / Int32 / Int64)	50

Add Property

Insert   Cancel

图 2-12　在 Entity 中加入 Property，输入 Property 名字，选择类型、长度等

给 PartitionKey、RowKey、Property 都输入一个值后，单击 Insert 按钮，就会出现如图 2-13 所示的界面。

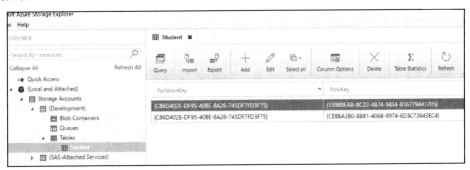

图 2-13　Entity 创建好之后的列表

注意，上述的 Time Stamp Property 也自动加入进来了。这个 Time Stamp 自动记录了 Entity 的最后创建或者修改时间。它默认的时间显示是按照 UTC 时间格式显示。这个 Entity 可以很方便地加 Property。比如开始的时候 Entity 只有几个 Property，后来由于业务需要，可以加入新的 Property，之前已经保存的 Entity 对应的新 Property 的值就为空。

Query 按钮可以查询 Entity,点开后出现如图 2-14 所示的界面。

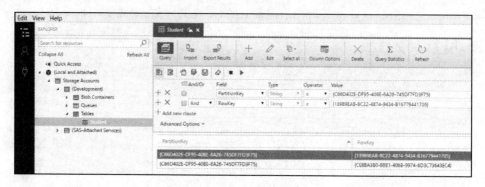

图 2-14 Query 界面

默认有两个条件:PartitionKey 和 RowKey。此两个条件是此表的主键,可以用最快的速度找到 Entity。如果需要按别的条件来查找 Entity,则可以将此两条件删除,再单击 Add new clause 按钮。新加的条件可以根据实际需要来改。

条件的比较符号有 6 种,如图 2-15 所示。

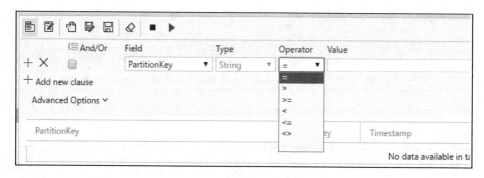

图 2-15 条件比较符

有等于、大于、大于或等于、小于、小于或等于、不等于几种。

除 PartitionKey 和 RowKey 之外的 Property 也可以加入。在窗口左边选择 And/Or 来指定条件的组合。当条件确定以后,点击三角形的按钮按此条件来查询 Entity。符合条件的 Entity 就会显示出来。

Export Results 按钮可以帮助把查询结果导出到数据文件中。导出的文件格式默认是 typed csv,即每个 Property 的格式都是有导出的。此例如图 2-16 所示。

```
1 PartitionKey,RowKey,Timestamp,name,name@type,address,address@type,wage,wage@type
2 {C86D402E-DF95-40BE-8A26-745DF7FD3F75},{139B9EAB-8C22-4874-9434-B16779441705},2018-04-05T13:11:59.400Z,test,Edm.String,,,
3 {C86D402E-DF95-40BE-8A26-745DF7FD3F75},{CE8BA3B0-8881-406B-9974-6D3C73643EC4},2018-04-05T13:13:38.833Z,sdf,Edm.String,nanshan district,Edm.String,,
4 {C86D402E-DF95-40BE-8A26-745DF7FD3F75},{DCD176EF-56ED-4932-977E-75190D6BD75F},2018-04-07T12:03:57.310Z,test,Edm.String,test,Edm.String,543.54,Edm.String
5
```

图 2-16 导出的 csv 文件格式

每一个自定义的 Property 后面都跟着一个 type 定义。这样可以明确每个 Property 的类型。这样数据导入、导出的时候格式没有丢失,但是其信息也比较多,当数据量大的时候会有问题。也可以选择没有 type 的 csv 格式,即那些@type 的字段都会去掉。这样 csv 文件体积相对会小很多。这样导入的时候就可能需要指定每个 Property 的类型。其格式如图 2-17 所示。

```
PartitionKey,RowKey,Timestamp,name,address,wage
{C86D402E-DF95-40BE-8A26-745DF7FD3F75},{139B9EAB-8C22-4874-9434-B16779441705},2018-04-05T13:11:59.400Z,test,,
{C86D402E-DF95-40BE-8A26-745DF7FD3F75},{CE8BA3B0-8861-406B-9974-6D3C73643EC4},2018-04-05T13:13:38.833Z,sdf,nanshan district,
{C86D402E-DF95-40BE-8A26-745DF7FD3F75},{DCD176EF-56ED-4932-977E-75190D6BD75F},2018-04-07T12:03:57.310Z,test,test,543.54
```

图 2-17 选择没有 type 的 csv 格式

如前所述,PartitionKey 是一个很关键的属性,Azure Table Storage 根据 PartitionKey 的值来决定存放的具体位置,这个功能类似之前关系型数据库的分区功能,需要合理地进行数据的分区。可以根据时间来分区,比如每年的数据一个分区,2017 年的数据一个分区,2018 年的数据一个分区;如果有特殊要求,还可以用其他属性来进行分区。这样就可以提高数据存储、查询的效率,当数据量特别大的时候,这些分区的策略帮助更大。

Import 按钮可以导入之前导出的 csv 文件。如果 PartitionKey 和 Row Key 重复,则会出现导入不成功。

Edit 按钮用来编辑某一个 Entity,单击 Edit 后,会弹出一个框来更改此 Entity 各 Property 的值。当保存后,其 Timestamp 会更新成最新的时间。这个 Timestamp Property 是系统控制的,不由我们控制。

Select All 按钮是选择所有 Entity。

Column Options 按钮是选择显示哪些 Column(字段)。

Delete 按钮是删除选定的那些 Entity。

Query Statistics 按钮是报告有多少个 Entity。

Refresh 按钮是用来刷新的。当多个客户端操作同一个表时,就需要此按钮。

之前介绍的都是本地模拟的存储,可以用 Storage Explorer 链接 Azure 平台中的 Storage。其方法是这样的:单击左边的电源插头按钮,弹出对话框,如图 2-18 所示。

如果有 Azure 账号,可以选择第一项,Azure environment 有 Azure、Azure China、Azure Germany、Azure US Government,一般地根据所在的环境来选择。如果拥有 Azure 账号的权限,那就有该账号的所有权限。可以看到该 Azure 账号所有的 Storage account,以及所有 Storage account 下的所有资源。这个是能看到的最全的信息。一般是 Azure 账号所有者才能做到。

第二项是 Storage 所有者分享出来的 Shared access signature,Shared access signature 是授权其他人访问的方式而不必告诉其他人 Account name 和 Account key。其他人只能访问授权的资源,比如某些表、某些 blob。

图 2-18　连接到 Azure Storage 的对话框

第三项是用 Storage name 和 Storage key，这个是 Storage account 所有者的信息。这样对此 Storage account 下的所有资源如 blob、file、queue、tables 都可以全权控制。

具体选哪一项，要看实际情况。一般开发的时候都是选第一项或者第三项。第二项也有可能，只是一般在开发的时候出现。这里以第三项为例说明，如图 2-19 所示。

图 2-19　使用 Account name 和 Account key 链接 Azure 中的 Storage 账号

输入 Account name 和 Account key,再看其他的输入项,一般是不需要改的,其他不需要改的话就单击 Next 按钮。如果是 Azure China 等,其域名后缀不同,则需要将 core.windows.net 改成对应的后缀域名,这点请注意。

下一个界面是确认界面,没有问题就单击 Connect 按钮。链接确认界面如图 2-20 所示。

![Connection Summary]

图 2-20　链接确认界面

加好 Storage Account 后,出现如图 2-21 所示界面。

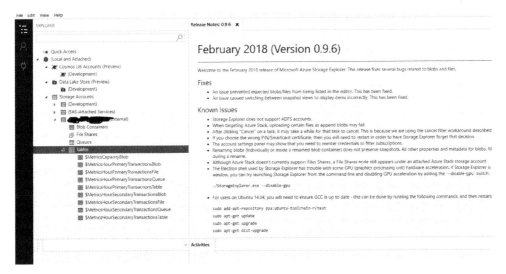

图 2-21　链接建立之后,显示 Azure 中 Storage 的资源信息

注意以 S 开始的那些 Table 是系统的表,不要动。在 portal.azure.com 的显示中是没有 Table 的,如图 2-22 所示。

这样就可以按照前面的方法来建立 Table、定义 Property 等,并进行数据输入、

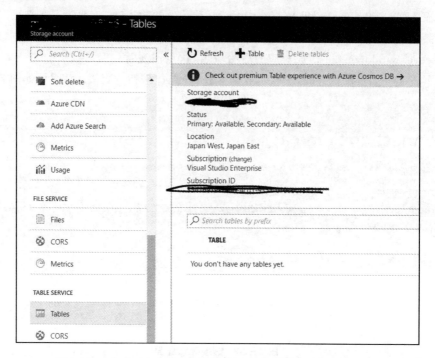

图 2-22 在 Azure Portal 中显示的 Storage Table 列表，此处为空

数据查询、导入导出等操作。这些操作与本机模仿出来的 Table Storage 是一样的。

3. 基于 Azure 账号访问 Azure 平台中的资源

单击左边的电源插头 按钮，弹出如图 2-23 所示界面。

图 2-23 使用 Azure Account 来链接

选择第一项 Add an Azure Account，然后 Azure environment 根据实际情况来选择 Azure（全球都有数据中心的）、Azure China（只在中国境内运营）等。之后弹出窗口，需要输入 Azure 平台的账号、密码，Azure 账号的资源就都被列出来了。如

图 2-24 所示,此 Azure 账号有两个 Storage Account,所以就都列出来了。这两个 Storage Account 可以分别管理不同的内容,互不干扰。所有的操作都是在其子树进行,其余的操作都与上述差不多。

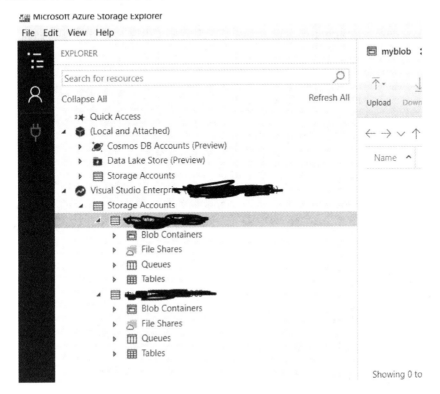

图 2-24 用 Azure Account 链接后显示的 Storage 资源

4. 基于 Shared Access Signature 访问 Azure 平台中的资源

上述根据 Storage Account 或者 Azure 账号来访问 Azure 平台中资源的方式,都是全部的权限。一旦将权限交给对方,对方就能全权控制所有的资源。比如 Azure 账号,是 Azure 平台最大的权限拥有者,可以看到所有此账号申请的资源,也包括 Storage。这样是不能接受的,需要一种能控制权限的访问。因此,Shared Access Signature 就出现了。

Shared Access Signature(SAS)分为几种。

(1) Service SAS

Service SAS 是针对某一个 Storage 资源来分配权限的,比如 blob、table、queue、file 之一,给一个读/写权限,或者只读权限;也可以更细,比如只是某个文件夹,或者某一些 Table、某一个具体的 queue,这样就可以给访问者需要的权限,不需要的权限一点也不多给。这样,相对权限控制比较精确、安全。

（2）Account SAS

Account SAS 是将 Storage Account 账号的权力代理给另外一个账号，如 Get/Set Service 属性、Get/Set Service Stats，以及 Storage 资源的读/写权限等都可代理给另外一个账号。Service SAS 所能给的权限 Account SAS 都包括。

Shared Access Signature 是一个签名了的 URL，由一些部分组成，通常其结构如图 2-25 所示。

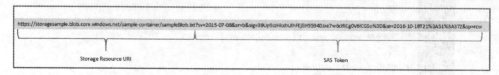

图 2-25 Shared Access Signature 的组成结构

其中的 sig 是签名，这个 sig 是用 Account Key 来签名的。Storage Resource URI 是具体资源在 container 里的路径。最前面的是资源 container 的名字，container 即容器的意思，先是容器名，然后才是资源在此容器里的路径（path）。SAS 的 URL 用来表示需要访问哪个资源，以及授权的签名，这样可以充分地证明客户端访问此资源的合法性。SAS 的 URL 包括如下一些 URL 参数（如图 2-26 所示，"*"号处是必须有的参数）。

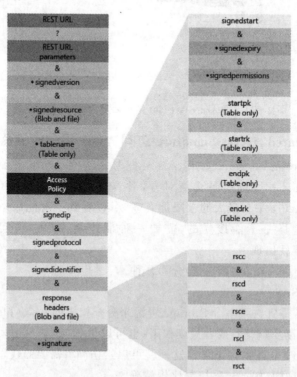

图 2-26 Shared Access Signature 的参数按颜色区分

signedversion 是一个必须有的参数,其 URL 缩写是 sv,凡 2012-02-12 之后的 Azure 版本都需要这个参数,用来表示 SAS 的版本。

signedresource 用来指定需要访问的资源类型。其 URL 缩写是 sr,可以是如下几个值:

b:表示需要访问 blob,允许访问 blob 的内容和元数据,只适用于 blob。

c:表示需要访问 container,允许访问容器中的元数据,或者 blob 的列表,只适用于 container。

f:表示需要访问 file,允许访问文件中的内容和元数据,只适用于 file service。

s:表示需要访问 share,允许访问此共享的内部所有文件和元数据,只适用于 share。

table name 的 URL 缩写是 tn,表示表名,即需要访问某个表,只适用于 Table storage。

signedexpiry 表示权限过期时间,用一个 ISO 8601 格式表示 UTC 的时间,形如 yyyy-MM-ddTHH:mm:ss.sssZ,例如:2017-09-01T03:09:12.678Z,这个 T 和 Z 是不能改的。

signedpermission 表示权限,其 URL 缩写是 sp,指读、写、删除的权限。

startpk 表示开始 partitionkey,其 URL 缩写是 spk。这个只适用于 Table。

startrk 表示开始 rowkey,其 URL 缩写是 srk。这个只适用于 Table。Startrk 必须和 startpk 一起使用。

endpk 表示结束 partitionkey,其 URL 缩写是 epk。这个只适用于 Table。

endrk 表示结束 rowkey,其 URL 缩写是 erk。这个只适用于 Table,只能和 Endpk 一起使用。

signature 是基于待签名的字符串,是用 HMAC-SHA256 算法签名出来的。其 URL 缩写是 sig。

如果是 blob 和 file service,则待签名的字符串构成是这样的:

```
StringToSign = signedpermissions + "\n" +
               signedstart + "\n" +
               signedexpiry + "\n" +
               canonicalizedresource + "\n" +
               signedidentifier + "\n" +
               signedIP + "\n" +
               signedProtocol + "\n" +
               signedversion + "\n" +
               rscc + "\n" +
               rscd + "\n" +
               rsce + "\n" +
               rscl + "\n" +
```

```
          rsct
```

如果是 Table，则待签名的字符串是这样的：

```
StringToSign = signedpermissions + "\n" +
               signedstart + "\n" +
               signedexpiry + "\n" +
               canonicalizedresource + "\n" +
               signedidentifier + "\n" +
               signedIP + "\n" +
               signedProtocol + "\n" +
               signedversion + "\n" +
               startingPartitionKey + "\n" +
               startingRowKey + "\n"
               endingPartitionKey + "\n" +
               endingRowKey
```

得到这个待签名字符串后，用 HMAC-SHA256 算法得到签名。这样整个 URL 就成了一个授权的 Service SAS。

Account SAS 的 URL 构成类似，有些参数与 Service SAS 一样，这里介绍不一样的参数：

signed services（缩写 ss）：可以是

-Blob（b）；

-Queue（q）；

-Table（t）；

-File（f）。

signed resource types（缩写 srt）：可以是

-Service（s）；

-Container（c）；

-Object（o）。

其用来产生签名的待签名字符串是这样构成的：

```
StringToSign = accountname + "\n" +
               signedpermissions + "\n" +
               signedservice + "\n" +
               signedresourcetype + "\n" +
               signedstart + "\n" +
               signedexpiry + "\n" +
               signedIP + "\n" +
               signedProtocol + "\n" +
               signedversion + "\n"
```

其同样也是用 HMAS-SHA256 算法来产生签名。

5. Table Storage 的设计思考

Azure 的 Table Storage 设计需要充分考虑不同的方面，如性能、可伸缩性、费用。特别是 PartitionKey、RowKey 的设计。Table Storage 会以 PartitionKey、RowKey 做一个聚集索引，合理地设计 PartitionKey、RowKey 会大大加速查询速度。先来看看 Table Storage 的限制，如表 2-1 所列。

表 2-1　Table Storage 的限制

项　目	内　容
一个 Storage Account 的所有空间限制	500 TB
表的个数	仅受限于此 Storage Account 的总空间
一个表分区个数限制	仅受限于此 Storage Account 的总空间
一个分区里 Entity 的个数限制	仅受限于此 Storage Account 的总空间
每个 Entity 的大小限制	最大 1 MB，最多 255 个 Properties，其中包括 PartitionKey、RowKey、Time Stamp
PartitionKey 的大小限制	最大 1 KB 的 String
RowKey 的大小限制	最大 1 KB 的 String
Entity group transaction 的大小限制	一个 Entity group transaction 最多包含 100 个 Entity，最多占用 4 MB 字节，并且只能更新同一个 Entity 一次，一个 Entity group transaction 不能跨 Partition

在设计表的时候，与关系型数据库的设计有所不同。为那些读压力大的应用设计查询、设计表的时候，针对读操作，需要考虑如下几点：

- 在查询语句里指定 PartitionKey 和 RowKey，这样就可以利用上其聚集索引，有利于提高查询速度。
- 可以考虑存储重复的 Entity，一定的数据重复，可以帮助提高查询的速度。Table Storage 是比较便宜的。例如可以存储 Summary Entity。
- 利用好 Compound Key（复合键），利用好 PartitionKey 和 RowKey 这个组合键，可以加速对 Entity 的访问。
- 尽量利用查询投影，即一次不要查询过多的 Property，只查询出需要的 Property 即可。只把此查询需要的 Property 查询出来，其余没有涉及的 Property 就不要查询出来。

针对写操作需要考虑如下几点：

- 不要创建热分区。合理地选择 PartitionKey，使数据分布在不同分区。
- 不要在写操作时中断。
- 不要为不同类型的 Entity 创建不同的表，可以考虑把它们放到同一个表中。

当一个查询语句进来时,通常会有如下几种操作方式:
- 若查询语句同时有 partitionkey=? 和 rowkey=?,那么就可以利用到索引,此时为 point query,可以很快就找到所在的分区和行块,这样最快。
- 若查询语句有 partitionkey=? 和 rowkey 的范围,那么就会部分用到索引,需要在此分区里搜索符合范围的 RowKey。此时相对第一种较慢。
- 若查询语句有 partitionkey=? 和另外一个 property=?,那么会部分用到索引,先找到分区,然后在此分区里逐条查找满足条件的 Entity。
- 最后一种是没有 PartitionKey 的查询,这样就会导致全表扫描,逐一检查每个 Entity 是否符合条件。这样就会非常地慢。

从上可知,如果要比较好地利用 Table Storage,应尽量用前三种操作方式。在建表的时候,PartitionKey 的选择就比较关键了。下面讲一下如何选择 PartitionKey。Entity 原本有若干个 Property,从中选择一些可以将数据分到不同区域的 Property 来作为 PartitionKey。首先得看数据的分布,有一些通常能将数据分布到不同分区的 Property,如日期、年份、小时、分钟。也有可能是别的 Property,需要对数据进行一些分析,应找到一个合理分区的 Property,将其作为 PartitionKey。

RowKey 的选择也很重要,即在同一个分区里,再进行一些小的分区,需要选择一些第二适合做分区的 Property 来作为 RowKey,这样就比较好地设计了表。当应用存储了相当大量的数据的时候,可以通过优化查询来加速数据的读/写速度。

6. Azure Table Storage 和 Azure SQL Server 的区别

Table Storage 更像 Nosql,它无事物控制,没有数据完整性检查,没有非聚集索引。Azure SQL Server 是一个关系型数据库,有事务控制,有数据完整性检查,除了聚集索引外,可以有非聚集索引。

样例代码下载地址:https://azure.microsoft.com/en-us/resources/samples/storage-table-dotnet-getting-started/。

代码里重要的点是 App.config 文件里的 Table Storage 配置:

```
<add key="StorageConnectionString" value="UseDevelopmentStorage=true;" />
<!-- <add key="StorageConnectionString" value="DefaultEndpointsProtocol=https;AccountName=[AccountName];AccountKey=[AccountKey]" /> -->
```

如上所述,其链接串可以是一个 Shared Access Signature,也可以是一个在 Azure 里的 Table Storage。

7. Common.cs 文件

```
using System;
using System.Threading.Tasks;
using Microsoft.Azure.CosmosDB.Table;
using Microsoft.Azure.Storage;
```

```csharp
using Microsoft.Azure;

namespace TableStorage
{
    public class Common
    {
        /// <summary>
        /// Validate the connection string information in app.config and throws an exception if
        /// it looks like the user hasn't updated this to valid values.
        /// </summary>
        /// <param name = "storageConnectionString"> Connection string for the storage service
        ///or the emulator </param>
        /// <returns> CloudStorageAccount object </returns>
        public static CloudStorageAccount CreateStorageAccountFromConnectionString(string storageConnectionString)
        {
            CloudStorageAccount storageAccount;
            try
            {
                storageAccount = CloudStorageAccount.Parse(storageConnectionString);
            }
            catch (FormatException)
            {
                Console.WriteLine(" Invalid storage account information provided. Please confirm the AccountName and AccountKey are valid in the app.config file - then restart the application.");
                throw;
            }
            catch (ArgumentException)
            {
                Console.WriteLine(" Invalid storage account information provided. Please confirm the AccountName and AccountKey are valid in the app.config file - then restart the sample.");
                Console.ReadLine();
                throw;
            }

            return storageAccount;
        }

        /// <summary>
        /// Create a table for the sample application to process messages in.
```

```csharp
/// </summary>
/// <returns> A CloudTable object </returns>
public static async Task <CloudTable> CreateTableAsync(string tableName)
{
    // Retrieve storage account information from connection string.
    CloudStorageAccount storageAccount = CreateStorageAccountFromConnectionString
(CloudConfigurationManager.GetSetting("StorageConnectionString"));

    // Create a table client for interacting with the table service
    CloudTableClient tableClient = storageAccount.CreateCloudTableClient();

    Console.WriteLine("Create a Table for the demo");

    // Create a table client for interacting with the table service
    CloudTable table = tableClient.GetTableReference(tableName);
    try
    {
        if (await table.CreateIfNotExistsAsync())
        {
            Console.WriteLine("Created Table named: {0}", tableName);
        }
        else
        {
            Console.WriteLine("Table {0} already exists", tableName);
        }
    }
    catch (StorageException)
    {
        Console.WriteLine("If you are running with the default configuration please make sure you have started the storage emulator. Press the Windows key and type Azure Storage to select and run it from the list of applications - then restart the sample.");
        Console.ReadLine();
        throw;
    }

    Console.WriteLine();
    return table;
}
```

CloudStorageAccount 方法用来创建 storage account，CreateTableAsync 方法用来创建 table。

8. SamplesUtils.cs 文件

```
using Microsoft.Azure;
using Microsoft.Azure.CosmosDB.Table;
using Microsoft.Azure.Storage;
using System;
using System.Threading.Tasks;
using TableStorage.Model;

namespace TableStorage
{
    class SamplesUtils
    {
        /// <summary>
        /// Demonstrate the most efficient storage query - the point query - where
        /// both partition key and row key are specified.
        /// </summary>
        /// <param name = "table"> Sample table name </param>
        /// <param name = "partitionKey"> Partition key - i.e., last name </param>
        /// <param name = "rowKey"> Row key - i.e., first name </param>
        /// <returns> A Task object </returns>
        public static async Task <CustomerEntity> RetrieveEntityUsingPointQueryAsync(CloudTable table, string partitionKey, string rowKey)
        {
            try
            {
                TableOperation retrieveOperation = TableOperation.Retrieve <CustomerEntity> (partitionKey, rowKey);
                TableResult result = await table.ExecuteAsync(retrieveOperation);
                CustomerEntity customer = result.Result as CustomerEntity;
                if (customer! = null)
                {
                    Console.WriteLine("\t{0}\t{1}\t{2}\t{3}", customer.PartitionKey, customer.RowKey, customer.Email, customer.PhoneNumber);
                }

                return customer;
            }
            catch (StorageException e)
            {
```

```csharp
            Console.WriteLine(e.Message);
            Console.ReadLine();
    throw;
        }
    }

    /// <summary>
    /// The Table Service supports two main types of insert operations.
    ///    1. Insert - insert a new entity. If an entity already exists with the same
    ///       PK + RK an exception will be thrown.
    ///    2. Replace - replace an existing entity. Replace an existing entity with a
    ///       new entity.
    ///    3. Insert or Replace - insert the entity if the entity does not exist, or
    ///       if the entity exists, replace the existing one.
    ///    4. Insert or Merge - insert the entity if the entity does not exist or, if
    ///       the entity exists, merges the provided entity properties with the already
    ///       existing ones.
    /// </summary>
    /// <param name = "table"> The sample table name </param>
    /// <param name = "entity"> The entity to insert or merge </param>
    /// <returns> A Task object </returns>
    public static async Task <CustomerEntity>
InsertOrMergeEntityAsync(CloudTable table, CustomerEntity entity)
    {
        if (entity == null)
        {
            thrownew ArgumentNullException("entity");
        }

        try
        {
            // Create the InsertOrReplace table operation
            TableOperation insertOrMergeOperation = TableOperation.InsertOrMerge(entity);

            // Execute the operation.
            TableResult result = await table.ExecuteAsync(insertOrMergeOperation);
            CustomerEntity insertedCustomer = result.Result as CustomerEntity;

            return insertedCustomer;
        }
```

```csharp
            catch (StorageException e)
            {
                Console.WriteLine(e.Message);
                Console.ReadLine();
                throw;
            }
        }


        /// <summary>
        /// Delete an entity
        /// </summary>
        /// <param name = "table"> Sample table name </param>
        /// <param name = "deleteEntity"> Entity to delete </param>
        /// <returns> A Task object </returns>
        public static async Task DeleteEntityAsync(CloudTable table, CustomerEntity deleteEntity)
        {
            try
            {
                if (deleteEntity == null)
                {
                    thrownew ArgumentNullException("deleteEntity");
                }

                TableOperation deleteOperation = TableOperation.Delete(deleteEntity);
                await table.ExecuteAsync(deleteOperation);
            }
            catch (StorageException e)
            {
                Console.WriteLine(e.Message);
                Console.ReadLine();
                throw;
            }
        }


        /// <summary>
        /// Check if given connection string is for Azure Table storage or Azure CosmosDB
        /// Table.
        /// </summary>
        /// <returns> true if azure cosmosdb table </returns>
        public static bool IsAzureCosmosdbTable()
        {
            string connectionString = CloudConfigurationManager.GetSetting("Storage-
```

```
ConnectionString");
            return !String.IsNullOrEmpty(connectionString) && connectionString.Contains("table.cosmosdb");
        }
    }
}
```

RetrieveEntityUsingPointQueryAsync 方法是用 point query 的方式取一个 Entity。InsertOrMergeEntityAsync 方法是插入或者更新一个 Entity。DeleteEntityAsync 是删除一个 Entity。

下面是从 AdvancedSamples.cs 出来的样例代码：

```
if (!SamplesUtils.IsAzureCosmosdbTable())
{
    // Create a SAS and try CRUD operations with the SAS.
    await AdvancedDataOperationsWithSasAsync(table);

    // Service Properties
    await ServicePropertiesSample(tableClient);

    // CORS
    await CorsSample(tableClient);

    // Service Stats
    await ServiceStatsSample(tableClient);

    // Table Acl
    await TableAclSample(table);

    // Create a SAS and try CRUD operations with the SAS and shared access policy on the table.
    await
    AdvancedDataOperationsWithSasAndSharedAccessPolicyOnTableAsync(table);
}
```

它演示了 Signed Access Signature、CRUD、Service Properties、CORS、Table Acl 等。

AdvancedSamples.cs 还有一个方法 GetTableSasUri，显示了如何创建 Shared Access signature。

```
/// <summary>
/// Returns a URI containing a SAS for the table.
/// </summary>
/// <param name = "table"> A CloudTable object. </param>
```

```csharp
/// <param name = "storedPolicyName"> A string containing the name of the stored access
///policy. If null, an ad-hoc SAS is created. </param>
/// <returns> A string containing the URI for the table, with the SAS token appended.
/// </returns>
private static string GetTableSasUri(CloudTable table, string storedPolicyName = null)
{
    string sasTableToken;

    // If no stored policy is specified, create a new access policy and define its
    // constraints.
    if (storedPolicyName == null)
    {
        // Note that the SharedAccessTablePolicy class is used both to define the
        // parameters of an ad-hoc SAS, and to construct a shared access policy
        // that is saved to the table's shared access policies.
        SharedAccessTablePolicy adHocPolicy = new SharedAccessTablePolicy()
        {
            // Permissions enable users to add, update, query, and delete entities
            // in the table.
            SharedAccessExpiryTime = DateTime.UtcNow.AddHours(24),
            Permissions = SharedAccessTablePermissions.Add |
            SharedAccessTablePermissions.Update |
            SharedAccessTablePermissions.Query |
            SharedAccessTablePermissions.Delete
        };

        // Generate the shared access signature on the table, setting the con-
        // straints directly
        // on the signature.
        sasTableToken = table.GetSharedAccessSignature(adHocPolicy, null);

        Console.WriteLine("SAS for table (ad hoc): {0}", sasTableToken);
        Console.WriteLine();
    }
    else
    {
        // Generate the shared access signature on the table. In this case, all of
        // the constraints for the shared access signature are specified on the
        // stored access policy, which is provided by name.
        // It is also possible to specify some constraints on an ad-hoc SAS and
        // others on the stored access policy.
        // However, a constraint must be specified on one or the other; it cannot be
        // specified on both.
```

```
            sasTableToken = table.GetSharedAccessSignature(null, storedPolicyName);

            Console.WriteLine("SAS for table (stored access policy): {0}", sasTable-
Token);

            Console.WriteLine();
        }

        // Return the URI string for the table, including the SAS token.
        return table.Uri + sasTableToken;
    }
```

此例的资源是一个 Table,那么这里得到的是一个 Service SAS,先要创建 SharedAccessTablePolicy,定义 Permission,然后再调用 Table 对象的 GetSharedAccessSignature 方法得到 Service SAS。

这个例子用的是 Table,其他如 Blob、File 等是与之类似的。

2.2　Azure Blob Storage

Azure Blob Storage 是一个存储二进制文件的地方。这有点类似 Oracle 的 Blob 字段,专用于存储二进制文件。需要建立 Blob Account 才能进行 Blob 的存储,如图 2-27 所示。

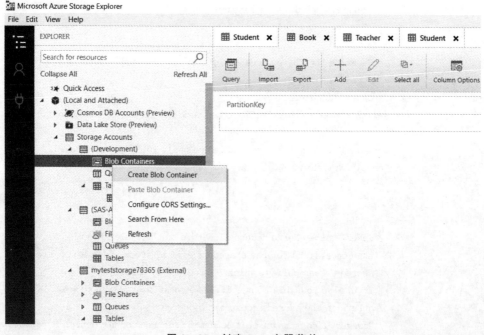

图 2-27　创建 Blob 容器菜单

右键单击 Blob Containers,出现菜单,选择 Create Blob Container,然后出现一个输入 Container 名字的框。输入名字即可,如图 2-28 所示。

图 2-28 创建 Blob 容器

Blob Containers 建立好之后,就可以往里面放文件了,如图 2-29 所示。

图 2-29 创建好的 Blob 容器

单击 Upload,出现如图 2-30 所示的下拉菜单。

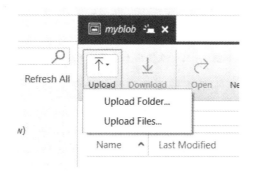

图 2-30 上传文件菜单

选择要上传的文件夹或者多个文件，假设选择上传文件夹，出现如图 2-31 所示的窗口。

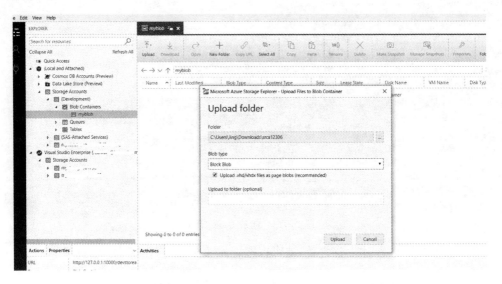

图 2-31　上传文件夹的窗口

选择一个文件夹之后，一般不需要改其他设置，单击 Upload 按钮，此文件夹就会被上传。可以点进去看文件夹中的文件，如图 2-32 所示。

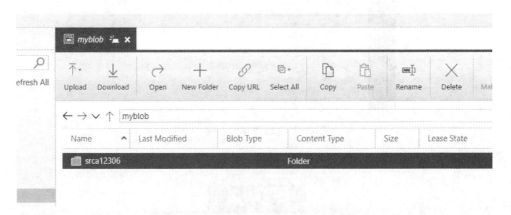

图 2-32　文件夹上传后的 Blob 容器

这个窗口里的按钮基本都跟资源管理器的操作类似，无非是文件操作，很好理解。差别比较大的是 Make Snapshot 和 Manage Snapshots 按钮。Make Snapshot 是创建快照，Manage Snapshots 是管理快照。所谓快照就相当于一个版本，Azure 云平台帮助建立一个快照，即把此文件的当前版本暂存起来，后来当需要恢复的时候可根据快照时间来恢复。这样，一个文件在 Azure 云平台里就可以有很多快照，做完快照后就可以根据实际需要来改文件，当需要恢复成某个版本的样子时，就按快照时间来选择恢复,如图 2-33 所示。

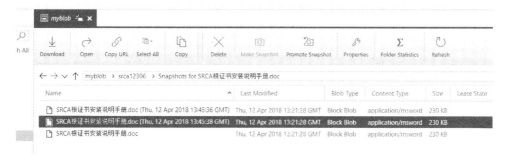

图 2-33 选择已保存的快照

单击一次 Make Snapshot,就会将文件进行一次快照,同时记录时间。

下面来看一个例子,可从 https://azure.microsoft.com/en-us/resources/samples/storage-blob-dotnet-getting-started/下载代码。

运行起来如图 2-34 所示。

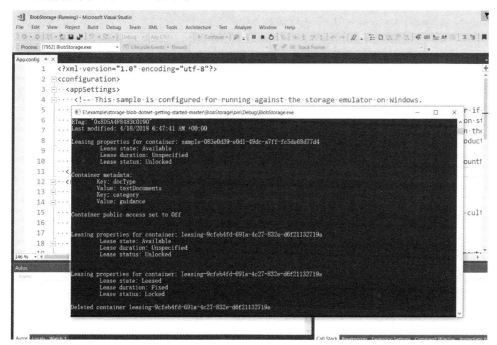

图 2-34 运行 Blob 容器演示程序

它自动创建 Container,上传或者创建 Block Blob、Page Blob、Block Directory,然后又删除。

下面是基本的 Block Blob 的操作,分别为创建 Container、上传 Block Blob,然后列 Blob 目录。

```csharp
/// <summary>
/// Basic operations to work with block blobs
/// </summary>
/// <returns> A Task object. </returns>
private static async Task BasicStorageBlockBlobOperationsAsync()
{
    const string ImageToUpload = "HelloWorld.png";
    string containerName = ContainerPrefix + Guid.NewGuid();

    // Retrieve storage account information from connection string
    CloudStorageAccount storageAccount = Common.CreateStorageAccountFromConnectionString();

    // Create a blob client for interacting with the blob service.
    CloudBlobClient blobClient = storageAccount.CreateCloudBlobClient();

    // Create a container for organizing blobs within the storage account.
    Console.WriteLine("1. Creating Container");
    CloudBlobContainer container = blobClient.GetContainerReference(containerName);
    try
    {
        // The call below will fail if the sample is configured to use the storage emula-
        // tor in the connection string, but the emulator is not running.
        // Change the retry policy for this call so that if it fails, it fails quickly.
        BlobRequestOptions requestOptions = new BlobRequestOptions() { RetryPolicy = new NoRetry() };
        await container.CreateIfNotExistsAsync(requestOptions, null);
    }
    catch (StorageException)
    {
        Console.WriteLine("If you are running with the default connection string, please make sure you have started the storage emulator. Press the Windows key and type Azure Storage to select and run it from the list of applications - then restart the sample.");
        Console.ReadLine();
        throw;
    }

    // To view the uploaded blob in a browser, you have two options. The first option is
    // to use a Shared Access Signature (SAS) token to delegate access to the resource.
    // See the documentation links at the top for more information on SAS. The second
    // approach is to set permissions to allow public access to blobs in this container.
    // Uncomment the line below to use this approach. Then you can view the image
```

```csharp
// using: https://[InsertYourStorageAccountNameHere].blob.core.windows.net/
// democontainer/HelloWorld.png await container.SetPermissionsAsync(new BlobCon-
// tainerPermissions { PublicAccess = BlobContainerPublicAccessType.Blob });

// Upload a BlockBlob to the newly created container
Console.WriteLine("2. Uploading BlockBlob");
CloudBlockBlob blockBlob = container.GetBlockBlobReference(ImageToUpload);

// Set the blob's content type so that the browser knows to treat it as an image.
blockBlob.Properties.ContentType = "image/png";
await blockBlob.UploadFromFileAsync(ImageToUpload);

// List all the blobs in the container.
/// Note that the ListBlobs method is called synchronously, for the purposes of the
/// sample. However, in a real-world application using the async/await pattern,
/// best practices recommend using asynchronous methods consistently.
Console.WriteLine("3. List Blobs in Container");
foreach (IListBlobItem blob in container.ListBlobs())
{
    // Blob type will be CloudBlockBlob, CloudPageBlob or CloudBlobDirectory
    // Use blob.GetType() and cast to appropriate type to gain access to properties
    // specific to each type
    Console.WriteLine("- {0} (type: {1})", blob.Uri, blob.GetType());
}

// Download a blob to your file system
Console.WriteLine("4. Download Blob from {0}", blockBlob.Uri.AbsoluteUri);
await blockBlob.DownloadToFileAsync(string.Format("./CopyOf{0}", ImageToUpload), FileMode.Create);

// Create a read-only snapshot of the blob
Console.WriteLine("5. Create a read-only snapshot of the blob");
CloudBlockBlob blockBlobSnapshot = await blockBlob.CreateSnapshotAsync(null, null, null, null);

// Clean up after the demo. This line is not strictly necessary as the container is
// deleted in the next call.
// It is included for the purposes of the example.
Console.WriteLine("6. Delete block blob and all of its snapshots");
await blockBlob.DeleteIfExistsAsync(DeleteSnapshotsOption.IncludeSnapshots, null, null, null);
```

```
            // Note that deleting the container also deletes any blobs in the container, and
            // their snapshots.
            // In the case of the sample, we delete the blob and its snapshots, and then the
            // container, to show how to delete each kind of resource.
            Console.WriteLine("7. Delete Container");
            await container.DeleteIfExistsAsync();
        }
```

下面是在 Storage Explorer 中看到的中间景象,如图 2-35 所示。

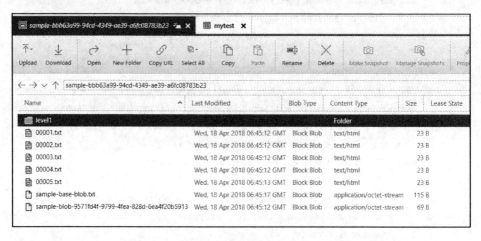

图 2-35 Blob 样例程序产生的中间结果

Advanced. cs 里演示了一些 Blob Storage 的高级用法,如 Container Metadata、Blob Metadata、SAS、CORS 等。

```
    /// <summary>
    /// Calls samples that demonstrate how to work with blobs.
    /// </summary>
    /// <param name = "container"> A CloudBlobContainer object. </param>
    /// <returns> A Task object. </returns>
    private static async Task CallBlobSamples(CloudBlobContainer container)
    {
        // Create a blob with a random name.
        CloudBlockBlob blob = await CreateRandomlyNamedBlockBlobAsync(container);

        // Get a reference to the blob created above from the server.
        // This call will fail if the blob does not yet exist.
        await GetExistingBlobReferenceAsync(container, blob.Name);

        // Create a specified number of block blobs in a flat structure.
        await CreateSequentiallyNamedBlockBlobsAsync(container, 5);
```

```
// List blobs in a flat listing
await ListBlobsFlatListingAsync(container, 10);

// Create a specified number of block blobs in a hierarchical structure.
await CreateNestedBlockBlobsAsync(container, 4, 3);

// List blobs with a hierarchical listing.
await ListBlobsHierarchicalListingAsync(container, null);

// List blobs whose names begin with "s" hierarchically, passing the container name
// as part of the prefix.
ListBlobsFromServiceClient(container.ServiceClient, string.Format("{0}/s", container.Name));

// List blobs whose names begin with "0" hierarchically, passing the container name
// as part of the prefix.
await ListBlobsFromServiceClientAsync ( container.ServiceClient, string.Format
("{0}/0", container.Name));

// Create a snapshot of a block blob.
await CreateBlockBlobSnapshotAsync(container);

// Copy a block blob to another blob in the same container.
await CopyBlockBlobAsync(container);

// To create and copy a large blob, uncomment this method.
// By default it creates a 100 MB blob and then copies it; change the value of the
// sizeInMb parameter to create a smaller or larger blob.
// await CopyLargeBlob(container, 100);

// Upload a blob in blocks.
await UploadBlobInBlocksAsync(container);

// Upload a 5 MB array of bytes to a block blob.
await UploadByteArrayAsync(container, 1024 * 5);
}
```

Azure Blob Storage 是一个相对比较便宜的文件存储，可以作为大量数据文件的存储。有的应用需要有大量的数据文件来交互，每天的数据文件产生很多，那么 Azure Blob Storage 就是一个很好的存储。其一个优点是安全，可以用 Shared Access Signature 控制各个访问者的权限，让各个访问者只能访问能访问的那部分 Blob。另一个优点是它有备份，即使发生硬盘损坏这种意外，也可很容易地恢复。还有就是它可以很好地跨平台，能允许各种编程语言的客户端连接、操作 Blob。

2.3　Azure Queue Storage

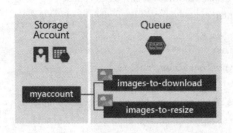

图 2-36　Azure Queue 的层级概念图

Azure Queue Storage 也是 Azure Storage 的一个组成部分。它可实现一些消息队列功能，可以允许建立消息队列，同时支持异步消息机制，允许多种客户端连接，不论客户端是 PC、服务器还是一个移动设备。

Azure Queue 的层级概念如图 2-36 所示。

这里 images-to-download 和 images-to-resize 都是 Queue 的名字。它们在 Storage Account 之下。

在 Storage Account 有了之后，就要创建 Queue，如图 2-37 所示。

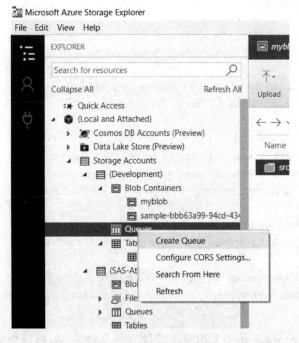

图 2-37　Create Queue 菜单

创建好后如图 2-38 所示。

这样就可以往 Queue 中加消息了。单击 Add Message，出现如图 2-39 所示界面。

图 2-38 创建好的 Queue

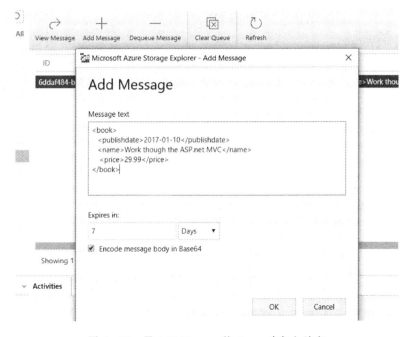

图 2-39 用 Add Message 往 Queue 中加入消息

消息是文本的,内容可以自己定。过期时间默认为 7 天。消息列表如图 2-40 所示。

Dequeue Message 将队列里第一个消息移出队列。

样例代码可下载：https://azure.microsoft.com/en-us/resources/samples/storage-queue-dotnet-getting-startcd/。

解压后如图 2-41 所示。

GettingStarted.cs 是一些使用 Azure Queue Storage 的基本例子。

图 2-40 Queue 中的消息列表

图 2-41 Azure Queue 的样例程序,需要修改连接串

```
/// <summary>
/// Test some of the file storage operations.
/// </summary>
public async Task RunQueueStorageOperationsAsync()
{
    // Create the queue name -- use a guid in the name so it's unique.
    string queueName = "demotest-" + System.Guid.NewGuid().ToString();

    // Create or reference an existing queue.
    CloudQueue queue = CreateQueueAsync(queueName).Result;

    // Demonstrate basic queue functionality.
    await BasicQueueOperationsAsync(queue);

    // Demonstrate how to update an enqueued message
    await UpdateEnqueuedMessageAsync(queue);

    // Demonstrate advanced functionality such as processing of batches of messages
    await ProcessBatchOfMessagesAsync(queue);

    // When you delete a queue it could take several seconds before you can recreate a
    // queue with the same name - hence to enable you to run the demo in quick succession
    // the queue is not deleted. If you want to delete the queue uncomment the line of
```

```csharp
    // code below.
    await DeleteQueueAsync(queue);
}
```

Advanced.cs 演示高级一些的 Azure Queue Storage 用法：

```csharp
/// <summary>
/// Test some of the queue storage operations.
/// </summary>
public async Task RunQueueStorageAdvancedOpsAsync()
{
    try
    {
        //***** Setup *****//
        Console.WriteLine("Getting reference to the storage account.");

        // Retrieve storage account information from connection string
        // How to create a storage connection string - http://msdn.microsoft.com/
        // en-us/library/azure/ee758697.aspx
        CloudStorageAccount storageAccount = Common.CreateStorageAccountFromConnectionString(CloudConfigurationManager.GetSetting("StorageConnectionString"));

        Console.WriteLine("Instantiating queue client.");
        Console.WriteLine(string.Empty);

        // Create a queue client for interacting with the queue service.
        CloudQueueClient cloudQueueClient = storageAccount.CreateCloudQueueClient();

        // List queues
        await ListQueuesSample(cloudQueueClient);

        // Service properties
        await ServicePropertiesSample(cloudQueueClient);

        // CORS Rules
        await CorsSample(cloudQueueClient);

        // Service Stats
        await ServiceStatsSample(cloudQueueClient);

        // Queue Metadata
        await QueueMetadataSample(cloudQueueClient);
```

```
        // Queue Acl
        await QueueAclSample(cloudQueueClient);
    }
    catch (Exception ex)
    {
        Console.WriteLine("Exception thrown. Message = {0}{1} Strack Trace = {2}", ex.
Message, Environment.NewLine, ex.StackTrace);
    }
}
```

其中,包括 List queue、Service properties、CORS Rules、Service Stats、Queue Metadata、Queue Acl。

Azure Queue Storage 和 Azure Service Bus Quene 的区别如下:

Azure Queue Storage 相对 Azure Service Bus Queue 来说是一个比较便宜的解决方案。另外,Azure Queue Storage 不支持事务机制,而 Azure Service Bus Queue 支持事务机制。我们需要知道什么时候使用 Azure Queue Storage,什么时候使用 Azure Service Bus Queue。

在下面的情况,选择 Azure Storage Queue:

① 队列成长超过 80 GB,最大容量不超过 200 TB,而且信息存在时间不超过 7 天。

② 单条信息的长度不超过 64 kbit。

在下面的情况,选择 Azure Service Bus Queue:

① 队列成长不会超过 80 GB,信息可以存在超过 7 天。

② 单条信息可以超过 64 kbit(但是不太可能接近 256 kbit 的限制)。

③ 需要保证绝对的 First-In-First-Out。

④ 需要和.NET Framework 中的 WCF 整合。

⑤ 需要批次发布及取用信息。

详情请查阅:https://docs.microsoft.com/en-us/azure/service-bus-messaging/service-bus-azure-and-service-bus-queues-compared-contrasted。

2.4　Azure File Storage

Azure File Storage 是一个通过 Server Message Block(SMB)协议提供云端文件共享的服务。通过 File Storage 共享的文件,能够被加载为云端或本地主机的磁盘,应用程序可以通过文件 API 像访问本地文件一样访问这些文件。

下面是 Azure File Storage 典型的应用场景:

● 轻松迁移那些有磁盘读/写操作的应用到云端。不用修改程序,只要通过

File Storage 加载相应的文件即可。
- 存放共享的应用程序配置文件。
- 存放日志等应用程序诊断数据。
- 存放管理员的常用工具。

图 2-42 描述了 File Storage 的基本组织结构。

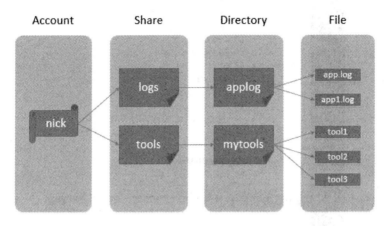

图 2-42 File Storage 的基本组织结构

Account 是最高一层。这个和 Blob 或者 Table 类似。

Share 是第二层,将某些资源共享出来,给这个共享起了一个名字。

在 Share 之下就是 Directory 和 File 了。这个就是树形结构,很好理解。

与 Blob Storage 相似,File 中的每个文件都可以通过 URL 来访问。URL 的详细格式如下:

https://<storage account>.file.core.windows.net/<share>/<directory/directories>/<filename>

注意,这个域名后缀 core.windows.net 是适用于 Azure 全球环境的。如果是别的环境,则需要改成相应的环境对应的域名。

这个 File Storage 在 Storage Emulator 中没有,只是在 Azure 平台中才有。

连接外部 Azure Storage 之后,右键单击 File Share,出现如图 2-43 所示的 Create File Share 菜单。

创建好之后出现如图 2-44 所示界面。

这与 Blob 很像,可以上传文件、文件夹,也有 Upload 按钮、Download 按钮,这两个按钮和 Blob 的类似,这里就不重复介绍了。其他按钮如 New Folder、Copy URL、Select All、Copy、Paste、Rename 都比较直观,从名字就可理解,略过。下面介绍 Connect VM 按钮,单击 Connect VM,出现如图 2-45 所示界面。

它给出一个命令,在对应的虚拟机中运行,就可将此 File Share 作为某个盘符来访问。这个 File Share 因为支持 SMB 协议,所以所有支持 SMB 的客户端都可以将

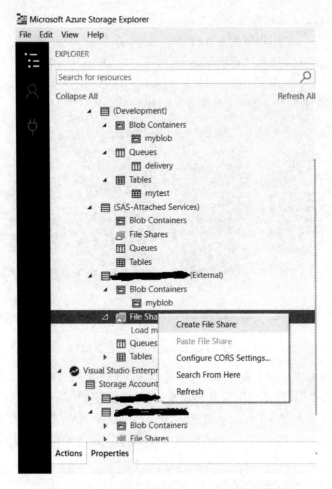

图 2-43　创建文件共享 Create File Share 菜单

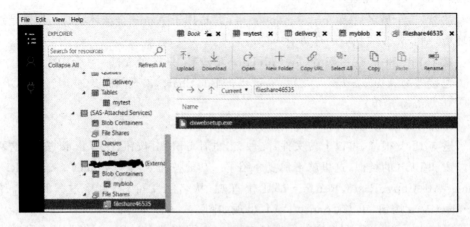

图 2-44　创建好的文件共享

图 2-45 将文件共享连接到一个虚拟机

File Share 作为一个远程文件系统来访问。现在支持 SMB 协议的客户端包括 Windows、Linux。

样例代码可从 https://azure.microsoft.com/en-us/resources/samples/aci-dot-net-manage-container-instances-2/下载。

代码文件 Program.cs 片段：

```
private async static void CreateFileShare(string saName, string storageAccountKey, string shareName)
        {
                CloudFileShare cloudFileShare = CloudStorageAccount.Parse(
$"DefaultEndpointsProtocol = https; AccountName = {saName}; AccountKey = {storageAccountKey}; EndpointSuffix = core.windows.net")
                .CreateCloudFileClient()
                .GetShareReference(shareName);
            await cloudFileShare.CreateAsync();
        }
```

创建 File Share。这个程序在创建 File Share 之后再去另外一个 Docker 虚拟机中把此 File Share 映射成一个盘。这样此 File Share 就能以与 Windows 文件系统差不多的形式访问了。

2.5 Azure Storage CDN

Azure Storage CDN 是将 Azure Storage 和 Azure CDN 结合起来。CDN 是 Content Delivery Network 的缩写，即服务商在用户附近的网络中心建立了就近的

服务器,如果将一些静态的内容和不常变的内容放到这些就近的服务器上,则当加载这些静态资源的时候就会比较快。将 Azure Storage 中的内容放到 Azure CDN 里时,Storage 中的内容就会被复制到离用户最近的服务器中去。用户每次访问的时候,就访问最近的服务器,其响应速度会比较快。Azure CDN 在背后做很多事情,帮助把内容更新传播到其他服务器,以便其他服务器也能看到更新。租用 Azure CDN,就会获得更快的速度,同时也得付出一些资费。

Azure CDN 需要先建立好 Storage account,然后才可设置 CDN,之前已经讲了如何建立 Storage account,这里就不重复了。在 Azure Portal 中打开此 Storage account,在左边菜单里找到 Azure CDN,如图 2-46 所示。

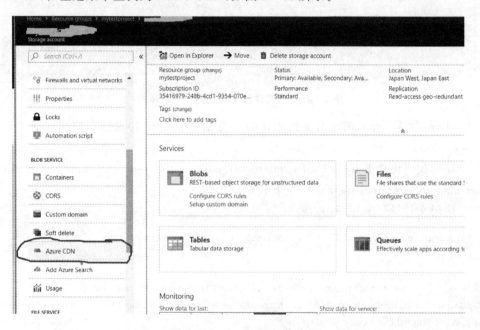

图 2-46 Azure CDN 菜单

单击 Azure CDN,右边会出现如图 2-47 所示的界面。

如果第一次创建 CDN,则最好创建新的 CDN profile;如果不是第一次,可以选择一个已经存在的 CDN profile,然后选择 Pricing tier。它有 Premium Verizon、Standard Verizon、Standard Akamai 三个选择,其价格 Premium 的会更高一些,Standard 的两个差不多。可以仔细阅读其价格说明,来选择一个。然后是输入 CDN endpoint name。这个应尽量想一个长一点的名字,免得跟其他人的重复。Origin hostname 是不需要改的。单击 Create 按钮,此 Endpoint 就会出现在 Endpoint 的列表中,如图 2-48 所示。

还需要将 CDN 设置一下才能让 CDN 用起来更好。不设置也可以,只是可能不是特别适用。

图 2 – 47　单击 Azure CDN 菜单后的界面

图 2 – 48　Azure CDN 创建好之后的界面

单击建立的 Endpoint，右边出现配置页面，如图 2-49 所示。

图 2-49　Azure CDN 的 Endpoint 设置界面

在此设置界面可以设置 GeoFilter、Compression（压缩）、Query String 等，也可以设置一个自定义的域名。

设置好以后，可以通过这样一个 URL 来访问 Azure Storage CDN：

http://<EndpointName>.azureedge.net/<myPublicContainer>/<BlobName>。

如果不需要 Azure Storage CDN，则可以将 CDN 的 Endpoint 删除。这样之后就没有 CDN 了。

第 3 章

App Service

3.1 App Service 简介

Azure App Service 是一个提供 Web App 的服务,可以支持各种编程语言构建 Web Services——微服务,当然,微软对自家开发的编程语言 C♯和. net framework 支持最好。App Service 底层有许多设施,如虚拟网络、虚拟的负载均衡、虚拟的防火墙、虚拟机,这些底层的资源共同在 App Service 的调度下,构成了一个提供 Web App 的服务。它可以供开发者部署不同类型的应用;同时,它默认具备基本的网络防火墙功能,具备基本的负载均衡功能。其目前支持的编程语言或者平台包括:.NET、.NET Core、Java、Ruby、Node. js、PHP、Python,可以方便地进行 DevOps 活动,如与 VSTS、Jenkins 等集成;与 Git 服务器、Docker 服务器等集成。App Service 可以是一个 Web App,也可以是一个 Mobile App,或者是一个微服务,具体的表现形式得看应用的要求而定。

3.2 用 ASP.NET Core 开发一个 App Service

用 VS 2017 安装程序安装 ASP.NET and web development 和 Azure development,如图 3-1 所示。

在 New Project 对话框中选择 Visual C♯→Web→ASP.NET Core Web Application,如图 3-2 所示。

选择 Web Application 模板,保证 Authentication 设置成 No Authentication。然后单击 OK 按钮,如图 3-3 所示。

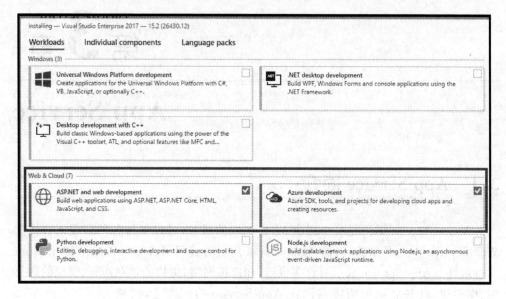

图 3-1　VS 2017 安装时须选择的部件

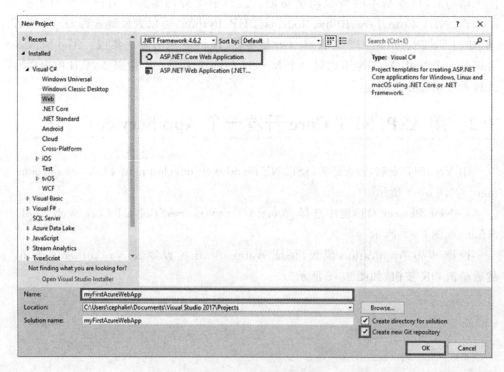

图 3-2　新建工程时选择 Web 类的工程

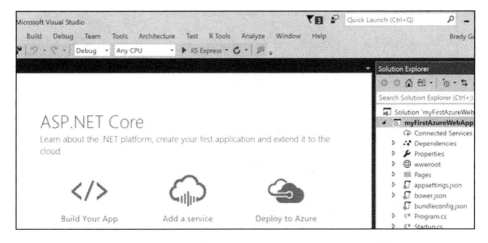

图 3-3 创建好 ASP.NET Core Web 应用后的工程

3.3 调 试

1. Program.cs 文件

```
namespace myFirstAzureWebApp
{
    public class Program
    {
        public static void Main(string[] args)
        {
            BuildWebHost(args).Run();
        }

        public static IWebHost BuildWebHost(string[] args) =>
            WebHost.CreateDefaultBuilder(args)
                .UseStartup<Startup>()
                .Build();
    }
}
```

BuildWebHost 方法用来创建一个 WebHost，并且定义一个起始配置(Startup 类)。BuildWebHost(args).Run()的 Run 方法是 IWebHost 类的方法。

2. Startup.cs 文件

```
namespace myFirstAzureWebApp
{
```

```
public class Startup
{
    public Startup(IConfiguration configuration)
    {
        Configuration = configuration;
    }

    public IConfiguration Configuration { get; }

    // This method gets called by the runtime. Use this method to add services to the
    // container.
    publicvoid ConfigureServices(IServiceCollection services)
    {
        services.AddMvc();
    }

    // This method gets called by the runtime. Use this method to configure the HTTP
    // request pipeline.
    public void Configure(IApplicationBuilder app, IHostingEnvironment env)
    {
        if (env.IsDevelopment())
        {
            app.UseBrowserLink();
            app.UseDeveloperExceptionPage();
        }
        else
        {
            app.UseExceptionHandler("/Error");
        }

        app.UseStaticFiles();

        app.UseMvc();
    }
}
```

Startup 类有这个 IConfiguration 参数，当运行创建此类的实例时，传入此参数。ConfigureServices 是用来设置服务的，运行时调用它。此方法在 ServiceCollection 中加入 Mvc 服务。之后在 Configure 中需要用 IApplicationBuilder.UseMvc 方法来启动 Mvc。env.IsDevelopment()是对运行环境进行检测，如果是开发环境，则可显示给开发员看的错误页面；如果是测试或者生产环境，则启用 Exception 路径/Error。

代码不多,可以在其中一些行打上断点。选菜单 Debug→Start Debugging,VisualStudio 会启动一个 IISexpress。这样是在本地运行。还可以一行一行代码地调试,如图 3-4 所示。

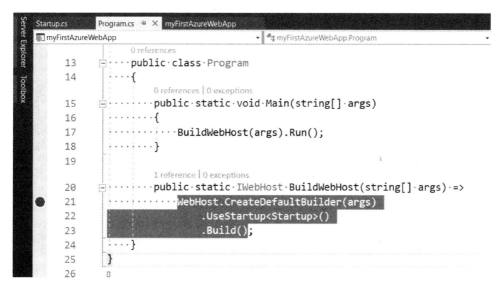

图 3-4　在 ASP.NET Core 程序中设置断点

运行后页面如图 3-5 所示。

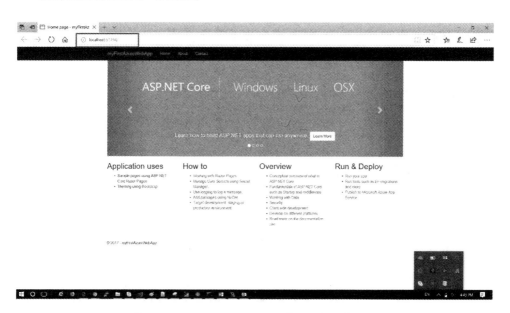

图 3-5　ASP.NET Core Web 应用运行后的页面

3.4 部　署

部署到 Azure 环境中去。在 Azure 环境中运行此应用程序，右键单击工程名，出现如图 3-6 所示的菜单。

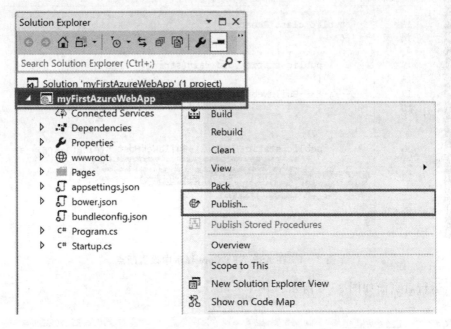

图 3-6　工程上的 Publish 菜单

选择 Publish 之后，出现如图 3-7 所示的界面。

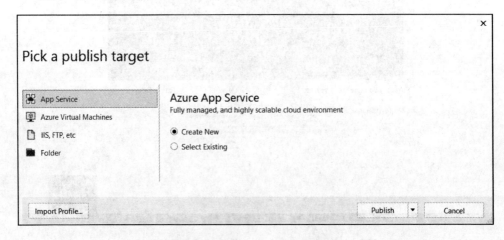

图 3-7　选择 Publish 后的对话框

单击 Publish 按钮，出现如图 3-8 所示的界面。

图 3-8 在 Azure 环境中创建新的 App Service

注意到此界面里有一个 Create a SQL Database 的链接，可以用来创建对应的 SQL Database。

单击 Create 按钮，系统需要一些时间。大约几分钟，这些资源在 Azure 中创建好之后，一个浏览器窗口会自动打开。注意到此时的 URL 变成了 http://myfirstazurewebapp20180424045034.azurewebsites.net/，它是一个部署在 Azure Web Site 中的应用。当然了，这个应用是根据 Visual Studio 模板创建的，相对比较简单。这就可以作为我们第一个部署的 App Service。之后可以继续修改，再单击 Publish 按钮。这样就把新版本发布到了 Azure 平台上。页面运行如图 3-9 所示。

回到 Visual Studio，可以看到此工程的 Publish 页面。刚刚在发布后建立了一个 Profile，所有部署的详细信息都已记录下来。这样下次就可以用已保存的 Publish Profile 文件。基于之前保存的 Publish Profile 文件，可以稍微修改后再进行下一次的发布，如图 3-10 所示。

打开 Azure Portal，查看刚刚创建的资源，如图 3-11 所示。

第一项就是 All services，单击它，查看详情，如图 3-12 所示。

微软 Azure 实战参考

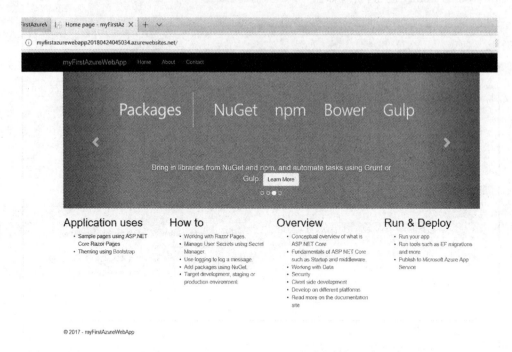

图 3-9　App Service 样例程序运行后的界面

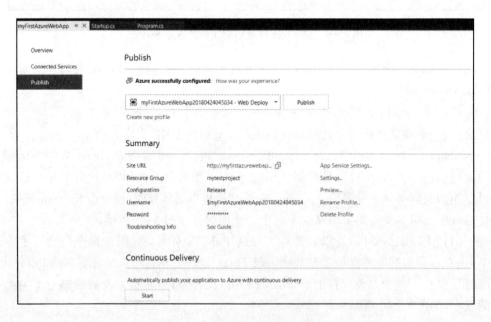

图 3-10　Publish Profile 的界面

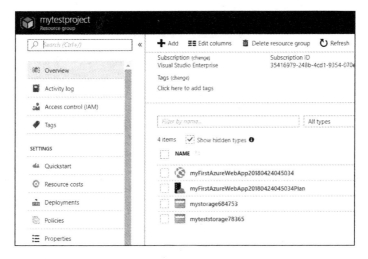

图 3-11 部署好的资源

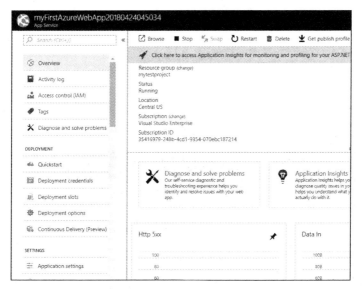

图 3-12 查看此 All services 的详情

进入详情页面。注意此菜单,如图 3-13 所示。

单击 Deployment credentials,这里设置部署用的 FTP 账号和密码,如图 3-14 所示。

选择 Deployment slots,默认是没有定义任何 Deployment slots。故创建一个,如图 3-15 所示。

单击 Add Slot,出现如图 3-16 所示的界面。

此处可选择已有的 Configuration Source。如果需要不同的 slot 的不同配置,则可另外建一个。部署槽建立好之后如图 3-17 所示。

图 3-13 此 App Service 的菜单

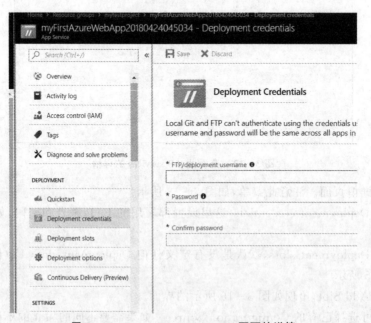

图 3-14 Deployment credentials 页面的详情

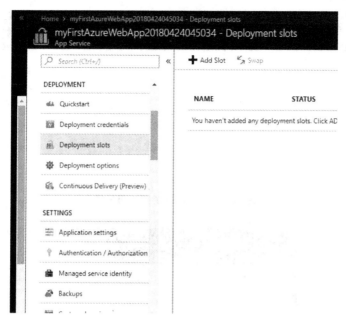

图 3-15　查看部署槽

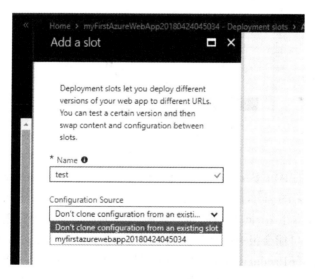

图 3-16　添加一个部署槽的界面

建立好两个以上的 slot 后,就可选择部署到哪个 slot 中;同时,这些 slot 还可以切换。可以部署不同的版本到不同的 slot 中。在生产环境中部署新版本,可以将新版本部署到一个 slot,然后切换作为当前起作用的 slot。单击 Swap,出现如图 3-18 所示的界面。

选一下 Source 和 Destination,然后单击 OK 按钮。目前可以看到建立了三个

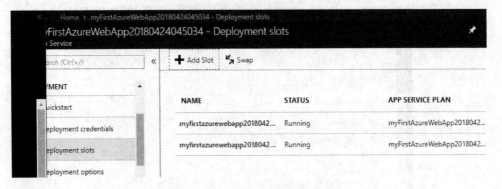

图 3-17 添加好了新的部署槽

图 3-18 切换部署槽

slot,在读者自己的 App Service 可以有两个、三个或者更多的 slot,这些 slot 可以做不同开发阶段部署的 slot,然后根据不同开发阶段来进行切换。这个 slot 切换可以快速地让某个版本生效,同时一般还不影响另外的版本。需要注意,这个 Preview Changes 部分,有可能会报一些警告。也可以选择 Swap with preview。一般是这样的模式:别的 slot→Production,Production 是默认就有的一个 slot,每创建一个 slot,就会有一个对应的 Web App 建立,如图 3-19 所示。

这样一个 Web App 保证了配置完全是分隔开的,版本之间切换是互不影响的。Application settings 菜单如图 3-20 所示。

这个菜单里可以设置链接串、路径,或者其他的一些配置信息。也可以打开或者关闭 Remote Debugging,如图 3-21 所示。

还有一些一般的 Settings,如 Java 版本、PHP 版本等。

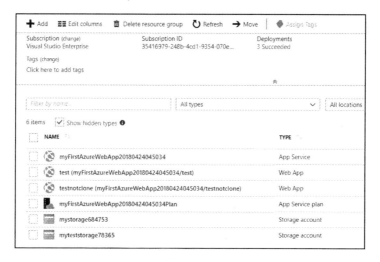

图 3-19 Production 槽和其他部署槽的区别

图 3-20 Application settings 菜单

Authentication/Authorization 菜单是用来做登录和应用中授权的,支持 Azure Active Directory、Google、Facebook 或者 Microsoft 账号,如图 3-22 所示。

这样 App Service 可以支持不同的登录方式,适应面还是比较广的。可以根据需要来选择一个登录方式。有了登录方式之后,就可以给每个登录账号分配权限,比如数据权限、菜单权限、工作流审批权限等。这些都可在应用中进行配置。

Custom domains 菜单如图 3-23 所示。

自定义域名可以让 App Service 以一个更容易记住的名字访问。如果没有域名,可以单击 Buy Domain 去买一个域名。

SSL settings 菜单如图 3-24 所示。

图 3-21　Application settings 页面中的一些设置

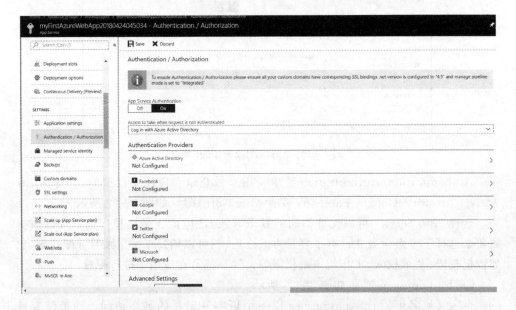

图 3-22　Authentication/Authorization 菜单所显示的相关外部 ID 提供者列表

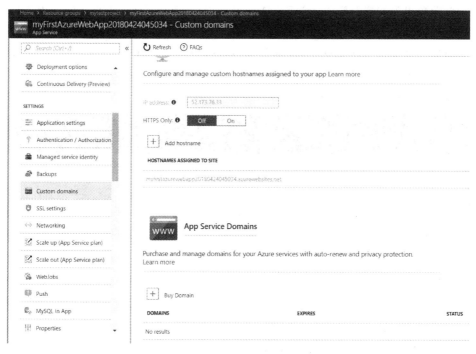

图 3-23　Custom domains 菜单

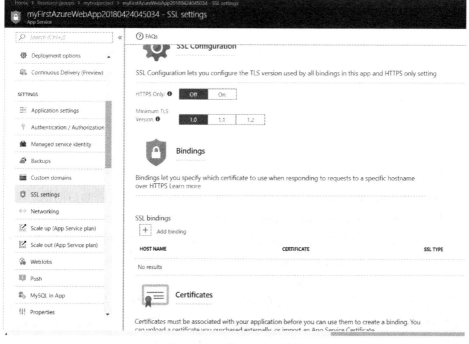

图 3-24　SSL settings 菜单

此菜单可设置安全证书,然后此 App Service 就可以用 https 来访问。由于很多手机都强制使用 https 协议,所以现在很多网站都必须安装一个安全证书。在这个页面可以管理买来的网站证书,同时设置 TLS 的版本,如 1.0、1.1、1.2。也可设置是否 https only,即是否只允许 https。

Networking 菜单如图 3-25 所示。

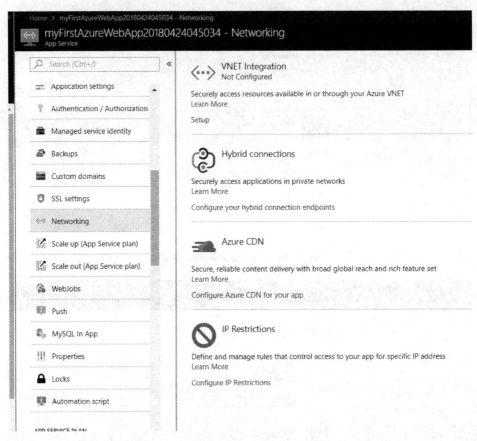

图 3-25　Networking 菜单,用来设置 App Service 的网络

在此也可以设置虚拟网络,比如此 App Service 是作为企业内部访问的,不对外公开,故可以建立 VPN,此 App Service 可以像企业内容网络里的服务器一样。还有可以将静态内容放置到 CDN,加速静态内容的访问。另外,可以用 IP 限制来阻止某些发起网络攻击的 IP 访问此 App Service。

Scale up 菜单如图 3-26 所示。

Scale up 菜单是当服务器不能负担全部的访问压力时使用的。首先可以考虑升级硬件,即换一个更高级的硬件、更高级的 CPU、更多的 CPU、更多的内存、更大更快的硬盘等。

WebJobs 菜单如图 3-27 所示。

App Service

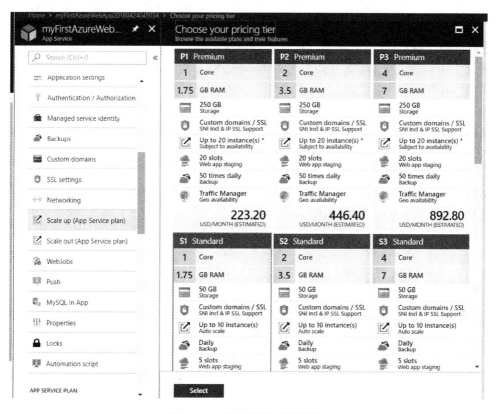

图 3-26 垂直扩展机器的界面

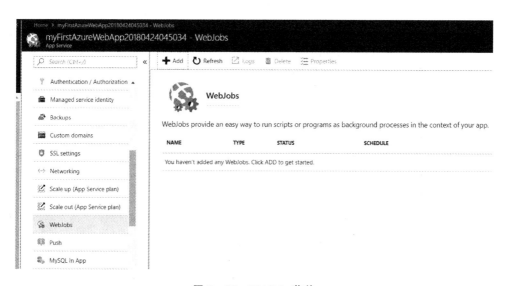

图 3-27 WebJobs 菜单

WebJobs 用来放置一些在 Web App 之外运行的任务，通常不需要用户界面，就一个命令行，这样可以设计一些数据处理的程序放到 WebJobs 中来运行。单击 Add 按钮，右边弹出一个窗口，如图 3-28 所示。

首先输入 WebJob 的名字（比如 test），然后选择一个执行文件，比如一个 .net 程序、一个 Python 程序、一个 Node.js 程序或者 Java 程序等。然后分别选择 Type 和 Scale。Type 的默认是 Continues，即一直运行。另外一种是触发，条件是时间点或者手动触发，如图 3-29 所示。

图 3-28　添加 WebJob

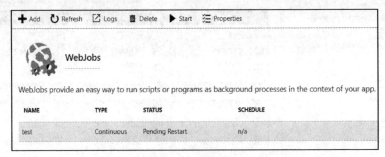

图 3-29　WebJobs 创建好之后的列表

触发条件可以选一个时间，此时间用一个类似 CRON 的格式来指定，如图 3-30 所示。

Visual Studio 可以很好地支持开发 WebJob 应用，如图 3-31 所示。

创建项目时，可以在 Cloud 项中找到 Azure WebJob(.NET Framework)。单击 OK 按钮后，项目即创建好。可以来看看代码：

```
//Program.cs 文件

using System;
using System.Collections.Generic;
using System.Linq;
using System.Text;
using System.Threading.Tasks;
using Microsoft.Azure.WebJobs;

namespace WebJob1
{
    // To learn more about Microsoft Azure WebJobs SDK, please see https://go.microsoft.
```

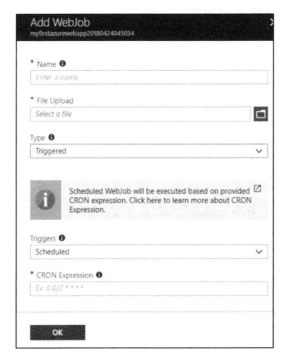

图 3-30　用 CRON 表达式触发的 WebJob

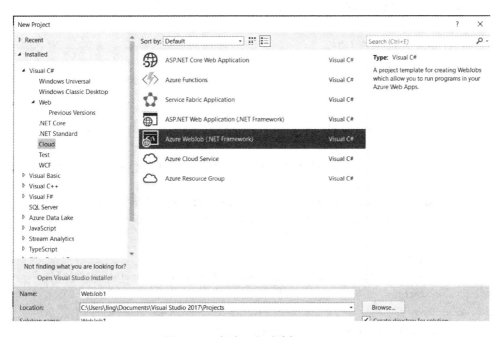

图 3-31　新建工程时选择 WebJob

```
// com/fwlink/? LinkID = 320976
    classProgram
    {
        // Please set the following connection strings in app.config for this WebJob to run:
        // AzureWebJobsDashboard and AzureWebJobsStorage
        static void Main()
        {
            var config = new JobHostConfiguration();

            if (config.IsDevelopment)
            {
                config.UseDevelopmentSettings();
            }

            var host = new JobHost(config);
            // The following code ensures that the WebJob will be running continuously
            host.RunAndBlock();
        }
    }
}
```

说明中要求设置 app.config 文件中的 Azure WebJobs Dashboard 和 Azure WebJobs Storage。

打开 app.config 看一下:

```xml
<?xmlversion = "1.0"encoding = "utf-8" ?>
<configuration>
    <connectionStrings>
        <!-- The format of the connection string is "DefaultEndpointsProtocol = https;AccountName = NAME;AccountKey = KEY" -->
        <!-- For local execution, the value can be set either in this config file or through environment variables -->
        <add name = "AzureWebJobsDashboard"connectionString = "" />
        <add name = "AzureWebJobsStorage"connectionString = "" />
    </connectionStrings>
    <startup>
        <supportedRuntime version = "v4.0"sku = ".NETFramework,Version = v4.6.1" />
    </startup>
    <runtime>
        <assemblyBinding xmlns = "urn:schemas-microsoft-com:asm.v1">
            <dependentAssembly>
                <assemblyIdentity name = "Microsoft.WindowsAzure.Storage"publicKeyToken =
```

```xml
                    "31bf3856ad364e35"culture = "neutral" />
            <bindingRedirectoldVersion = "0.0.0.0 - 7.0.0.0"newVersion = "7.0.0.0" />
        </dependentAssembly>
        <dependentAssembly>
            <assemblyIdentity name = "Newtonsoft.Json"
publicKeyToken = "30ad4fe6b2a6aeed"culture = "neutral" />
            <bindingRedirectoldVersion = "0.0.0.0 - 6.0.0.0"newVersion = "6.0.0.0" />
        </dependentAssembly>
    </assemblyBinding>
    </runtime>
</configuration>
```

Azure WebJobs Dashboard 指的是 Azure SDK 记录 WebJobs Dashboard 的 log 的一个 Storage Account。

Azure WebJobs Storage 指的是 Azure SDK 记录 WebJobs 上传文件到 Blob Storage，处理 Queue Storage 的一个 Storage Account。

```csharp
//Functions.cs 文件

using System;
using System.Collections.Generic;
using System.IO;
using System.Linq;
using System.Text;
using System.Threading.Tasks;
using Microsoft.Azure.WebJobs;

namespace WebJob1
{
    public class Functions
    {
        // This function will get triggered/executed when a new message is written
        // on an Azure Queue called queue.
        public static void ProcessQueueMessage([QueueTrigger("queue")] string message, TextWriter log)
        {
            log.WriteLine(message);
        }
    }
}
```

这个 QueueTrigger 的 Attribute 帮助 WebJob 监听某一个特定的 Queue Storage，然后就可以处理此 Queue 上新进入的消息；同时，也提供了 TextWriter log 对

象作为参数，WebJob 就可通过此对象来写 Log 到 Storage 中。此代码是最简单的原型，还可以根据业务需求来加入更多的代码。如前所述，WebJob 是批量处理数据的一个好工具，和 App Service 配合起来使用，就可以更加灵活地完成复杂业务场景的任务。

3.5 在大流量下的垂直扩展和水平扩展

此功能是 Scale out 菜单，它与 Scale up 的区别是：Scale up 是升级单台服务器的硬件，而 Scale out 是将服务扩展到多台服务器上。在此界面中可以设置有多少个 Scale up 指定的服务器来作为此 App Service 的服务器集群，然后可以设置根据什么条件来启动一个集群中的新空闲服务器，比如 CPU 比例达到 80%，内存占用达到 80% 等。当设置了这些以后，Azure 会自动将 App Service 的程序和配置文件以及配置项都自动复制到其他服务器上，在默认情况下，这些集群里的服务器是被 Azure 平台静置的，只有设定的启动新服务器的条件达到，比如 CPU 比例达到 80%，才会从服务器集群中启动一个新的服务器，然后 Azure 平台自动将新的请求导入到此新启动的服务器，由新启动的服务器来服务用户。这个相当于传统分布式服务器中可伸缩的 Web 层，唯一不同的是我们不需要具体地管理这些集群中的服务器，这些集群中的服务器，包括网络负载均衡等，都是由 Azure 平台来管理的。

默认此菜单中是没有 Enable autoscale 的，即没有自动伸缩特性。需要单击 Enable autoscale 按钮，然后才会出现如如图 3-32 所示的菜单。

图 3-32 如何设置 Scale out

Scale Mode 可以根据一个 metric 来伸缩,所谓 metric 即某些参数,如 CPU 比例、内存比例、IO、磁盘等,metric 由 Rule 来定义;也可以根据一个服务器实例个数来伸缩。如果那些参数低于某个值,则可以将某些服务器暂停,因为停止的服务器基本不需要付钱,或者需要付较少的钱,所以这样的模式可以帮助 Azure 的用户来省钱,比较灵活。业务量大的时候就启动足够的服务器来响应用户,如果没有那么多的用户时,就让相应数量的服务器停止,还可以省去不少服务器的费用。

Rules 用来表示启动一个新空闲服务器的规则,包含 CPU 比例、内存比例、IO、磁盘等。选择 Add a rule 后会出现最右边的小窗口(见图 3-32)。可根据实际需要来设定。

Instance limit 用来表示集群中需要的服务器个数(包括最小值,最大值及默认值)。一般来说,默认个数是等于最小个数的。可以根据需要来设定。一般没有启动的服务器不需要付费。注意,最大实例个数是 10。

如图 3-33 所示是一个例子。

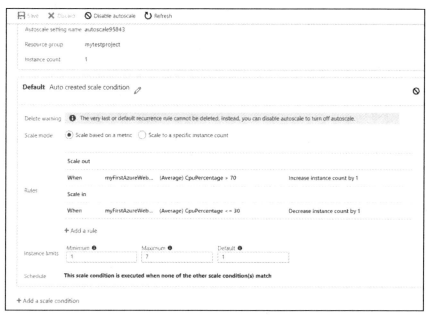

图 3-33 进行 Scale out 的例子

CPU 比例大于或等于 70%、持续 10 分钟就启动一个新实例,直到最大实例的个数达到。最大允许的实例个数是 10。有读者会问,如果每个实例都用了最强的服务器类型,同时此实例个数也是 10(最大值),如果压力还是超过此 App Service 能负载的范围,那么可以再从业务上垂直分割,分割出多个子网站来。每个子网站有子域名,不同的子域名对应不同的 App Service,这样,就可以无限制地扩展了。详见以下模式:

假设顶级域名后缀是：某公司.com.cn；

子域名1：某公司.com.cn对应一个App Service；

子域名2：某公司.com.cn对应一个App Service；

子域名3：某公司.com.cn对应一个App Service；

子域名4：某公司.com.cn对应一个App Service；

子域名5：某公司.com.cn对应一个App Service。

因为子域名的个数完全就没有限制，所以这样的方式可以说是无限制地伸缩的。当发展过程中又出现了某一个子域名对应的App Service不堪压力的情况时，可以再对此App Service进行类似于上面的垂直分割，再分出一些子网站来，域名就得再来一次更新和整理。

另外，如果一个集群CPU比例小于或等于设定值的30%，则持续10分钟，就减少一个实例。这是当访问量下降时就减少服务器的实例，这样可以节省一些费用。这种设定是一个非常灵活的集群设置。不必像传统的IT那样，采购了多少服务器就是多少服务器，不管访问量是大还是小，其数量一般都是比较稳定的，无法根据需要动态调整。空闲的时候，多余的服务器就浪费了；忙的时候，可能现有服务器数量还不够。Azure云平台可以根据实际需求，动态启动足够数目的服务器实例来服务用户；不用的时候又可以暂停服务器，暂停的服务器就不用付费了。

如果需要开发一个可分布式存储Session数据的应用，则需要在Startup.cs中写上这样的代码：

```
// This method gets called by the runtime. Use this method to add services to the container.
publicvoid ConfigureServices(IServiceCollection services)
{
    services.AddMvc();
    services.AddDistributedRedisCache(options =>
    {
        options.Configuration = "localhost";
        options.InstanceName = "SampleInstance";
    });
}
```

Azure平台有Redis分布式服务。可以申请一个Redis缓存服务，然后ASP.net core使用此Redis服务。Session数据都缓存到Redis中，这样可以实现可伸缩性。

3.6 创建Redis缓存

Redis是一个在内存中存储键值对的快速缓存服务。它可以做分布式的安装。微软Azure提供了一个分布式的Redis服务，它是兼容开源版本的。在All Services里搜索Redis，会出来一个Redis Cache，选择它，就到了Redis服务的页面。需要单击 Create Redis Cache 按钮来创建第一个 Redis 服务，出现如图 3-34 所示的界面。

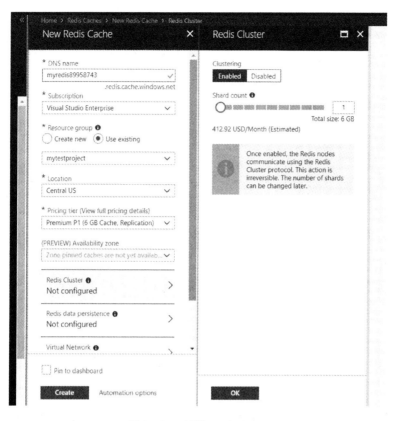

图 3-34 创建 Redis Cache

Pricing Tier 是选择服务级别，有 Basic 系列、Standard 系列、Premium 系列，如图 3-35 所示。

它们的价格也不一样。从图中可以看到其大致的差别：容量不一样，能否复制备份也不一样。需要提到的一点是，Premium 级别的可以提供 Redis Cluster（集群）服务。这是与 Standard 级别的主要差别。如图 3-35 所示，在选了 Premium 之一后，就可以在 Redis Cluster 中设置 Cluster 的参数。首先启用，然后选择 Shard 数目。系统会给出一个估计的费用，Shard 为 1 的时候大约一个月 412 美金。选好之后单击 Create 按钮，就开始运行。等一段时间后 Redis Cache 就建立好了，如

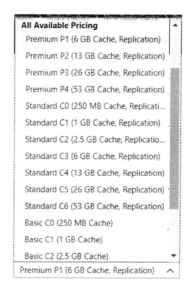

图 3-35 Redis Cache 的 Pricing Tier

图 3-36 所示。

图 3-36 创建好的 Redis Cache

注意到图 3-36 中右边有一个 Keys,选择 Show access keys,然后拷贝出 Primary Connection String：

myredis4385092. redis. cache. windows. net:6380, password = j/1UFDEtcxV-8TVxrw/Evf6/YbMkA/MRjNCjGM/6IZQ8=, ssl=True, abortConnect=False

这个链接串可以给客户端如. net、Java 或者 PHP 等用。客户端只要连上之后就可以链接,然后存储数据到 Redis 中。

在 Visual Studio 的 nuget 管理界面中安装两个包,首先是 asp. net core 的 session 库,如图 3-37 所示。

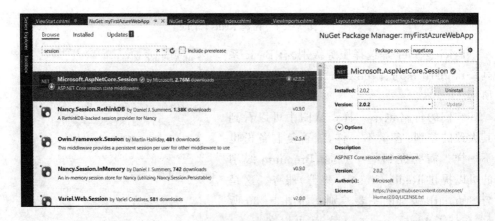

图 3-37 用 nuget 下载所需的引用库

然后是微软提供的 Redis 库。Microsoft. Extensions. Caching. Redis. Core 库目前是 1.0.3 版。此库是开源的,源码在下述地址可以访问到：https://github.com/aspnet/Caching。它提供了三种 Distributed Cache 的实现：In Memory、Redis、Sql Server,如图 3-38 所示。

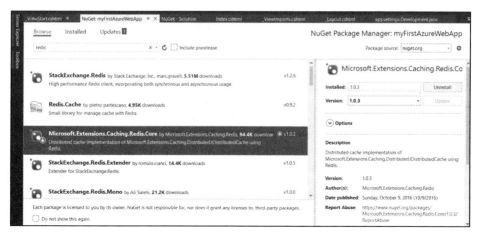

图 3-38 在 nuget 中下载 Microsoft.Extensions.Caching.Redis.Core 库

这样程序代码就要改一下：

```
//Startup.cs 文件
using System;
using System.Collections.Generic;
using System.Linq;
using System.Threading.Tasks;
using Microsoft.AspNetCore.Builder;
using Microsoft.AspNetCore.Hosting;
using Microsoft.Extensions.Configuration;
using Microsoft.Extensions.DependencyInjection;

namespace myFirstAzureWebApp
{
    public class Startup
    {
        public Startup(IConfiguration configuration)
        {
            Configuration = configuration;
        }

        public IConfiguration Configuration { get; }

        // This method gets called by the runtime. Use this method to add services to the
        // container.
        publicvoid ConfigureServices(IServiceCollection services)
        {
            services.AddMvc();
```

```
            services.AddSession();
            services.AddDistributedRedisCache(options =>
            {
                options.Configuration = Configuration.GetConnectionString("RedisConnection");
                options.InstanceName = "master";
            });
        }

        // This method gets called by the runtime. Use this method to configure the HTTP
        // request pipeline.
        public void Configure(IApplicationBuilder app, IHostingEnvironment env)
        {
            if (env.IsDevelopment())
            {
                app.UseBrowserLink();
                app.UseDeveloperExceptionPage();
            }
            else
            {
                app.UseExceptionHandler("/Error");
            }

            //app.UseStaticFiles();

            app.UseSession();
            app.UseMvcWithDefaultRoute();
        }
    }
}
```

Asp.net core 使用 Redis 的 Session，需要三步：

第一步：在 Startup.cs 中，使用 Configure Services 方法时要调用以下两个方法（如灰底纹处字）：

```
services.AddSession();
services.AddDistributedRedisCache(options =>
{
    options.Configuration = Configuration.GetConnectionString("RedisConnection");
    options.InstanceName = "master";
});
```

第二步：在 Configure 方法中调用以下方法（如灰底纹处字）：

```
app.UseSession();
```

以下是 Redis 链接串的设置文件：

```
//appSettings.json
{
"ConnectionStrings": {
"RedisConnection": " myredis4385092.redis.cache.windows.net:6380,password = j/1UFDEtcxV8TVxrw/Evf6/YbMkA/MRjNCjGM/6IZQ8 = ,ssl = True,abortConnect = False"
    },
"Logging": {
"IncludeScopes": false,
"LogLevel": {
"Default": "Warning"
      }
    }
}
```

第三步：设置 MVC controller 的代码。

```
//KeepController.cs
using System;
using System.Collections.Generic;
using System.Linq;
using System.Threading.Tasks;
using Microsoft.AspNetCore.Http;
using Microsoft.AspNetCore.Mvc;
using Microsoft.Extensions.Caching.Distributed;

namespace myFirstAzureWebApp.Controllers
{
    public class KeepController : Controller
    {
        private IDistributedCache _cache;

        public KeepController(IDistributedCache cache)
        {
            _cache = cache;
        }

        // GET: Keep
        [HttpGet]
        public ActionResult Index()
        {
```

```
                string value = _cache.GetString("CacheTime");

                if (value == null)
                {
                    value = DateTime.Now.ToString();

                    var options = new DistributedCacheEntryOptions();
                    options.SetSlidingExpiration(TimeSpan.FromMinutes(1));
                    _cache.SetString("CacheTime", value, options);
                }
                ViewData["CacheTime"] = value;
                ViewData["CurrentTime"] = DateTime.Now.ToString();
            return View();
        }

    }
}
```

注意这几个类基本都有一个 interface 为参数的构造函数。如：

Startup(IConfiguration configuration);

KeepController(IDistributedCache cache);

即现在 asp.net core 已经把 Dependency Injection(依赖注入)实现了,所以很多 asp.net core 本身的一些接口和类都默认有依赖注入机制。所以将这些接口作为一个类的构造函数的参数,会自动有容器为此接口提供一个实例化的对象,这些对象是容器管理的对象,不必担心此对象的生命周期,不需要去管理此对象的回收。

调用 _cache.SetString 存放的数据可以用别的 Redis 客户端查看,如图 3-39 所示。

可以看到上述代码存放了 CacheTime 数据到 Azure 的 Redis 服务器上。

加入访问 Session 的代码,如下(见灰色的代码行):

```
//KeepController.cs 文件
using System;
using System.Collections.Generic;
using System.Linq;
using System.Threading.Tasks;
using Microsoft.AspNetCore.Http;
using Microsoft.AspNetCore.Mvc;
using Microsoft.Extensions.Caching.Distributed;

namespace myFirstAzureWebApp.Controllers
{
    public class KeepController : Controller
```

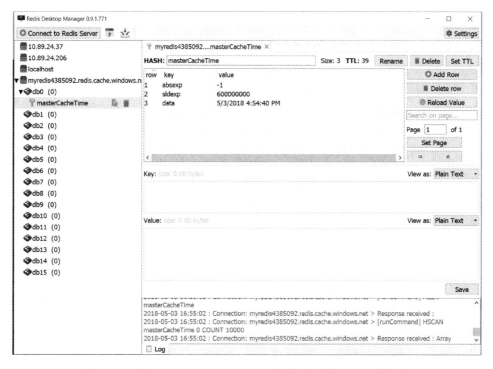

图 3-39　用 Redis Desktop Manager 查看用程序代码写入的数据

```
{
    private IDistributedCache _cache;

    public KeepController(IDistributedCache cache)
    {
        _cache = cache;
    }

    // GET: Keep
    [HttpGet]
    public ActionResult Index()
    {
        string value = _cache.GetString("CacheTime");
        string sessionValue = HttpContext.Session.GetString("sessionTime");

        if (value == null)
        {
            value = DateTime.Now.ToString();

            var options = new DistributedCacheEntryOptions();
            options.SetSlidingExpiration(TimeSpan.FromMinutes(1));
```

```
                _cache.SetString("CacheTime", value, options);
            }
            if (sessionValue == null)
            {
                HttpContext.Session.SetString("sessionTime", DateTime.Now.ToString());
            }
            else
            {
                ViewData["sessionTime"] = sessionValue;
            }
            ViewData["CacheTime"] = value;
            ViewData["CurrentTime"] = DateTime.Now.ToString();
            return View();
        }

    }
}
```

注：此代码第一次调用的时候去取 Session 中的数据，如果发现没有，就给 Session 中存一个名为 sessionTime 的字符串。再次刷新页面的时候，还是去 Session 中名为 sessionTime 的字符串，这次就存在了，取出来后放到 ViewData 中。然后页面里就显示：

/Views/Keep/Index.cshtml 文件（这是一个 Razor view）

@*

    For more information on enabling MVC for empty projects, visit https://go.microsoft.com/fwlink/? LinkID = 397860

*@
@{
}

keep page loaded

cached time: @ViewData["CacheTime"]

current time: @ViewData["CurrentTime"]

session time: @ViewData["sessionTime"]

可以看到 session 的数据也放到了 Redis 中，如图 3-40 所示。

加了几行代码：

```
        string cart = null;
        HttpContext.Session.SetString("cart", "{\"cart\":\"test\"}");
        cart = HttpContext.Session.GetString("cart");
```

也可以发现，Session 数据还是存在同一个键值下，即这个 Session 对应的就是 GUID：868dcdcc-4009-ace1-3467-bb69aed772a2，如图 3-41 所示。

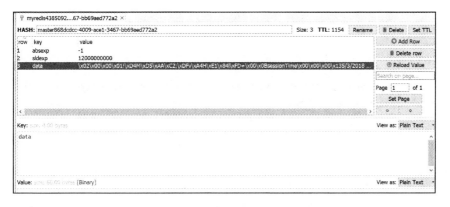

图 3 – 40　查看 Redis 中存入的 Session 数据

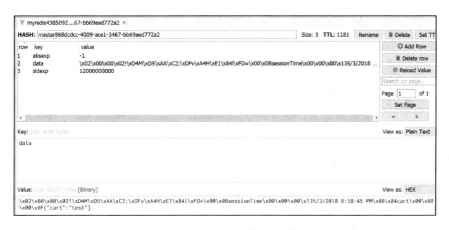

图 3 – 41　往 Session 中存入更多数据,其键值还是同一个

有一个 sessionTime 的数据,它是以一个二进制的形式存储的;也就是说这个 Redis 服务可以为 asp. net core 的服务来存取,asp. net core 自行进行 Session 数据的管理。在 2002 年到 2008 年的 asp. net web form 时代,曾经有一个 asp. net session 不能同时写的问题,即一个页面对 Session 写并且持续时间长的话,其他 asp. net 页面就不能写,还会影响到其他 asp. net 页面的读,因为它们被一定程度地阻止住了。所以那时开发 asp. net web form 的应用多数性能都不高。后来有一些改进方案,如 asp. net mvc,但并不是很彻底的改进。所以很多 asp. net mvc 的应用都是关闭 Session Module 的(即不用 asp. net 本身的 Session 机制)。现在 asp. net core 对 asp. net 是一个比较彻底的改进,它的整个页面渲染管线(pipe line)都重新设计了,以便 asp. net core 能在不同操作系统平台上运行。在 Session 中的改进就是 Session 不再是一个阻塞性的操作,多个线程可以同时对 Session 进行存取,不论采用 In-Memory、Redis 或者 Sql Server 做存储 Session 的地方,不同 Session 间都不会互相影响,同一个 Session 的多个线程都可以对 Session 数据进行操作,Session 机制保证写操作不会被丢失或者引起冲突。

在开发 asp.net core 应用程序的时候是选择 Session，还是选择如 Redis 这样的缓存呢？此处建议还是选择 Session 机制，因为 Session 机制由 asp.net core 来管理。其在 Redis 中的 Key 是根据 Sesison ID 来区分开的。其有效期可以由 Session 的 Option 来管理。只要 Session 还处于有效期，Session 中存储的值就还是可以访问的。

什么样的数据适合放到 Session 中呢？

首先是要能序列化的数据。有一些复杂的对象，如果不能序列化成一个字符串，就不能存入 Session。一些复杂的对象，比如打开文件的句柄、多媒体打开的句柄等，都属于不能序列化的对象，这样的数据是不能直接存入 Session 的。

其次就是尺寸比较大的数据最好不要放进 Session 中，比如一个体量比较大的文件，或者是一个完整的比较大的订单信息就不要放入。Session 中并非不能放比较大的数据，只是这样做的效率不高，网络传输量也比较大。序列化、网络传输、反序列化等都要耗费不少 CPU、内存，这样很不划算。

3.7 用 Asp.net core 开发 Azure App Service 架构需考虑的一些要点

架构师要考虑的输入包括功能性需求以及非功能性需求。功能性需求与具体的领域有关，比如石油行业有石油行业的功能性需求，这个就不具体讲了。非功能性需求通常包括：系统响应时间、数据处理时间、页面加载时间、数据容量、系统抗压能力、异常情况处理能力、安全性、可扩展性、可维护性以及成本。为了保证系统处理时间，App Service 的 Pricing Tier 就得很好地选择，既要成本能接受，同时服务器又要能抗住访问压力。当一台服务器不足以抗住访问压力时，就要能伸缩到多台服务器上。Azure 平台默认提供了可伸缩的模式。应用程序也要配合 Azure 可伸缩的模式来编写。为了能配合可伸缩的模式，需要做到以下几点。

1. App Service 的服务最好是无状态的

无状态就是 Web 服务器中不记录任何状态信息，这样做的好处是万一发生单点失败，此服务器实例上没有任何状态信息，那么集群中另外一台 Web 服务器立即可以捡起客户的请求来处理。而客户感觉不到集群中发生了一个单点失败。

2. 即使需要有状态，状态数据也最好存到 Azure Redis 服务中

有些应用的场合，必须要有状态数据，比如网上购物，购物车中放了一些商品，是用户选择的，但是还没有买单。这样就需要为用户保存这个状态信息。这样的场合就必须要有状态数据。可以将状态数据放到 Redis 服务器中。因为 Asp.net core 提供了 Session 机制，Session 是 Asp.net core 提供的异于其他编程语言的一种存储状态数据的方式。Session 最终也是将状态数据存储到 Redis 中，但是 Session 机制进行了一些管理。如果用的是 Redis 的 List 或者 Set，那么可以有 $2^{32}-1$ 个元素的容量。每个元素可以是一个最长512字节的字符串。所以在考虑用Session的时候，也不要什么数据都往Session中放，比如商品详情信息，就不要往Session中放。只有那些简单的id、key之类的数据才往Session中放。只要这些放到Session中的数据可以用来找到其他剩余的数据就可以了。

前面讲过Azure Redis可以有Cluster，还有异地灾备(需选择Premium类的Pricing tier)。Redis可以比较安全地存储大量的key、value数据。

3. 应用可以按层来划分，即表示层、商务逻辑层、数据库层等

应用分层，便于建立不同的层，各层之间降低耦合性，在各层建定义调用合同，比如REST API，或者WCF服务调用之类的调用合同。这样，各层可以相对独立地变化。只要遵守调用合同，就不会有问题。一般地，表示层是Web层，有前端代码，如angular js、typescript、jquery+HTML5；有的是纯前端代码，即没有razor之类的引擎，页面代码都是用纯前端代码写的。基于前端框架的特点，可以用一些面向对象的编程。纯前端代码的表示层的可单元测试性相对比较好。可以利用如jasmine之类的javascript单元测试框架来进行单元测试，以提高前端代码的单元测试覆盖率。表示层也可以是如asp.netrazor的技术，页面代码是通过razor引擎生成的。前端代码和razor代码是混合的，相对来讲其可单元测试性要比纯前端代码差一些。所以在表示层这里，尽量采用纯前端的技术，如angular、typescript、jquery+HTML5，可以利用html5引入的browser的local storage、session storage、cookie等。至于Java Script面向对象编程，也是成熟技术，针对目标领域的那些对象进行分析和编程就可以了。单元测试就可以用jasmine、mocha之类的框架。

商务逻辑层和表示层得有一个调用合同。比较流行的是商务逻辑层用微服务的形式以及REST API接口。当然也可以是别的形式，如WCF，或者Xml Web Service等形式，只要定义妥当即可。商务逻辑层最好采用一些设计模式来增加可维护性、可扩展性。另外，各商务逻辑层要设计安全性，所有的登录、权限等都是在商务逻辑层这里设计，安全性是更重要的。安全性的保障与选用的登录系统有关系，登录系统要能像LDAP那样支持目录树结构，然后支持针对某些组进行权限分配。前面提到像Google、Facebook、Amazon那样的登录系统，也有Azure本身提供的登录系统。本书之后的章节会谈到Azure AD。

数据库层也需要良好的设计。可以选用nosql的Mongo DB、Azure Table Storage，也可以选用Azure SQL Server。Mongo DB支持很大量的数据。Azure Table Storage也可支持很大量的数据，同时比较便宜。前面的章节谈到了如何设计Azure Table Storage的PartitionKey和RowKey。Azure SQL Server是集群的。如果选用Premium以上的Pricing tier，则可以享受集群、灾备等高可用服务。应很好地设计每一个表，一般地要满足一定的数据库范式，如bcnf范式。只有在特别极端的情况下才考虑降低范式。同时，要设计好表的聚集索引以及非聚集索引。需要根据高频查询类进行设计。聚集索引的候选字段通常是短的、单向递增的，如时间。然后是

最高频率的查询。可以按照此方法来设计聚集索引。非聚集索引根据那些次高频率的查询来设计。非聚集索引不要求单向递增,但是同样也要求候选字段是短的。写查询语句尽量避免输出很大量的数据,比如查出整个表的所有字段、所有记录,这样,需要在网络上传输的数据量就很大。另外,要避免在记录数很多的表上进行表扫描。尽量让查询语句命中索引,不管是聚集索引还是非聚集索引。数据库层的事务机制需要进行良好的设计。一般可以利用数据库本身的事务机制,定义好数据库事务分离级别。默认是 Read committed,还可以选用别的事务分离级别。同时,要尽量优化查询语句,让每一个事务尽量运行时间短,就可大大减小和其他数据库事务重叠的概率,也可大大增加数据库系统的吞吐量。

3.8 用 Java 开发一个 Azure App Service

Azure 平台可以支持多种编程语言,如 Java、Python、PHP、Node.js 等。用 Java 开发 Azure App Service 需要一些先决条件。

需要 Eclipse IDE for Java EE Developers,同时安装 Azure Toolkit for Eclipse。需要打开 Eclipse Market,搜索 Azure,Azure Toolkit for Eclipse 就出现了,点击安装即可,如图 3-42 所示。

图 3-42　安装 Azure Toolkit for Eclipse

然后新建 Dynamic Web Project，File→New→Dynamic Web Project，如图 3-43 所示。

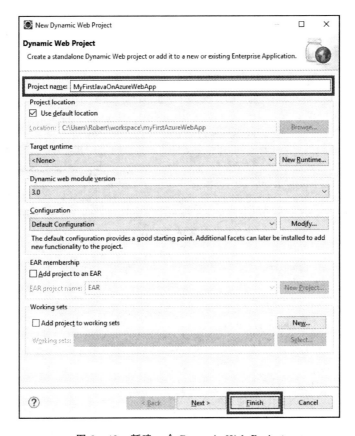

图 3-43　新建一个 Dynamic Web Project

输入名字 MyFirstJavaOnAzureWebApp，单击 Finish 按钮。然后创建一个 JSP 页面，如图 3-44 所示。

右键单击 WebContent，选择 New→JSP File，如图 3-45 所示。

输入名字 index.jsp，然后单击 Finish 按钮。

打开 index.jsp，输入如下内容：

\<body\>

\<h1\> \<% out.println("Hello Azure!"); %\> \</h1\>

\</body\>

注：会有一个错误信息，缺少一个 Javsservlet 类。可以忽略它，如图 3-46 所示。

右键单击 Java 工程，然后选择弹出菜单 Azure→Publish as Azure Web App，如图 3-47 所示。

然后出现如图 3-48 所示对话框。

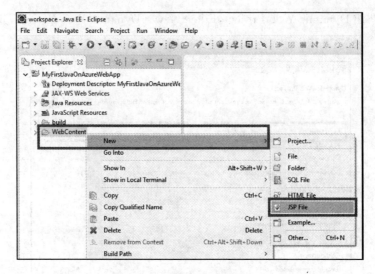

图 3-44　新建一个 JSP 文件

图 3-45　输入文件名并单击 Finish 按钮

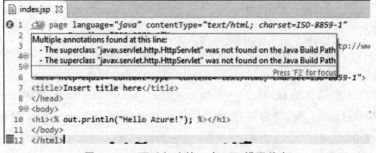

图 3-46　可以忽略的一个 JSP 错误信息

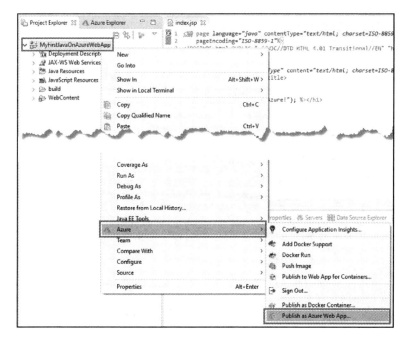

图 3-47 用菜单将此工程部署到 Azure 上

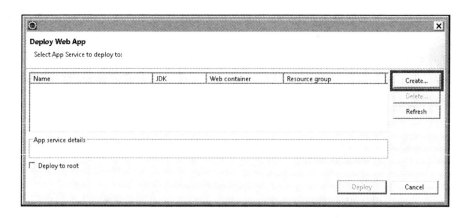

图 3-48 部署 Java 工程到 Azure 的对话框

在 Deploy Web App 对话框中单击 Create 按钮,出现如图 3-49 所示对话框。

在此对话框中,注意数字可能与此图中的不同。Location 就从下拉框中选择一个,如 East US、West US、Central US、Australia Contral 之类的 Region。

Pricing tier 是服务器级别,其收费级别也不同,可以仔细看看。下面这个链接是关于 App Service 的价格信息:https://azure.microsoft.com/en-us/pricing/details/app-service/? ref=microsoft.com&utm_source=microsoft.com&utm_medium=docs&utm_campaign=visualstudio。

图 3-49 创建部署的 Profile

Create new 表示创建一个新的。如果选择 Use existing,则表示覆盖现有的一个 App Service,用新版本替代旧版本。

打开 Resource group 页面,根据实际需要选择。要么创建新的 Resource group,要么用已有的 Resource group,都只是把资源放到某个 group 里管理,所以只要方便管理即可,如图 3-50 所示。

图 3-50 针对资源组的设置

打开 JDK 页面，可以选 JDK 版本和 Tomcat 版本。可以根据需要来选择，如图 3-51 所示。

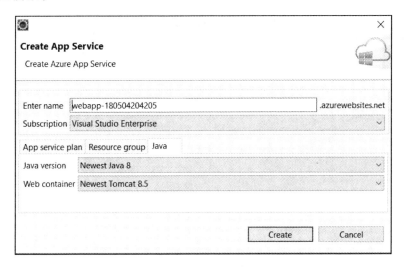

图 3-51　选择运行时的 JDK 版本和 Web 容器

完成各页面的设置后单击 Create 按钮。等一下，会出现如图 3-52 所示的界面。

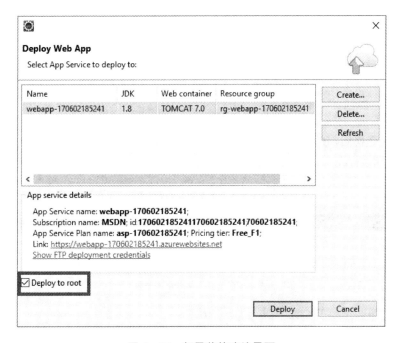

图 3-52　部署前的确认界面

该界面是部署之前的确认页。勾选 Deploy to root，确认无误后单击 Deploy

按钮。

随后打开 Azure Portal,就可以看到已部署的 App Service,如图3-53所示。

图3-53 部署好的 Java 的 App Service

根据所显示的 URL,链接访问,则出现如图3-54所示的界面。

图3-54 应用 Java 的页面部署后的效果

这个就是我们加的 index.jsp 的内容。

打开 Advanced Tools 菜单,再打开 Process Explorer,可以看到它上面运行的进程,如图3-55所示。

图3-55 在 Process Explorer 查看其中运行的 Java 进程

这个是一个 Tomcat 进程。

对于那些使用 Spring Boot、Spring Cloud 的用户，Spring Boot 支持默认的 tomcat、jetty，也可以在没有默认容器的情况下启动内置的 tomcat 来作为容器。如果要将 Spring Boot 部署到 Azure 平台的 Tomcat 中，则需要再做一些事情，比如配置 tomcat 运行。

设置为 war 文件，于是第一步就需要修改项目的类型为 war，修改 pom.xml 配置文件。将打包的类型设置为 war：

```xml
<plugin>
    <artifactId>maven-war-plugin</artifactId>
    <version>2.5</version>
    <configuration>
        <failOnMissingWebXml>false</failOnMissingWebXml>
        <archive>
            <manifest>
                <mainClass>${start-class}</mainClass>
                <addDefaultImplementationEntries>true</addDefaultImplementationEntries>
            </manifest>
        </archive>
    </configuration>
</plugin>
```

然后对 Springboot 启动类做个简单处理。

```java
package com.somecompany;

import org.springframework.boot.SpringApplication;
import org.springframework.boot.autoconfigure.SpringBootApplication;
import org.springframework.boot.builder.SpringApplicationBuilder;
import org.springframework.boot.context.embedded.EmbeddedServletContainer;
import org.springframework.boot.web.support.SpringBootServletInitializer;
@SpringBootApplication
public class SpringBootStarter extends SpringBootServletInitializer{
    @Override
    protected SpringApplicationBuilder configure(SpringApplicationBuilder builder) {
        return builder.sources(SpringBootStarter.class);
    }
    public static void main(String[] args) {

        SpringApplication.run(SpringBootStarter.class, args);
    }
}
```

这样，此工程就可以在 Tomcat 这样的容器里运行，然后配合 Azure 发布程序来把 Publish Java 工程加到 Azure App Service 中。

3.9 用 Python 开发一个 Azure App Service

Python 近年来变得越来越热了，排名渐渐地靠前了。TensorFlow 机器学习框架就是用 Python 写的。很多开源代码也是用 Python 写的，所以用 Python 写应用程序也比较常见。Python 也可以用来开发 Azure App Service。此处需要两个前提，一个是安装 Git 软件，方便下载开源 Python 代码，如果是 Windows，可下载 Git for Windows；如果是 Linux，就下载 Git。另一个是 Python 软件，用来运行 Python 程序。

用此命令下载样例程序：

git clone https://github.com/Azure-Samples/python-docs-hello-world

下载后其目录结构如图 3-56 所示。

图 3-56 Python 样例程序的文件目录

首先关注一下 requirements.txt：

//requirements.txt 文件内容
Flask == 0.12.1

注意到它是要求用 Flask 0.12.1 版本。用 pip 命令安装的时候会用到此文件。它会安装上 Flask 库。

//main.py 文件内容
from flask import Flask
app = Flask(__name__)

```
@app.route('/')
def hello_world():
    return 'Hello, World! '

if __name__ == '__main__':
    app.run()
```

//ptvs_virtualenv_proxy.py 文件内容

```
# ##############################################################
#
# Copyright (c) Microsoft Corporation.
#
# This source code is subject to terms and conditions of the Apache License, Version 2.0. A
# copy of the license can be found in the License.html file at the root of this distribution.
# If you cannot locate the Apache License, Version 2.0, please send an email to
# vspython@microsoft.com. By using this source code in any fashion, you are agreeing
# to be bound by the terms of the Apache License, Version 2.0.
#
# You must not remove this notice, or any other, from this software.
#
# ##############################################################

import datetime
import os
import sys
import traceback

if sys.version_info[0] == 3:
    def to_str(value):
        return value.decode(sys.getfilesystemencoding())

    def execfile(path, global_dict):
        """Execute a file"""
        with open(path, 'r') as f:
            code = f.read()
        code = code.replace('\r\n', '\n') + '\n'
        exec(code, global_dict)
else:
    def to_str(value):
        return value.encode(sys.getfilesystemencoding())
```

```python
def log(txt):
    """Logs fatal errors to a log file if WSGI_LOG env var is defined"""
    log_file = os.environ.get('WSGI_LOG')
    if log_file:
        f = open(log_file, 'a+')
        try:
            f.write('%s: %s' % (datetime.datetime.now(), txt))
        finally:
            f.close()


ptvsd_secret = os.getenv('WSGI_PTVSD_SECRET')
if ptvsd_secret:
    log('Enabling ptvsd...\n')
    try:
        import ptvsd
        try:
            ptvsd.enable_attach(ptvsd_secret)
            log('ptvsd enabled.\n')
        except:
            log('ptvsd.enable_attach failed\n')
    except ImportError:
        log('error importing ptvsd.\n')


def get_wsgi_handler(handler_name):
    if not handler_name:
        raise Exception('WSGI_ALT_VIRTUALENV_HANDLER env var must be set')

    if not isinstance(handler_name, str):
        handler_name = to_str(handler_name)

    module_name, _, callable_name = handler_name.rpartition('.')
    should_call = callable_name.endswith('()')
    callable_name = callable_name[:-2] if should_call else callable_name
    name_list = [(callable_name, should_call)]
    handler = None
    last_tb = ''

    while module_name:
        try:
            handler = __import__(module_name, fromlist=[name_list[0][0]])
            last_tb = ''
            for name, should_call in name_list:
```

```
                    handler = getattr(handler, name)
                    if should_call:
                        handler = handler()
                break
            except ImportError:
                module_name, _, callable_name = module_name.rpartition('.')
                should_call = callable_name.endswith('()')
                callable_name = callable_name[:-2] if should_call else callable_name
                name_list.insert(0, (callable_name, should_call))
                handler = None
                last_tb = ': ' + traceback.format_exc()

    if handler is None:
        raise ValueError('"%s" could not be imported%s' % (handler_name, last_tb))

    return handler


activate_this = os.getenv('WSGI_ALT_VIRTUALENV_ACTIVATE_THIS')
if not activate_this:
    raise Exception('WSGI_ALT_VIRTUALENV_ACTIVATE_THIS is not set')

def get_virtualenv_handler():
    log('Activating virtualenv with %s\n' % activate_this)
    execfile(activate_this, dict(__file__=activate_this))

    log('Getting handler %s\n' % os.getenv('WSGI_ALT_VIRTUALENV_HANDLER'))
    handler = get_wsgi_handler(os.getenv('WSGI_ALT_VIRTUALENV_HANDLER'))
    log('Got handler: %r\n' % handler)
    return handler

def get_venv_handler():
    log('Activating venv with executable at %s\n' % activate_this)
    import site
    sys.executable = activate_this
    old_sys_path, sys.path = sys.path, []

    site.main()

    sys.path.insert(0, '')
    for item in old_sys_path:
        if item not in sys.path:
            sys.path.append(item)
```

```python
    log('Getting handler %s\n' % os.getenv('WSGI_ALT_VIRTUALENV_HANDLER'))
    handler = get_wsgi_handler(os.getenv('WSGI_ALT_VIRTUALENV_HANDLER'))
    log('Got handler: %r\n' % handler)
    return handler
```

这个是在 App Service 运行中需的文件，定义了一些虚拟环境下的 handler。

```xml
//web.2.7.config
<?xml version="1.0"?>
<configuration>
    <appSettings>
        <add key="WSGI_ALT_VIRTUALENV_HANDLER" value="main.app" />
        <add key="WSGI_ALT_VIRTUALENV_ACTIVATE_THIS"
            value="D:\home\site\wwwroot\env\Scripts\activate_this.py" />
        <add key="WSGI_HANDLER"
            value="ptvs_virtualenv_proxy.get_virtualenv_handler()" />
        <add key="PYTHONPATH" value="D:\home\site\wwwroot" />
    </appSettings>
    <system.web>
        <compilation debug="true" targetFramework="4.0" />
    </system.web>
    <system.webServer>
        <modules runAllManagedModulesForAllRequests="true" />
        <handlers>
            <remove name="Python27_via_FastCGI" />
            <remove name="Python34_via_FastCGI" />
            <add name="Python FastCGI"
                path="handler.fcgi"
                verb="*"
                modules="FastCgiModule"
scriptProcessor="D:\Python27\python.exe|D:\Python27\Scripts\wfastcgi.py"
                resourceType="Unspecified"
                requireAccess="Script" />
        </handlers>
        <rewrite>
          <rules>
            <rule name="Static Files" stopProcessing="true">
                <conditions>
                    <add input="true" pattern="false" />
                </conditions>
            </rule>
            <rule name="Configure Python" stopProcessing="true">
```

```
                <matchurl = "(.*)"ignoreCase = "false" />
                <conditions>
                    <addinput = "{REQUEST_URI}" pattern = "^/static/.*" ignoreCase = "true" negate = "true" />
                </conditions>
                <actiontype = "Rewrite"
                    url = "handler.fcgi/{R:1}"
                    appendQueryString = "true" />
            </rule>
        </rules>
    </rewrite>
</system.webServer>
</configuration>
```

这个是当运行环境为 Python 2.7 时的配置文件。还有一个是当运行环境为 Python 3.4 时的配置文件,文件内容类似,这里就不再列了。

运行此命令:cd Python-docs-hello-world。

运行此命令:pip install -r requirements.txt。

运行此命令后会安装上 Flask 库,需要有互联网连接。

运行命令 python main.py,就开始运行样例程序。然后打开 http://localhost:5000/,此程序监听 5 000 个端口,页面如图 3 - 57 所示。

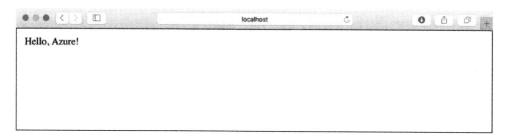

图 3 - 57 开始运行样例程序

回到运行命令的窗口,按 Ctrl + C 键终止程序运行。

下面的命令需要先安装 Azure CLI,这个链接可以告诉我们怎么安装:https://docs.azure.cn/zh-cn/cli/install-azure-cli?view=azure-cli-latest。

安装后才可以用 Azure CLI 命令。

运行 Azure 命令:

```
az login
```

注:在 Azure China 中使用 Azure CLI 2.0 之前,请首先运行 az cloud set -n AzureChinaCloud 更改云环境。如果要切换回全局 Azure,请再次运行 az cloud set -n AzureCloud。

用此命令来创建部署用户:

az webapp deployment user set --user-name <username> --password <password>

密码长度必须至少为8个字符,其中包含以下3种元素中的2种:字母、数字、符号。

在Web应用中进行FTP和本地Git部署时需要一个部署用户。用户名和密码都为账户级别。它们与Azure订阅账号不同。

如果收到"'Conflict'. Details:409"错误提示,请更改用户名。如果收到"'Bad Request'. Details:400"错误提示,请使用更强的密码。

用此命令查一下location列表:

az appservice list-locations

用此命令创建resource group:

az group create --name myResourceGroup --location <从前一个命令中返回的location>

用此命令创建appservice plan:

az appservice plan create --name myAppServicePlan --resource-group myResourceGroup --sku FREE

注:sku FREE是Pricing tier的一种,请查阅Azure App Service的资料。选择一个正确的Pricing。

创建好之后会有类似如下所列的一些输出信息:

{
"adminSiteName": null,
"appServicePlanName": "myAppServicePlan",
"geoRegion": "China North",
"hostingEnvironmentProfile": null,
" id ": "/subscriptions/0000 - 0000/resourceGroups/myResourceGroup/providers/Microsoft.Web/serverfarms/myAppServicePlan",
"kind": "app",
"location": "China North",
"maximumNumberOfWorkers": 1,
"name": "myAppServicePlan",
" <""JSON",
"data""removed",
"for""brevity.",
"> ": null,
"targetWorkerSizeId": 0,
"type": "Microsoft.Web/serverfarms",

```
"workerTierName": null
}
```

然后用此命令创建 AppService：

```
az webapp create -- resource - group myResourceGroup -- plan myAppServicePlan -- name <app_name> -- runtime "python|3.4" -- deployment - local - git
```

注：runtime 版本可以切换成 2.7，运行完后会出现类似于如下所列的信息：

```
Local git is configured with url of 'https://<username>@<app_name>.scm.chinacloudsites.cn/<app_name>.git'
{
    "availabilityState": "Normal",
    "clientAffinityEnabled": true,
    "clientCertEnabled": false,
    "cloningInfo": null,
    "containerSize": 0,
    "dailyMemoryTimeQuota": 0,
    "defaultHostName": "<app_name>.chinacloudsites.cn",
    "deploymentLocalGitUrl": "https://<username>@<app_name>.scm.chinacloudsites.cn/<app_name>.git",
    "enabled": true,
    "<""JSON",
    "data""removed",
    "for""brevity.",
    ">"
}
```

注：此信息是 Azure China 中的信息，与 Azure Global 略有不同。

至此，已创建了一个空的新 Web 应用并启用了 Git 部署。

Git 远程部署的 URL 将显示在 deploymentLocalGitUrl 属性中，其格式为 https://<username>@<app_name>.scm.chinacloudsites.cn/<app_name>.git。保存此 URL，因为稍后将需要它。如果是 Azure Global，则其 URL 是这样的：https://<username>@<app_name>.scm.azurewebsites.net/<app_name>.git。注意，其后缀名不同，一个是 chinacloudsites.cn 结尾，一个是 azurewebsites.net 结尾。

访问刚部署的 http://<app name>.azurewebsites.net 或者 http://<app name>.chinacloudsites.cn，出现如图 3-58 所示的界面。

用此命令：

```
git remote add azure <前一步里获得的部署 gitURL>
```

即 https://<username>@<app_name>.scm.azurewebsites.net/<app_name>.git 或者 https://<username>@<app_name>.scm.chinacloudsites.cn/<app_

图 3-58　部署后的 Python 程序

name＞.git，然后需要输入之前建立的部署用户的账号和密码。

创建好这个 Azure Remote 之后，就可以改本地代码，当达到可以部署的程度时，就可以用此命令来把 Python 代码部署到 Azure 服务上：git commit -am "updated output"。此命令是提交到本地计算机的 repository；同时，写上注释信息，让别人也能看懂代码。

git push azure master

此命令是将本地代码改变推送到远端名为 Azure 的 Master 分支上。前面我们定义过的 Azure 远端库就是之前建立的 App Service 的部署 git URL。所以这就是在进行部署。

访问刚部署的 http://＜app name＞.azurewebsites.net 或者 http://＜app name＞.chinacloudsites.cn，应该会看到代码的变化或者部署的结果。

第 4 章

Azure SQL Database

Azure SQL Database 是 Azure 平台提供的关系型数据库服务。它支持更多类型的数据,如关系型数据、JSON、Spatial 和 XML。SQL 数据库提供托管的单一 SQL 数据库、弹性池中的托管 SQL 数据库,以及 SQL 托管实例(公共预览版)。它可以实现动态可伸缩性能,并提供列存储索引(用于极端分析和报告)和内存中 OLTP(用于极端事务处理)等选项。微软可无缝处理 SQL 代码库的所有修补和更新,并避开底层基础结构的所有管理。

Azure SQL Database 与 MS SQL Server Database Engine 共享代码。根据微软云优先的策略,最新的功能特性都是先在 Azure SQL Database 中出现,然后才会在 MS SQL Server 中出现。这种策略让用户可以使用到最新的功能特性,并且不用花精力去升级和打补丁,因为这些最新功能特性都在上百万个数据库上测试过,是成熟可靠的。下面这个链接可以找到一些最新的功能特性和后面要实现的一些功能特性:

https://azure.microsoft.com/roadmap/?category=databases。

Azure SQL Database 在多个服务级别上提供可预测的性能,有的服务级别可以提供动态可伸缩性,全年无下线时间,内带的智能优化,全球的动态伸缩和高可用性,高级的安全控制,而且基本不需要进行管理。这些能力让用户可以聚焦到应用的开发,缩短提交到市场的时间,而不是花很多时间来管理虚拟机、基础架构。Azure SQL Database 部署在全球 38 个数据中心,随着更多数据中心的建设,可以选择一个离用户更近的数据中心来架设 Azure SQL Database。

4.1 Azure SQL Database 对比 SQL Server

既然 Azure SQL Database 与 MS SQL Server Database Engine 共享代码,那么这两者之间大体上是相同的。Azure SQL Database 是 Azure 平台提供的一个类似之前 MS SQL Server 的服务。在 Azure Portal 可以创建 Azure SQL Server 的实例,同时可以像之前一样地写 tsql,其语法基本是一致的。其基本对象模型,比如表、索引、视图、存储过程、触发器、事务机制等都和 MS SQL Server 差不多。当然此服务与之前的 MS SQL Server 也有一定的差别。

Azure SQL Database 最近有一个还在预研中的新特性：Azure SQL Database Managed Instance,这个新特性的好处是 100% 兼容独立的 MS SQL Server(Enterprise Edition)。

Azure SQL Database 与 Managed Instance 的异同点如表 4-1 所列。

表 4-1　Azure SQL Database 与 Managed Instance 的异同点

SQL	Azure SQL Database	Managed Instance（预研）
Always Encrypted	Yes—see Cert store and Key vault	Yes—see Cert store and Key vault
Always On Availability Groups	High availability 包含在每一个数据库中。请参考这个链接查看灾备恢复：https://docs.microsoft.com/en-us/azure/sql-database/sql-database-business-continuity	High availability 包含在每一个数据库中。请参考这个链接查看灾备恢复：https://docs.microsoft.com/en-us/azure/sql-database/sql-database-business-continuity
Attach a database	No	No
Application roles	Yes	Yes
Auditing	Yes	Yes
Automatic backups	Yes	Yes
Automatic tuning (plan forcing)	Yes	Yes
Automatic tuning (indexes)	Yes	No
BACPAC file (export)	Yes—see SQL Database export https://docs.microsoft.com/en-us/azure/sql-database/sql-database-export	Yes
BACPAC file (import)	Yes—see SQL Database import https://docs.microsoft.com/en-us/azure/sql-database/sql-database-import	Yes
BACKUP command	No—see only system-initiated automatic backups-see https://docs.microsoft.com/en-us/azure/sql-database/sql-database-automated-backups	System-initiated automated backups and user initiated copy-only backups-see Backup differences
Built-in functions	Most—see individual functions	Yes—see Stored procedures, functions, triggers differences
Change data capture	No	Yes
Change tracking	Yes	Yes
Collation statements	Yes	Yes

续表 4-1

SQL	Azure SQL Database	Managed Instance（预研）
Columnstore indexes	Yes—https://docs.microsoft.com/sql/relational-databases/indexes/columnstore-indexes-overview	Yes
Common language runtime (CLR)	No	Yes—see CLR differences
Contained databases	Yes	Yes
Contained users	Yes	Yes
Control of flow language keywords	Yes	Yes
Cross-database queries	No—see Elastic queries	Yes—see plus Elastic queries
Cross-database transactions	No	Yes—see Linked server differences
Cursors	Yes	Yes
Data compression	Yes	Yes
Database mail	No	Yes
Data Migration Service (DMS)	Yes	Yes
Database mirroring	No	No
Database configuration settings	Yes	Yes
Data Quality Services (DQS)	No	No
Database snapshots	No	No
Data types	Yes	Yes
DBCC statements	Most—see individual statements	Yes—see DBCC differences
DDL statements	Most—see individual statements	Yes—see T-SQL differences
DDL triggers	Database only	Yes
Distributed partition views	No	Yes
Distributed transactions MS DTC	No—see Elastic transactions	No—see Elastic transactions
DML statements	Yes	Yes

续表 4-1

SQL	Azure SQL Database	Managed Instance（预研）
DML triggers	Most—see individual statements	Yes
DMVs	Most—see individual DMVs	Yes—see T-SQL differences
Dynamic data masking	Yes	Yes
Elastic pools	Yes	Built-in-a single Managed Instance can have multiple databases that share the same pool of resources
Event notifications	No—see Alerts	Yes
Expressions	Yes	Yes
Extended events	Some—see https://docs.microsoft.com/en-us/azure/sql-database/sql-database-xevent-db-diff-from-svr	Yes—see Extended events differences
Extended stored procedures	No	No
Files and file groups	Primary file group only	Yes
Filestream	No	No
Full-text search	第三方的分词工具不支持	第三方的分词工具不支持
Functions	Most—see individual functions	Yes—see Stored procedures, functions, triggers differences
Geo-restore	Yes	No—you can restore COPY_ONLY full backups that you take periodically—see Backup differences and Restore differences
Geo-replication	Yes	No
Graph processing	Yes	Yes
In-memory optimization	Yes—Premium and Business Critial tiers only	No
JSON data support	Yes	Yes
Language elements	Most—see individual elements	Yes—see T-SQL differences
Linked servers	No—see Elastic query	Only to SQL Server and SQL Database
Log shipping	High availability is included with every database. Disaster recovery is discussed in Overview of business continuity with Azure SQL Database	High availability is included with every database. Disaster recovery is discussed in Overview of business continuity with Azure SQL Database

续表 4-1

SQL	Azure SQL Database	Managed Instance（预研）
Master Data Services（MDS）	No	No
Minimal logging in bulk import	No	No
Modifying system data	No	Yes
Online index operations	Yes	Yes
OPEN DATA SOURCE	No	Yes—see T-SQL differences
OPENJSON	Yes	Yes
OPEN QUERY	No	Yes—see T-SQL differences
OPEN ROW SET	No	Yes—see T-SQL differences
OPEN XML	Yes	Yes
Operators	Most—see individual operators	Yes—see T-SQL differences
Partitioning	Yes	Yes
Point in time database restore	Yes—see SQL Database recovery	Yes—see SQL Database recovery
Polybase	No	No
Policy-based management	No	No
Predicates	Yes	Yes
R Services	Preview release—see What's new in machine learning	No
Resource governor	No	Yes
Restore statements	No	Yes—see Restore differences
Restore database from backup	From automated backups only—see SQL Database recovery	From automated backups—see SQL Database recovery and from full backups—see Backup differences
Row Level Security	Yes	Yes
Semantic search	No	No
Sequence numbers	Yes	Yes
Service Broker	No	Yes—see Service Broker differences
Server configuration settings	No	Yes—see T-SQL differences

续表 4-1

SQL	Azure SQL Database	Managed Instance（预研）
Set statements	Most—see individual statements	Yes—see T-SQL differences
SMO	Yes	Yes
Spatial	Yes	Yes
SQL Data Sync	Yes	No
SQL Operations Studio	Yes	Yes
SQL Server Agent	No—see Elastic jobs	Yes—see SQL Server Agent differences
SQL Server Analysis Services (SSAS)	No—see Azure Analysis Services	No—see Azure Analysis Services
SQL Server Auditing	No—see SQL Database auditing	Yes—see Auditing differences
SQL Server Data Tools (SSDT)	Yes	Yes
SQL Server Integration Services (SSIS)	Yes, with a managed SSIS in Azure Data Factory (ADF) environment, where packages are stored in SSISDB hosted by Azure SQL Database and executed on Azure SSIS Integration Runtime	Yes, with a managed SSIS in Azure Data Factory (ADF) environment, where packages are stored in SSISDB hosted by Managed Instance and executed on Azure SSIS Integration Runtime
SQL Server Management Studio (SSMS)	Yes	Yes
SQL Server PowerShell	Yes	Yes
SQL Server Profiler	No—see Extended events	Yes
SQL Server Replication	Transactional and snapshot replication subscriber only	No
SQL Server Reporting Services (SSRS)	No—see Power BI	No—see Power BI
Stored procedures	Yes	Yes
System stored functions	Most—see individual functions	Yes—see Stored procedures, functions, triggers differences
System stored procedures	Some—see individual stored procedures	Yes—see Stored procedures, functions, triggers differences
System tables	Some—see individual tables	Yes—see T-SQL differences
System catalog views	Some—see individual views	Yes—see T-SQL differences
Temporary tables	Local and database-scoped global temporary tables	Local and instance-scoped global temporary tables

续表 4-1

SQL	Azure SQL Database	Managed Instance（预研）
Temporal tables	Yes	Yes
Threat detection	Yes	Yes
Trace flags	No	No
Variables	Yes	Yes
Transparent data encryption（TDE）	Yes	No—not in Public Preview
VNet	Partial—see VNET Endpoints	Yes—Resource Manager model only
Windows Server Failover Clustering	High availability is included with every database. Disaster recovery is discussed in Overview of business continuity with Azure SQL Database	High availability is included with every database. Disaster recovery is discussed in Overview of business continuity with Azure SQL Database
XML indexes	Yes	Yes

4.2 创建一个 SQL Azure 服务器

在 Azure Portal 界面，选左边的 SQL Database，再单击 Create SQL Database，出现如图 4-1 所示的界面。

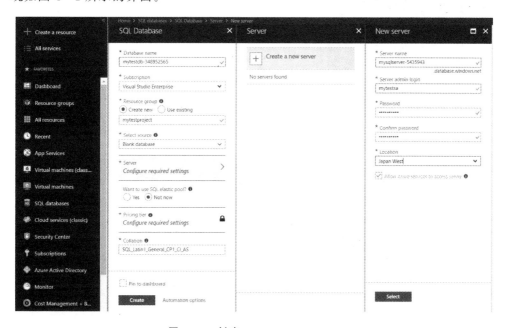

图 4-1　创建 Azure SQL Database

输入数据库的名字,选择 Subscription,创建一个 Resource groups,然后选择 Blank database 的 Source。其他的 Source 尽量不要选,除非需要 Sample 数据库。然后就是 Server。Server 还没有建立,点击 Server 处,会在右边出现如图 4-1 所示的创建 Server 的界面。输入 Server 名字、admin 登录名、密码,再选择 location,此例中选择了 Japan West,这个 location 相对比较近,速度会快点,如图 4-2 所示。都输入后单击 Select。等待一段时间,此 Server 就创建好了。

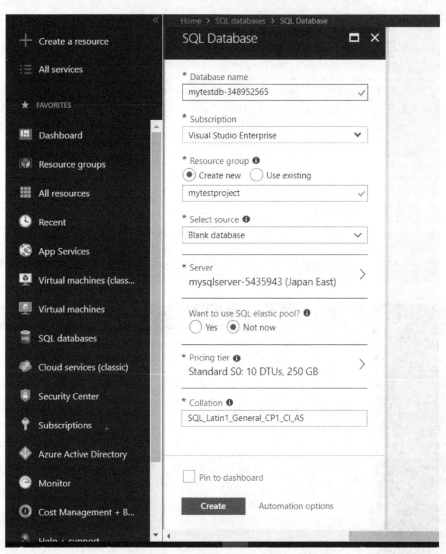

图 4-2 创建 SQL Server 选择的选项

可见到 Server 名字已经填写到此界面中。然后就是选择 Pricing tier,如图 4-3 所示。

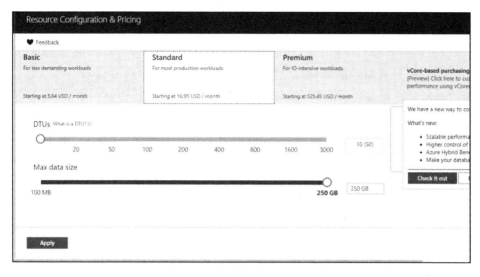

图 4-3 选择 Pricing tier

主要是 Basic、Standard、Premium 的选择,根据需要选择一个。测试环境用一个 Standard 就可以了,开发环境可以用 Basic,生产环境就用 Premium。除此之外,还可以用一种新的方式来选择 Pricing tier,如图 4-3 中最右边的那个弹出框中所说,Azure 可以按 Scala performance solutions,更强地控制 storage 花销,还有混合 SQL Server 的安装方式,以及异地灾备措施等都可以用新的方式来选择,单击 Check it out 按钮,出现如图 4-4 所示的界面。

图 4-4 用新的方式选择 Pricing tier

在此界面里,可以选择 CPU Core 的个数、最大数据大小、备份数据几天,还有 Scala Performance Solutions、更强的控制 storage 花销、混合 SQL Server 的安装方式、异地灾备措施等。如果觉得这样选麻烦,可以选择 Looking for basic,则 Standard、Premium 回到刚才选择 Basic、Standard、Premium 的界面,这是个套餐,三个级别可以照顾到绝大多数的情况。刚刚说的新的方式可以更加定制化需求,可以更精准地控制费用,根据使用情况来调整。

最后一个要关注的是 Collation,这是定义 SQL Database 的默认字符集。这个得根据需要来定。如果是中文,则需要选用 Unicode 的字符集,或选用支持中文的字符集。如果是开发别的语言的应用,则需要选用对应的字符集。

Collation 定义的是字符集。一个字符可以是日文、韩文、俄文、中文、英文等,其字符的存储是占用一定的字节的,比如 3 个字节或者 4 个字节;有的字符如英文字符,其存储值需要一个字节就够了。国际标准定义了字符集,给每个字符集定义了名字,比如:Chinese_PRC_CI_AI_WS、Latin1_General_100_CI_AS_SC、Japanese_Bushu_Kakusu_100_CI_AS_SC,可以看到其中有 CI、AI、WS 等,它们代表什么意思呢?

- _CI(CS) 表示是否区分字母大小写,CI 是不区分,CS 是区分。如果区分大小写,那么排序的时候小写字母的排在大写字母的前面;如果不区分大小写,那么排序的时候视大小写字母相同。
- _AI(AS) 表示是否区分重音,AI 是不区分,AS 是区分。如果不区分重音,那么排序的时候视"a"和"?"为相同字符。
- _KI(KS) 表示是否区分假名类型,KI 是不区分,KS 是区分,在日语中应用。
- _WI(WS) 表示是否区分全半角,WI 是不区分,WS 是区分。半角是单字节,全角是双字节。

Collation 一共有四个级别,分别是 Server-level、Database-level、Column-level 和 Expression-level,在此处创建 SQL Database 是 Server-level。Server-level 可以不影响 Database-level,Database-level 可以不影响 Column-level,即 Server-level collation 可以和 Database-level 的 Collation 不同,Database-level 的 Collation 可以和 Column-level 的 Collation 不同,但是这样不是很方便,最好是 Collation 到处都一致。所以在创建 SQL Database 时就要选择正确的 Collation。

选择完成就单击 Create 按钮。等一段时间后,SQL Database 就创建完成了。在 Dashboard 页面选择此新建的 Azure SQL Database,进入其详情页面,如图 4-5 所示。

Azure SQL Database

图 4-5 创建完成的 Azure SQL Database 的详情页面

4.3 创建一个 SQL Azure 数据库

首先复制此数据库服务器的名字,看到图 4-6 中的 Click to copy。

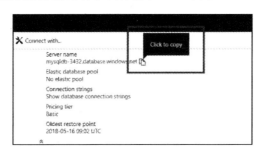

图 4-6 单击 Click to copy 按钮来复制服务器名

在 Azure Portal 界面中,可以建立数据库。
在 Azure Portal 左边主菜单中,点击 SQL databases,出现如图 4-7 所示界面。
单击 Add 按钮,出现如图 4-8 所示界面。
这个界面需要输入数据库名,选择 Pricing tier 和 Collation,这些都在前文中介

图 4-7 SQL databases 菜单

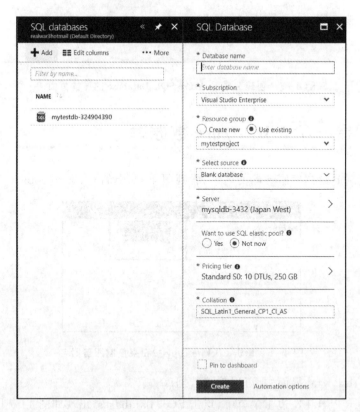

图 4-8 创建一个 SQL databases 的界面

绍过。确定好后,单击 Create 按钮,就创建了一个新的数据库。Server 还是前面建的那个 Server。这个界面比在 MS SQL Server Management Studio 中创建数据库时的选项要多一些,可以选择 Pricing tier 等。

4.4 用 SQL Server Management Studio 链接 Azure SQL Database

除了可以在 Azure Portal 中管理 Azure SQL Database,也可以在 SQL Server Management Studio 中管理 Azure SQL Database。可以用 MS SQL Server Management Studio 链接此数据库。建立链接,如图 4-9 所示。

图 4-9 用 MS SQL Server Management Studio 链接 Azure SQL Database

输入创建时给的 Admin 用户名和密码,因为是第一次访问此服务器,相应的防火墙规则还没有建立,所以会出现如图 4-10 所示的界面。

图 4-10 新建一个防火墙规则

可以将此链接的外网 IP 加入到防火墙规则中，即选第一项。有时用户所在的公司网络有多个出口 IP，而且 IP 会变，所以选第二项也是可以的。不过建议一般情况下选择第一项比较保险。防火墙规则开的口子越小，越有可能减少攻击的机会。好，就选择第一项，Add my client IP（实际 IP），这样就进入了此服务器，如图 4-11 所示。

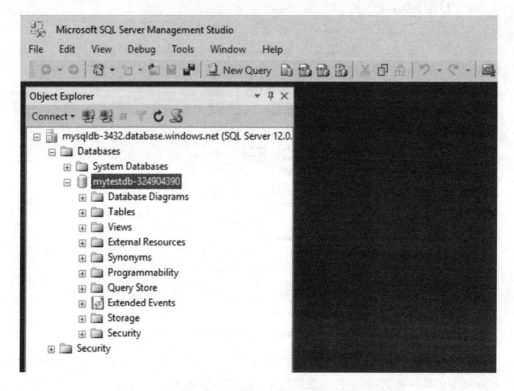

图 4-11　链接上 Azure SQL Database，显示出其内部资源

默认此 SQL Database 创建了一个数据库。如此例中的 mytestdb-324904390，也可以再创建新的数据库，这个数据库是默认创建的，有时应用中需要的数据库不止一个，需要将数据放到多个数据库中进行管理。

右键单击 Databases，如图 4-12 所示。

选择 New Database，输入名字就可以了，单击 OK 按钮，如图 4-13 所示。

等一会儿之后，此数据库就创建好了。可以看到，它默认选择的是 Standard 的 Pricing tier，如图 4-14 所示。

Azure SQL Database

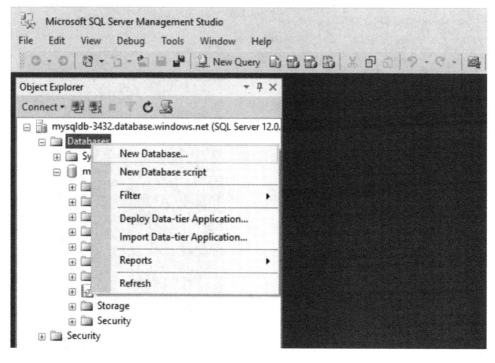

图 4-12 新建 Databases 的菜单

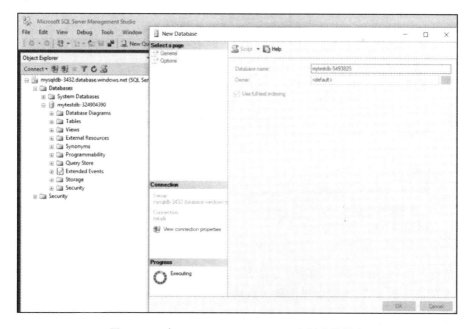

图 4-13 在 MS Management Studio 中新建数据库

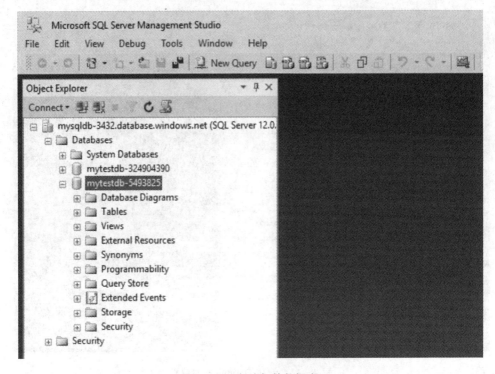

图 4-14 创建好的数据库

4.5 在 Azure SQL Database 中执行的 TSQL

这里将介绍如何使用 SSMS 在 SQL Azure 中执行 TSQL 语句。现在创建一个表。

```
create table dbo.Book
(
    bookindex nvarchar(100) not null ,
    name nvarchar(200) not null,
    publishdate datetime not null,
    author nvarchar(100) not null
)
```

执行此语句,显示成功,如图 4-15 所示。

现在是一个空表,需要往里面插入数据,用 INSERT 语句。

```
INSERT INTO [dbo].[Book]
            ([bookindex]
            ,[name]
```

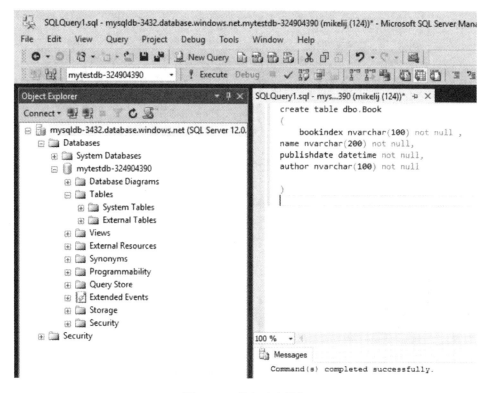

图 4-15　执行建表语句

```
,[publishdate]
,[author])
VALUES
('9787111525516'
,'电子与嵌入式系统设计丛书'
,'2016-04-01'
,'李庆华')
GO
```

执行成功,如图 4-16 所示。

看一下数据,如图 4-17 所示。

看一下这个表的设计界面,如图 4-18、图 4-19 所示。

这四个字段都是按之前建表语句建立的。再看主键,目前是没有的。索引也是没有的。对于一个比较小的表,没有主键、没有索引还可以接受,如果是生产环境的数据,很大可能是数据比较多的。一般地都要建立主键,还要有若干个索引。

这里要解释几个概念:

1. 主键(primary key)

主关键字(primary key,简称主键):是表中的一个或多个字段,它的值用于唯一

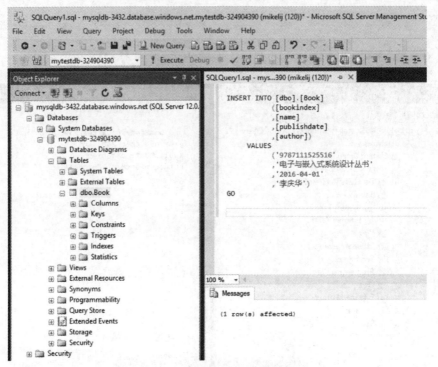

图 4-16 在空表中插入数据

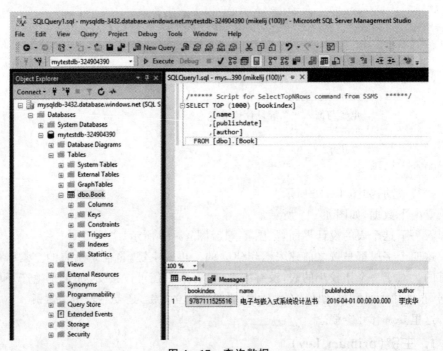

图 4-17 查询数据

Azure SQL Database

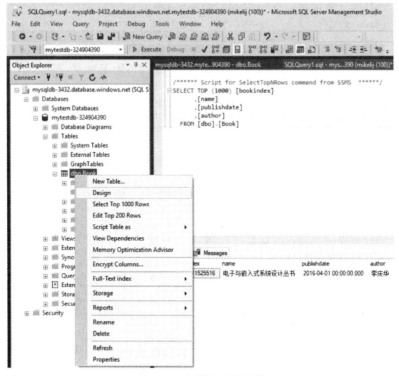

图 4-18 设计表的菜单

图 4-19 设计表的字段

地标识表中的某一条记录。在两个表的关系中,主关键字用来在一个表中引用来自于另一个表中的特定记录。主关键字是一种唯一关键字,是表定义的一部分。一个表的主键可以由多个关键字共同组成,并且主关键字的列不能包含空值。主关键字是可选的,并且可在 CREATE TABLE 或 ALTER TABLE 语句中定义。

主关键字是被挑选出来,作为表的行的唯一标识的候选关键字。一个表只有一个主关键字。主关键字又可以称为主键。主键可以由一个字段,也可以由多个字段组成,分别称为单字段主键或多字段主键,又称主码。它可以唯一确定表中的一行数据,或者唯一确定一个实体。

主键的作用如下:

① 保证实体的完整性;

② 加快数据库的操作速度;

③ 在表中添加新记录时,数据库系统会自动检查新记录的主键值,不允许该值与其他记录的主键值重复;

④ 数据库系统自动按主键值的顺序显示表中的记录。如果没有定义主键,则按输入记录的顺序显示表中的记录。

2. 外键(foreign key)

简单地说,外键就是别的表的主键,只是在当前表中进行引用。既然是别的表的主键,外键就可以在被引用表中唯一标识一个记录,相当于一个指针从当前表指向了被引用表。

外键的作用如下:

① 保证数据的完整性、一致性。

② 主要目的是控制存储在外键表中的数据。

③ 使两张表形成关联,外键只能引用外表中的列的值或使用空值。

3. 聚集索引(clustered index)

聚集索引是指数据库表行中数据的物理顺序与键值的逻辑(索引)顺序相同。一个表只能有一个聚集索引,因为一个表的物理顺序只有一种情况,所以,对应的聚集索引只能有一个。如果某索引不是聚集索引,则表中的行的物理顺序与索引顺序不匹配,与非聚集索引相同,聚集索引有着更快的检索速度。

因为记录的物理排列顺序就是按照聚集索引键值的顺序来排的,所以若查询语句是按聚集索引来搜索,则会直接利用聚集索引来进行查找,这样速度会比较快。聚集索引可以由一个或者多个字段构成,而且通常聚集索引会作为主键来使用,就是此聚集索引可以用来标识一个记录,以示与本表中其他记录的区别。

聚集索引适用的情况如下:

① 含有大量非重复值的列。

② 使用 BETWEEN、>、> =、<、<= ,返回一个范围值的列。

③ 被连续访问的列。
④ 返回大型结果集的查询。
⑤ 经常被使用链接或 GROUP BY 子句查询访问的列。

表 4-2 总结了何时使用聚集索引或非聚集索引。

表 4-2 何时使用聚集索引或非聚集索引

动作描述	使用聚集索引	使用非聚集索引
列经常被分组排序	应	应
返回某范围内的数据	应	不应
一个或极少的不同值	不应	不应
小数目的不同值	应	应
大数目的不同值	不应	应
频繁更新的列	不应	应
外键列	应	应
主键列	应	应
频繁修改索引的列	不应	应

聚集索引的设计应符合一定的规则：

① 应是单向递增的，如日期时间，它是单向递增的；或者自增的 identity 字段。这个要求是因新增记录的时候，数据库引擎只需在本表的结尾处添加新的记录，按照聚集索引的特性，物理顺序和聚集索引的顺序一致，如果一个记录需要插入到之前的物理位置，则需要很多的 IO 开销来物理移动相应的记录集合，才能将记录插入，这样的设计将引起插入记录时 IO 开销很高。如果聚集索引能保证单向递增的特性，则在插入新记录时，新记录的值刚好是最大的，因为它刚好符合单向递增的特性，那么只需要在所有现存物理记录的最后位置写入新记录就可以了。这样带来的 IO 开销就比较小。

② 应是比较短的数据，如日期、比较短的字符串等。如果聚集索引中有比较长、大的数据，则搜索比较的时候效率就比较差，耗 CPU 更多。聚集索引的字节数限制是最大 900 字节，如果将几个比较长的字段放入聚集索引，则有可能超出聚集索引最大 900 字节的限制而导致索引失败。

③ 查询语句需要经常对聚集索引的字段进行搜索，如单个记录的搜索，或者范围搜索，即聚集索引的字段被查询的概率最大或者很高。

④ 聚集索引中的字段越少越好，刚好够用来标识一个记录就行了。不要出现一些可有可无的字段。所谓可有可无的字段是指，有些字段即使参与进聚集索引，对唯一标识一个记录也没有关系，不能用来唯一标识记录。这样的字段就没有必要放进聚集索引中。

⑤ 一旦创建后,聚集索引字段的值不需要再改动,这个特性符合的话,聚集索引就可以用来做主键。只有那些不在聚集索引中的字段需要改动,这是必须遵循的。设想如果聚集索引的字段需要改动,则会引起其物理位置的改动,移动一条记录的位置,需要将所有受影响的记录都进行移动,所涉及的IO开销与插入时不遵循单向递增的情况下一样大,这将是不可忍受的。如果这个聚集索引字段频繁被改,则会频繁引起大量IO,CPU也会很高。所以,这个特性是必须遵循的。

4. 非聚集索引(non clustered index)

非聚集索引是指除聚集索引之外的索引。其物理顺序与聚集索引的顺序是不同的。非聚集索引需要用数据行之外的空间来进行存储。非聚集索引存储了索引键值,每一个索引键值包括一个指向数据行的指针。这个指向数据行的指针被称作行定位器。行定位器的结构取决于数据页是存储在一个堆上还是一个聚集索引上,如果数据是存储在一个堆上,则行定位器是一个指向数据行的指针;如果数据是存储在一个聚集索引上,则行定位器是一个指向聚集索引Key的指针。

可以包括非键的字段到叶子级别的非聚集索引,来避免受索引长度的限制(非聚集索引的最大字节数限制是1 700字节,而聚集索引的最大字节数限制是900字节),并且执行带索引的查询。建立一个带Included Column的非聚集索引的步骤如下:

```
-- Creates a nonclustered index on the Person.Address table with four included (non-key) columns.
-- index key column is PostalCode and the nonkey columns are
-- AddressLine1, AddressLine2, City, and StateProvinceID.
CREATE NONCLUSTERED INDEX IX_Address_PostalCode
ON Person.Address (PostalCode)
INCLUDE (AddressLine1, AddressLine2, City, StateProvinceID);
GO
```

注意到它有Include子句,将四个字段包含在索引上。

用在此处的样例表,其语句如下:

```
CREATE NONCLUSTERED INDEX IX_BookName
ON dbo.Book(name)
INCLUDE (publishdate, author);
GO
```

则此索引的建立如图4-20所示。

可以看到其Included Column。这些Included Column不是键的一部分,所以不受1 700字节的限制。

以上是几个概念。下面接着介绍使用SQL Server Management Studio执行TSQL。

Azure SQL Database

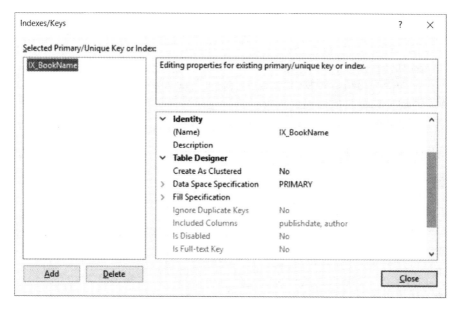

图 4-20 设计一个包含 Included Column 的非聚集索引

创建表的时候可以指定 Primary Key，如：

/****** Object: Table [dbo].[Book] Script Date：5/24/2018 1:50:06 PM ******/
SET ANSI_NULLS ON
GO

SET QUOTED_IDENTIFIER ON
GO

CREATETABLE [dbo].[Book](
　　[bookindex] [nvarchar](100)NOT NULL,
　　[name] [nvarchar](200)NOT NULL,
　　[publishdate] [datetime] NOT NULL,
　　[author] [nvarchar](100)NOT NULL,
CONSTRAINT [PK_Book] PRIMARY KEY CLUSTERED
(
　　[bookindex] ASC
)WITH (STATISTICS_NORECOMPUTE = OFF,IGNORE_DUP_KEY = OFF)ON [PRIMARY]
)ON [PRIMARY]
GO

注意其中的 Primary Key 关键字和 Clustered 关键字，即这个表是以 bookindex 作为主键，同时它也是聚集索引。

如果建表的时候没有创建主键，则可以用如下语句来创建主键：

```
ALTER TABLE dbo.Book ADD CONSTRAINT
    PK_Book PRIMARY KEY CLUSTERED
    (
    bookindex
    )WITH(STATISTICS_NORECOMPUTE = OFF,IGNORE_DUP_KEY = OFF,ALLOW_ROW_LOCKS = ON,ALLOW_PAGE_LOCKS = ON)ON [PRIMARY]
```

如前所述,主键可以有一个或者多个字段,要看实际情况。

另外再根据查询的需要,创建一些非聚集索引。

```
CREATE NONCLUSTERED INDEX IX_BookName ON dbo.Book
    (
    name
    )WITH(STATISTICS_NORECOMPUTE = OFF,IGNORE_DUP_KEY = OFF,ALLOW_ROW_LOCKS = ON,ALLOW_PAGE_LOCKS = ON)ON [PRIMARY]
    GO
```

比如按书名查找,则

```
CREATE NONCLUSTERED INDEX IX_Author ON dbo.Book
    (
    author
    )WITH(STATISTICS_NORECOMPUTE = OFF,IGNORE_DUP_KEY = OFF,ALLOW_ROW_LOCKS = ON,ALLOW_PAGE_LOCKS = ON)ON [PRIMARY]
    GO
```

比如按作者查找,可以在 author 字段上创建一个索引。

加入一个字段的 TSQL:

```
ALTER TABLE dbo.Book ADD price float(53)NULL
```

删除一个字段的 TSQL:

```
ALTER TABLE dbo.Book DROP COLUMN price
```

修改表的时候需要检测一个字段是否已经存在,或者此字段是否已经删除,即让脚本可以重复运行。

```
if exists(select * from sys.syscolumns where name = 'price'and id = object_id('dbo.Book'))
        print' 存在 '
else
    begin
        print' 不存在,增加 price 字段 ';
        ALTER TABLE dbo.Book ADD price float(53)NULL;
    end;
```

即在准备加字段之前检测一下字段是否已经存在。同理,在删除字段的时候,可以检测一下字段是否存在,若存在,则运行删除语句删除之;如果不存在,就不再进行删除。

```
if exists(select * from sys.syscolumnswhere name = 'price'and id = object_id('dbo.Book'))
    begin
    print' 存在,删除 price 字段 ';
        ALTER TABLE dbo.Book DROP COLUMN price;
    end;
else
    print' 已删除 ';
```

4.6 用 Visual Studio 开发应用链接 Azure SQL Database

可以用 Visual Studio 2017 来新建一个项目,如图 4-21 所示。

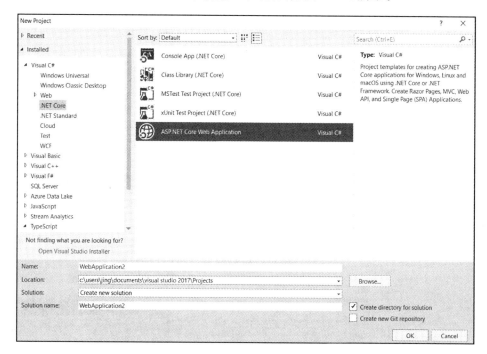

图 4-21 创建一个应用来访问 Azure SQL Database

选择 ASP. NET Core Web Application,然后在下一个界面里选择 API,如图 4-22 所示。

这是要创建一个 ASP. NET Core API 类型的应用。当然,也可以选择其他类型的应用。API 类型的应用可以方便建立 RESTFul 的 API。其他类型如 Web Appli-

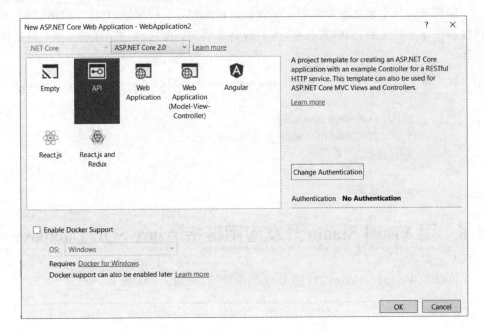

图 4-22 创建一个 API 类型的应用

cation,或者 Web Application(Model-View-Controller)、Angular 等都有相应的后台代码,只要有后台代码,就可以写建立到 Azure SQL Database 的代码。

这里选择之后,项目就建立好了。项目建立好之后如图 4-23 所示。

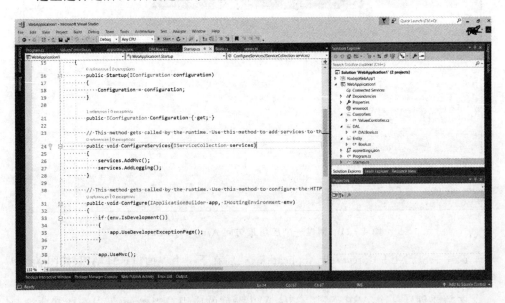

图 4-23 创建好的 API 类型的 ASP.NET Core 应用

代码如下:

```csharp
//Program.cs
using System;
using System.Collections.Generic;
using System.IO;
using System.Linq;
using System.Threading.Tasks;
using Microsoft.AspNetCore;
using Microsoft.AspNetCore.Hosting;
using Microsoft.Extensions.Configuration;
using Microsoft.Extensions.Logging;

namespace WebApplication1
{
    public class Program
    {
        public static void Main(string[] args)
        {
            BuildWebHost(args).Run();
        }

        public static IWebHost BuildWebHost(string[] args) =>
            WebHost.CreateDefaultBuilder(args)
                .UseStartup<Startup>()
                .Build();
    }
}
```

这个和之前的差不多,要有 Build WebHost、Setup 环境。

```csharp
//Startup.cs
using System;
using System.Collections.Generic;
using System.Linq;
using System.Threading.Tasks;
using Microsoft.AspNetCore.Builder;
using Microsoft.AspNetCore.Hosting;
using Microsoft.Extensions.Configuration;
using Microsoft.Extensions.DependencyInjection;
using Microsoft.Extensions.Logging;
using Microsoft.Extensions.Options;

namespace WebApplication1
```

```
{
    public class Startup
    {
        public Startup(IConfiguration configuration)
        {
            Configuration = configuration;
        }

        public IConfiguration Configuration { get; }

        // This method gets called by the runtime. Use this method to add services to the
        // container.
        public void ConfigureServices(IServiceCollection services)
        {
            services.AddMvc();
            services.AddLogging();
        }

        // This method gets called by the runtime. Use this method to configure the HTTP
        // request pipeline.
        publicvoid Configure(IApplicationBuilder app, IHostingEnvironment env)
        {
            if (env.IsDevelopment())
            {
                app.UseDeveloperExceptionPage();
            }

            app.UseMvc();
        }
    }
}
```

这个类和之前的类似,提供了 MVC 和 Logging 的服务。同样,IConfiguration 是构造函数输入的参数。

```
//Book.cs
using System;
using System.Collections.Generic;
using System.Linq;
using System.Threading.Tasks;

namespace WebApplication1.Entity
{
```

```csharp
public class Book
{
    public Int64 bookindex { get; set; }
    public string name { get; set; }
    public string author { get; set; }
    public DateTime publishDate { get; set; }
    public double price { get; set; }
    public string ISBN { get; set; }
    public string publisher { get; set; }
}
```

定义一个实体类,用来传数据。注意其字段都是可以序列化的。没有不能序列化的字段。

```csharp
//DALBook.cs
using System;
using System.Collections.Generic;
using System.Data.SqlClient;
using System.Linq;
using System.Threading.Tasks;
using Microsoft.Extensions.Configuration;
using Microsoft.Extensions.Logging;
using WebApplication1.Entity;

namespace WebApplication1.DAL
{
    public class DALBook
    {
        public DALBook(IConfiguration configuration, ILogger logger)
        {
            Configuration = configuration;
            _logger = logger;
        }
        private readonly ILogger _logger;
        public IConfiguration Configuration { get; }

        public List<string> GetBookNames()
        {
            SqlConnection connection = new SqlConnection(Configuration.GetConnectionString("myConnection"));
            try
            {
```

```csharp
            connection.Open();
            SqlCommand sqlCommand = new SqlCommand("select name from dbo.Book", connection);
            SqlDataReader reader = sqlCommand.ExecuteReader();
            List <string> bookNameList = new List <string>();
            while (reader.Read())
            {
                bookNameList.Add(reader.GetString(0));
            }
            reader.Close();
            connection.Close();
            return bookNameList;
        }
        catch(SqlException sqlException)
        {
            _logger.LogError(sqlException, sqlException.Message, null);
        }
        catch (Exception exception)
        {
            _logger.LogError(exception, exception.Message, null);
        }
        returnnull;
    }

    public List <Book> GetBookList()
    {
        SqlConnection connection = new SqlConnection(Configuration.GetConnectionString("myConnection"));
        try
        {
            connection.Open();
            SqlCommand sqlCommand = new SqlCommand("select bookindex, name, publishdate,author,price,publisher,ISBN from dbo.Book", connection);
            SqlDataReader reader = sqlCommand.ExecuteReader();
            List <Book> bookList = new List <Book>();
            while (reader.Read())
            {
                Book book = new Book();
                book.bookindex = reader.GetInt64(0);
                book.name = reader.GetString(1);
                book.publishDate = reader.GetDateTime(2);
                book.author = reader.GetString(3);
```

```csharp
                book.price = reader.GetDouble(4);
                book.publisher = reader.GetString(5);
                book.ISBN = reader.GetString(6);
                bookList.Add(book);
            }
            reader.Close();
            connection.Close();
            return bookList;
        }
        catch (SqlException sqlException)
        {
            _logger.LogError(sqlException, sqlException.Message, null);
        }
        catch (Exception exception)
        {
            _logger.LogError(exception, exception.Message, null);
        }
        returnnull;
    }

    public Book GetBookById(Int64 id)
    {
        SqlConnection connection = new SqlConnection (Configuration.GetConnectionString("myConnection"));
        try
        {
            connection.Open();
            SqlCommand sqlCommand = new SqlCommand("select bookindex,name,publishdate,author,price,publisher,ISBN from dbo.Book where bookindex = ?", connection);
            SqlDataReader reader = sqlCommand.ExecuteReader();
            sqlCommand.Parameters.Add(id);
            Book book = new Book();
            while (reader.Read())
            {
                book.bookindex = reader.GetInt64(0);
                book.name = reader.GetString(1);
                book.publishDate = reader.GetDateTime(2);
                book.author = reader.GetString(3);
                book.price = reader.GetDouble(4);
                book.publisher = reader.GetString(5);
                book.ISBN = reader.GetString(6);
            }
```

```
                reader.Close();
                connection.Close();
                return book;
            }
            catch (SqlException sqlException)
            {
                _logger.LogError(sqlException, sqlException.Message, null);
            }
            catch (Exception exception)
            {
                _logger.LogError(exception, exception.Message, null);
            }
            returnnull;
        }
    }
}
```

此为访问数据的类。其构造函数的参数是由 Controller 传入的。而 Controller 是从容器中获得的。Asp.net core 默认带一个容器,这个容器类似于过去的 Unity framework,可以用来管理一些对象的生命周期,管理对象的创建、销毁。此类用的是 ado.net 的方式访问数据库。GetBookNames 方法返回书名列表。GetBookList 返回书的列表,此列表包括 Book 表的几个字段。GetBookById 根据 ID 返回特定的 Book 数据,返回的都是 json 格式的数据。如 Book 列表就是一个 Book 的数组。

运行起来后,访问 http://localhost:55903/api/values,此 API 返回如下的数据:

```
[{
    "bookindex": 1,
    "name": "面向对象设计",
    "author": "李先生",
    "publishDate": "2017-05-01T00:00:00",
    "price": 60.0,
    "isbn": "54880594354",
    "publisher": "北航出版社"
},
{
    "bookindex": 2,
    "name": "页面设计",
    "author": "章和都",
    "publishDate": "2016-08-09T00:00:00",
    "price": 40.0,
    "isbn": "43546554654",
    "publisher": "电子工业出版社"
```

}]

最外面是中括号,表明它是一个数组。里面是数组元素,用一对大括号{}括起来。{}里的都是一个Book对象的几个字段。这段代码访问的表有两个:Book和Booknumber,这是Book表:

```
CREATE TABLE [dbo].[Book](
    [bookindex] [bigint] IDENTITY(1,1) NOT NULL,
    [name] [nvarchar](200) NOT NULL,
    [publishdate] [datetime] NOT NULL,
    [author] [nvarchar](100) NOT NULL,
    [price] [float] NULL,
    [ISBN] [nvarchar](100) NULL,
    [publisher] [nvarchar](100) NULL,
CONSTRAINT [PK_Book] PRIMARY KEY CLUSTERED
(
    [bookindex] ASC
) WITH (STATISTICS_NORECOMPUTE = OFF, IGNORE_DUP_KEY = OFF) ON [PRIMARY]
) ON [PRIMARY]
GO


ALTER TABLE [dbo].[Book] WITH CHECK ADD CONSTRAINT [FK_Book_BookNumber]
FOREIGN KEY([ISBN])
REFERENCES [dbo].[BookNumber]([ISBN])
ON UPDATE CASCADE
ON DELETE CASCADE
GO
```

注意看到它有一个外键,引用了BookNumber的ISBN字段。
BookNumber表是这样的:

```
CREATE TABLE [dbo].[BookNumber](
    [ISBN] [nvarchar](100) NOT NULL,
    [count] [int] NULL,
CONSTRAINT [PK_BookNumber] PRIMARY KEY CLUSTERED
(
    [ISBN] ASC
) WITH (STATISTICS_NORECOMPUTE = OFF, IGNORE_DUP_KEY = OFF) ON [PRIMARY]
) ON [PRIMARY]
GO


//ValuesController.cs
using System;
```

```csharp
using System.Collections.Generic;
using System.Linq;
using System.Threading.Tasks;
using Microsoft.AspNetCore.Mvc;
using Microsoft.Extensions.Configuration;
using Microsoft.Extensions.Logging;
using WebApplication1.DAL;
using WebApplication1.Entity;

namespace WebApplication1.Controllers
{
    [Route("api/[controller]")]
    publicclassValuesController : Controller
    {
        private readonly ILogger _logger;
        public IConfiguration Configuration { get; }
        DALBook dalBook;
        public ValuesController(IConfiguration configuration, ILogger <ValuesController> logger)
        {
            Configuration = configuration;
            _logger = logger;
            dalBook = new DALBook(Configuration, _logger);
        }
        // GET api/values
        [HttpGet]
        public IEnumerable <Book> Get()
        {
            return dalBook.GetBookList().ToArray();
        }

        // GET api/values/5
        [HttpGet("{id}")]
        public Book Get(int id)
        {
            return dalBook.GetBookById(id);
        }

        // POST api/values
        [HttpPost]
        public void Post([FromBody]string value)
        {
```

```
        }

        // PUT api/values/5
        [HttpPut("{id}")]
        public void Put(int id, [FromBody]string value)
        {
        }

        // DELETE api/values/5
        [HttpDelete("{id}")]
        public void Delete(int id)
        {
        }
    }
}
```

这个是 Controller 类,提供了 api/values,目前实现了 api/values 和 api/values/{id}这两个 api。

[Route("api/[controller]")]

注解表明了其 route 是 api/…,接着的是 controller 的名字。这里有一个 Values Controller,那么就有一个 api/values 的可访问路径。[controller]代表的是 controller 的名字。其构造函数参数 IConfiguration configuration, ILogger <ValuesController> logger 是要求容器提供这两个对象。容器建立了这两个对象,所以可以提供这两个对象。这些 API 都是 json 格式的消息。输入消息和返回消息都是 json 格式的。返回对象都必须是能序列化的。系统会自动将对象序列化成 json 消息。

此处用到的配置文件如下:

```
//appsettings.json
{
    "Logging": {
        "IncludeScopes": false,
        "Debug": {
            "LogLevel": {
                "Default": "Warning"
            }
        },
        "Console": {
            "LogLevel": {
                "Default": "Warning"
            }
        }
```

        },
        "ConnectionStrings": {
            "myConnection": "Data Source = mysqldb - 3432.database.windows.net;Initial Catalog = mytestdb - 324904390;uid = mikelij;pwd = fyt6&G%;"
        }
    }

4.7 把本地的 SQL Server 数据库迁移到 Azure SQL 云数据库上

把本地的 SQL Server 数据库迁移到 Azure SQL Database 上是一种比较顺利的过程。微软官方提供了一些工具和方法，帮助我们完成迁移。提供的工具包括 Microsoft® Data Migration Assistant，方法主要是两种：事务复制和有 downtime 的迁移。Data Migration Assistant 工具可以检测现有数据库中的所有对象是否有迁移兼容性问题，因为有一些部分支持功能和函数，所以只要用到了这些部分支持的功能和函数，就会存在迁移兼容性问题。

1. 有迁移兼容性问题的功能和函数

TSQL 语法有部分不兼容的地方：

Create 和 Alter Database 语句有很多选项。此语句包括文件位置、FILESTREAM，还有 Service Broker 的选项，这些选项只能适用于 SQL Server。如果在迁移前先用手工操作的方式创建 Azure SQL Database，那么这些选项没有关系。如果是用脚本的方式来进行数据库的迁移，则需要仔细地审阅这些 Create 或者 Alter Database 的语句，以确保所有的选项都与 Azure SQL Database 兼容。如果有一些不兼容的地方，则需要进行语句的修改。此处提到的选项是不兼容的。所以提到的这些选项都不能直接在 Azure SQL Database 上运行，需要修改脚本，去除这些选项后才能在 Azure SQL Database 上运行。Azure SQL Database 有 Service Objective 和 Elastic Scale 选项只适用于 Azure SQL Database。这个语句可以根据实际需要来选择是否加上 Service Objective 和 Elastic Scale 的选项。当然这两个选项可以不必在创建时就选，可以在稍后的时候再选择。可以通过 Azure Portal 界面来进行 Azure SQL Database 的配置，给它加上 Service Objective 和 Elastic Scale 的选项。

Create 和 Alter Login 的语句也是部分支持，因为此语句的一些选项在 Azure SQL Database 中不支持。为了保证数据库更具兼容性，Azure SQL Database 鼓励尽量使用内含的数据库用户，而不是用登录用户名。

2. 在 Azure SQL Database 中不支持的 TSQL 语法

前文已对比了 SQL Server 和 Azure SQL Database 的差别。除了那些之外，

TSQL 也有一些在 Azure SQL Database 中不支持的地方：

Systemobjects 不支持 Collation。

与连接相关的：Azure SQL Database 不支持 Windows Authentication，但是支持 Azure ADauthentication，可以支持最新的 SQL Server Management Studio。

3. 用三个或者四个部分名字的跨数据库的查询(只读查询语句只能用在 elasticpoolquery)

跨数据库的拥有者链，TRUSTWORTHY 设置如下：

EXECUTE AS LOGIN 需要改成 'EXECUTE AS USER'。

除了一些可扩展的 Key 管理，加密是支持的。

事件机制：事件、事件的通知、查询的通知。

文件的存放位置：那些关于文件位置的语法不再支持，因为 Azure SQL Database 管理文件的存放位置。

高可用性的设置是由 Azure SQL Database 来管理的，之前备份的语法、恢复的语法、AlwaysOn 语法、数据库镜像、Logshipping 的机制、恢复模式的语法，都不再需要。

Log 读取语法如 Push Replication、Change Data Capture 等不再在 Azure SQL Database 中支持。Azure SQL Database 可以作为 Push Replication 的订阅，这些都由 Azure SQL Database 管理。

这几个功能无法在 Azure SQL Database 中使用：fn_get_sql,fn_virtualfilestats,fn_virtualservernodes。

关于硬件的语法：如内存、工作者线程数、CPU、Trace 标记都不在 Azure SQL Database 中支持，需要改成 Service Level。

还有更多的不支持信息，请查阅：

https://docs.microsoft.com/en-us/azure/sql-database/sql-database-transact-sql-information。

https://docs.microsoft.com/en-us/azure/sql-database/sql-database-features。

方法一：有 Downtime 的迁移。

用此方法迁移一个或者集群化的 SQL Server 数据库需要有一些 Downtime。或者需要进行一些迁移后的针对生产环境的测试。下面介绍一些一般化的步骤来迁移 SQL Server 到 Azure SQL Database 或者 Managed Instance，如图 4-24 所示。

用最新版的 Data Migration Assistant(DMA)访问数据库并检测数据库的兼容性，如果存在兼容性问题，则 DMA 会列出来。

针对这些兼容性问题，准备一些必要的解决方法(fix，一些 TSQL 脚本)。

做一个事务级别完整的备份或者停止新的事务进入数据库，这样方便进行数据库的迁移。具体方法包括关闭客户端的连接，或者是创建数据库的一个快照(snapshot)。迁移后，可以用事务复制来更新迁移的数据库。

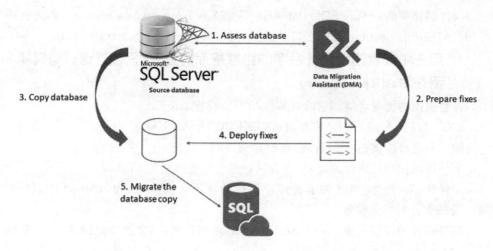

图 4-24 将 SQL Server 实例迁移到 Azure SQL Database

运行 TSQL 脚本来解决之前发现的兼容性问题,然后用 DMA 迁移数据库到一个新的 Azure SQL Database。

迁移完后,可以用事务复制来同步那些做快照后新产生的事务。

在迁移的时候有可能有性能问题,以下是一些建议,可以减少迁移时的性能问题。

- 选择预算允许的最高的服务级别和性能层来最大化数据传输的性能。在迁移完成后可以再改成相对便宜的服务级别和 Pricing tier,这样可以省点费用。
- 尽量缩短 BACPAC 文件和目的数据中心的物理距离。
- 取消自动进行统计计算的活动。
- 将表分区,并且建立必要的索引。
- 删除索引的视图,并且在迁移后再创建它们。
- 移除很少用的历史数据到别的数据库并迁移这些历史数据到一个其他的 Azure SQL Database。可以用 Elastic 查询这些历史数据。
- 在迁移后再允许全部重新统计计算。

方法二:用事务复制的方法进行迁移。

在不能停机(因为停机会带来服务的暂停、用户的失望)的情况下,只能使用事务复制的方法。要使用事务复制的方法,源数据库必须符合事务复制方法的要求并且兼容 Azure SQL Database,即没有前文所述的那些不兼容的问题,才能使用事务复制的方法进行迁移。如果有不符合事务复制方法的问题,必须先解决这些问题,修改之后才能采用事务复制的方法。

这种方法是将 Azure SQL Database 作为源数据库 SQL Server 的一个事务复制订阅者，然后让事务复制运行，将所有对数据的改动都复制到 Azure SQL Database，等完成了所有的事务复制，数据和最新改动都已复制到 Azure SQL Database，就可以将应用程序的链接串改到 Azure SQL Database。这样 Azure SQL Database 就成了新的生产环境。而 SQL Server 就可以处理掉了，同时事务复制也可以关掉了。

一个小帖士：可以只复制一个 SQL Server 上的一部分数据库，或者更少，比如一部分表，或者是部分数据，但这要根据实际情况来操作和选择。

当然，微软也提供了另外一个选择，叫作 Managed Instance。这个 managed instance 可以 100% 兼容本地的 SQL Server。有兴趣的读者可以参考，详见：https://docs.microsoft.com/en-us/azure/sql-database/sql-database-managed-instance。

4.8 Azure SQL Database Data Sync 数据同步功能

1. Data Sync 介绍

Data Sync 数据同步功能是 Azure SQL Database 提供的一个在多个 Azure SQL Database 和 SQL Server 同步数据的服务。

Data Sync 基于 Sync Group 的概念。Sync Group 可以将一些需要同步的数据库放到一个 Group 里来，然后 Sync 同步机制保证新产生的改动会双向同步到其他数据库。

一个 Sync Group 有如下一些属性：

Sync Schema 指的是哪个数据被同步。

Sync Direction 可以是双向，或者单向。可以是 Hubto Member，或者 Memberto Hub，或者是双向。

Sync Interval 表示同步多久发生一次。

Conflict Resolution Policy 指的是当有冲突时，如何解决冲突。可以是 Hubwins 表示采用 Hub 的数据，或者 Memberwins 表示采用 Member 的数据。

Data Sync 定义一个 Hub 和一些 Member，然后由一些拓扑来同步数据。只能有一个数据库作为 Hub，其余的作为 Member，只能 Azure SQL Database 作为 Hub，其余的如本地 SQL Server Database 或者在 Azure 虚拟机里的 SQL Server 可以作为 Member。当然 Azure SQL Database 也可以作为 Member。如果有两个以上 Azure SQL Database，则一个做 Hub，一个做 Member。

需要有一个数据库作为 Sync Database，Sync Database 包含 Metadata 和同步的 Log。Sync Database 必须是与 Hub 的 Azure SQL Database 同一个 Region 的 Azure SQL Database。

如图 4-25 所示是 Sync Group 的示意图。

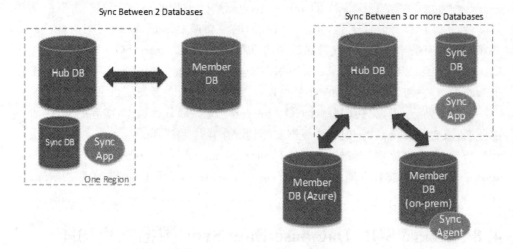

图 4-25 Sync Group 的示意图

2. Data Sync 的应用

① 混合式的应用（Hybrid Data Synchronization）：采用 Data Sync，可以在 on-premise 数据库和 Azure SQL Database 之间同步数据，并允许混合式的应用。这个能力对考虑将应用移动到 Azure 中的客户有吸引力。

② 分布式的应用（Distributed Applications）：在许多情况下，将压力分到多个数据库是有好处的。例如：有一个大的生产环境数据库，同时需要基于这些数据运行报表或者数据分析，这样就会有额外的压力，安排第二个数据库会对这些额外的压力有帮助。这样可以将额外压力造成的性能影响减到最小，可以用 Data Sync 来保持两个数据库同步。

③ 全球分布式的应用（Globally Distributed Applications）：许多商业应用扩展到好几个区域甚至好几个国家。为了最大化地减小网络延迟，最好将数据放到比较近的数据中心。有 Data Sync，可以让多个区域的数据库保持同步。

Data Sync 不适合如下几种情况：

① 灾难备份。

② 可伸缩的读。

③ ETL(OLTP to OLAP)。

④ 从 SQL Server 迁移到 Azure SQL Database。

3. Data Sync 如何工作

① 跟踪数据变动：用 Insert、Update、Delete 触发器来检测数据变动。这些变动被记录到一个另外的表中。

② 同步数据：Data Sync 被设计成一个 Hub 和 Spoke 模型。此 Hub 与每个 Member 进行单独同步。变动从 Hub 下载到 Member，同时变动也从 Member 上传

到 Hub。

③ 解决冲突：Data Sync 提供了两种方法来解决数据冲突：Hub 赢，或者 Member 赢。如果选项是 Hub 赢，那么 Hub 的变动总是会覆盖 Member 的变动。如果选项是 Member 赢，那么 Member 的变动总是覆盖 Hub 的变动。如果存在多个 Member，则最终的结果取决于哪个 Member 最先进行同步。

4. Data Sync 的需求和限制

(1) 最终的数据一致性

既然 Data Sync 是基于触发器的，故它不能完全符合事务型的一致性。但是微软保证数据变动都会被复制到其他数据库服务器，并且不会有数据丢失。

(2) 对性能的影响

Data Sync 用 Insert、Update、Delete 触发器来跟踪变动，创建另外的表来记录变动，对性能是有一些影响的。可以检查 Azure 服务的级别，如果需要升级就升级。

(3) 基本要求

① 每个表必须有一个主键(Primary Key)，每个主键不会被改动。

② Snapshot 的分离级别必须打开。

(4) 限　制

① 表中不能有不是主键的 Identity 字段。

② 主键不能有 Datetime 的字段。

③ 对象(数据库、表、视图、字段等)的名字不能有不能打印的字符，如""""[、]"，Azure AD Authentication 不支持。

(5) 不支持的数据类型

① File Stream 文件流；

② SQL/CLRUDT、SQL/CLR 自定义类型；

③ XML Schema Collection；

④ Cursor、Timestamp、Hierarchyid。

5. 创建 Sync Group

打开 Azure Portal，单击左边的 SQL databases 菜单，如图 4-26 所示。

在列出的所有数据库列表中选择需要作为 Hub 的数据库。这个作为 Hub 的数据库是作为中心点的，其他所有的 Member 数据库都是要跟这个 Hub 数据库进行同步的。

找到 Sync to other databases 菜单，如图 4-27 所示。

接着单击 New Sync Group，它分三步。

第一步是 Create sync group，在 Sync Group Name 框中输入名字。Sync Metadata Database 有两个选项，一个是新建一个 Metadata Database，另一个是利用已有的 Metadata Database，一般是选择新建一个，如图 4-28 所示。

图 4-26　单击 SQL databases 菜单

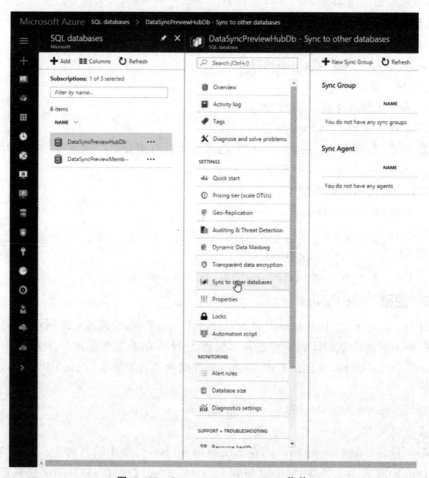

图 4-27　Sync to other databases 菜单

Azure SQL Database

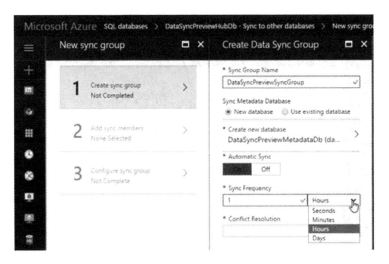

图 4-28 Great sync group

如果新建,则需要输入新 Metadata Database 的名字,配置好 Service Tier 等。然后单击 OK 按钮。

在 Automatic Sync 处,根据需要来选择 On 或者 Off。

如果选择 On,则需要在 Sync Frequency 下的框中选择一个时间间隔,如 1 Hours(小时),还有如图中显示的 Seconds(秒)、Minutes(分钟)、Days(天)。

在 Conflict Resolution 框中可选择 Hub Win(Hub 赢)或者 Member Win(Member 赢)。Hub Win 即以 Hub 数据库为准,若有冲突则按照 Hub 的数据来,丢弃 Member 中有冲突的数据。Member Win 即以 Member 数据库为准,如有冲突,则丢弃 Hub 中有冲突的数据。如果有多个 Member,则最先进行同步的 Member 对结果有影响。

第二步是 Add sync members,如图 4-29 所示。

Member 可以是 Azure SQL Database,也可以是 SQL Server。可以选择 Add an Azure Database,也可以选择 Add an On-Premises Database,应根据实际需要而定。这里先讲如何加入一个 Azure SQL Database。选择 Add an Azure Database,然后 Configure Azure Database 页面就出现了,如图 4-30 所示。

输入 Sync Member Name,然后选择一个 Subscription,如果只有一个 Subscription,就默认已选好了。然后就选择一个 Azure SQL Server,再就选择一个此 Azure SQL Server 的 Azure SQL Database,最后是选择同步方向。如果是双向(Bi-directionSync),则或者是往 Hub 同步(To the Hub),或者是从 Hub 到 Members(From the Hub),可根据实际需要来选择。

然后输入此 Azure SQL Server 的用户名和密码。这用来同步时能登录此数据库。

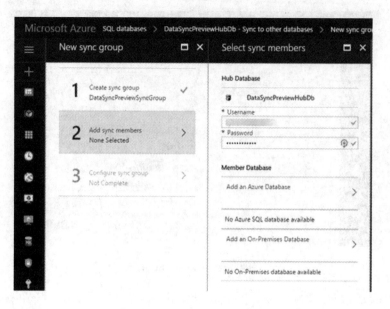

图 4-29 Add sync members

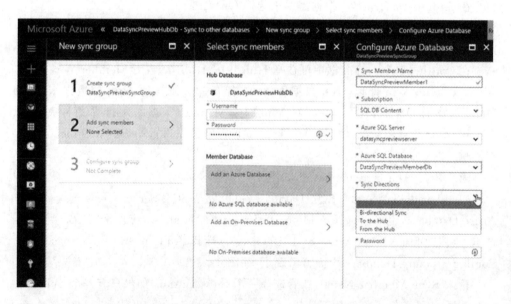

图 4-30 选择 Sync 类型

最后单击 OK 按钮。

可以选择 Add an On-Premises Database 来加一个 MS SQL Server 到此 Sync Group。Configure On-Premises 页面出来了,如图 4-31 所示。

一般需要创建一个 Sync Agent。

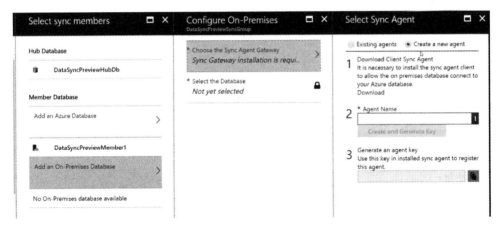

图 4 - 31　选择 Sync Agent

第一步是下载一个 Sync Agent 的客户端软件。这个客户端软件需要安装到 SQL Server。

第二步是输入一个 Agent 的名字。

第三步是产生一个 Agent Key，安装 Sync Agent 的时候用。详见图 4 - 32。

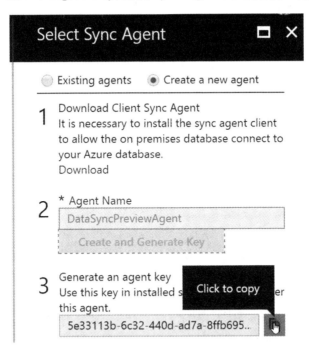

图 4 - 32　选择已有的 Sync Agent 或者建立新的 Sync Agent

在 SQL Server 的服务器上，运行 Client Sync Agent 程序，如图 4 - 33 所示。
先选择 Submit Agent Key，出现如图 4 - 34 所示界面。

图 4-33 运行 Client Sync Agent 程序

图 4-34 SQL Data Sync 2.0 界面

　　将前面产生的 Agent Key 粘贴到这里的 Agent Key 中，然后输入 Azure SQL Database 的 Metadata Database 的用户名和密码。可以单击 Test Connection 按钮测试链接。如果成功，则单击 OK 按钮。这里有可能遇到防火墙的问题，现象是连接不上，原因是 Azure SQL Database 默认都有防火墙防护安全的，你的外网 IP 没有添加到防火墙规则中。可以用 SSMS 连接一下作为 Hub 的 Azure SQL Database。这样就需要登录 Subscription 账号。接着就会问是否将外网 IPxx.xx.xx.xx 添加到 Azure SQL Database 的防火墙规则里，添加之后外网 IP 就可以通过防火墙了，这样就可以建立连接了。单击 OK 按钮，就会顺利连接。

　　设置好 Agent Key 之后，就可以注册一些 SQL Server 了。Register/Unregister 按钮就可以点击了。

单击 Register 按钮,出现如图 4-35 所示的界面。

图 4-35 建立到 SQL Database 的链接

输入 SQL Server 的 IP 或者域名以及数据库名字。然后点击 SQL 或者 Windows 登录方式,这可根据实际情况选择。如果有 SQL 登录账号,就可以选择它;如果有 Windows 的登录账号,也可以选择 Windows 登录账号。这个账号只在 Azure SQL Agent 中使用,用来访问 SQL Server。没有硬性规定,两种方法都可以。输入了 SQL/Windows 的登录账号之后,选择 Test Connection,成功后单击 Save 按钮。

然后回到 Azure Portal,在 Configure On-Premises 页面,因为之前的设置,可以选择之前设置的 Azure SQL Agent,见图 4-36 中第二个下拉框。然后就可以选择同步方向。双向(Bi-direction Sync)要求的网络流量比较大一些,改动会发生在 Hub 和 Members,万一发生了冲突,就会用到设置的冲突解决策略 Hub Win 或 Member Win。如果没有这么复杂,比如改动只发生在 Members,需要同步到 Hub,则可选 To The Hub;如果是改动只发生在 Hub,需要同步到 Members,则可以选择 From the Hub。详见图 4-36。

单击 OK 按钮关闭 Select Database 页面,单击 OK 按钮关闭 Configure On-Premises 页面。等这个新的 Sync Member 建立后,单击 OK 按钮关闭 Select Sync-Members 页面,如图 4-37 所示。

链接到 SQL Data Sync 和本地 Agent,将用户名加到 Role DataSync_Executor,Data Sync 在 SQL Server 创建这个 Role Data Sync_Executor。

然后就是配置 Sync group(ConfigureSyncGroup),这时要选择哪些表需要 Sync,选择 Configure Sync Group 之后,出现选择 Tables 页面,如图 4-38 所示。

上面有一个 Select a database,选择需要同步的数据库。选择了一个数据库名字,会刷新此数据库的表结构。Table 的列表会刷新。选择需要进行同步的表。将其左边的勾勾选上,如图 4-39 所示。

图 4-36 选择 Sync 数据库

图 4-37 选择 sync members

图 4-38 选择 Tables 页面

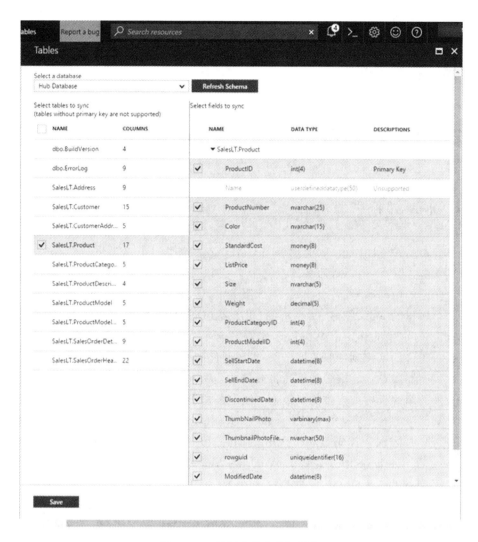

图 4-39 选择字段来进行同步

其右边会出现选择字段的页面,默认是全部字段。如果有特殊需要,则可以选择某些字段进行同步。最后单击 Save 按钮。

Azure SQL Data Sync 最小同步频率是 5 min。

第 5 章

Azure Active Directory

5.1 简 介

  Azure Active Directory(简称 Azure AD)是一个类似于 Windows Active Directory 的服务。在 Windows 服务器系统中有一个 Active Directory 的服务，它负责管理一个企业的内部账号、各级组织、各个服务器，以及一定数量的 PC，域管理员可以管理整个域里的资源、账号、共享资源、服务器访问权限、域的 Policy 等。这些都是在 Windows 域中的。一个 Windows 域需要有一个主域控制器服务器，比较大的企业还会有备份的域控制器服务器，万一出现硬件问题，可以由备份域控服务器来顶替主域控制器服务器。域管理界面将域里的资源显示成一个树形结构，其属性结构是按企业的组织架构来安排的。最上一级是公司最上层的组织。中间的是各层的组织。某某部门之下有某某子部门。Azure Active Directory 与 Windows Active Directory 的服务有一些区别。首先 Azure Active Directory 是基于微软云平台 Azure 的，而 Windows Active Directory 是基于 Windows Server 搭建的网络。其次它们提供的 API 是不同的，如 Azure Active Directory 的 API 可以是 RESTFul 的，要调用其 API，客户端可以是任何编程语言和操作系统平台，只要能支持发出 RESTFul 的调用请求就可以了。而 Windows Active Directory 的 API 是基于 Windows dll 的，还有基于.net dll 的，客户端必须得是能调用 Windows dll 的程序，或者能调用.net dll 的程序。通常来说 Linux 下是不支持运行 Windows dll 或者.net dll 的。这样对客户端的平台有一定要求。还有另外一个区别，在 Azure Active Directory 中，认证后的客户端拥有一个 JWT token，是一个字符串形式的。而在 Windows Active Directory 中，或者认证后的客户端，只是在内存里有一个 System.Directory.Services.Account Management.User Pricipal 类型。Azure Active Directory 调用的是 Microsoft.Identity Model.Clients.Active Directory 名称空间的类型，而 Windows Active Directory 调用的是 System.Directory Services.Account Management 名称空间的类型。

  Azure Active Directory 是微软开发的支持多个租户(tenant)、基于 Azure 云平台的账号管理服务，将核心目录服务、应用访问权限服务、账号保护等功能集成至一个解决方案。Azure Active Directory 提供了一个功能强大、基于标准的平台，允许开发者对他们的应用开发进行访问和控制,其基于中心化的政策(policy)和规则(rule)。

1. Azure Active Directory 的可靠性

Azure Active Directory 有多租户、地理位置分布式的、高可用性的设计。因此,用户可以在很多关键商业应用中信赖它,在全球 28 个数据中心中运行,并且有自动的 failover,假设一个数据中心出现问题,至少会有两个以上另外的数据中心拥有所有的目录树结构,这些数据中心都有不同的地理位置,不会受到影响,因此应用程序还是可以继续访问 Azure Active Directory 的数据,不受一个数据中心问题的影响。

2. Azure Active Directory 的版本

Azure Active Directory 有一个免费版本,此免费版本可以帮助用户管理账号和组,与 On-Premise 的目录同步,获得在 Azure、Office 365 的单点登录的功能,还有诸如 Salesforce、Workday、Concur、DocuSign、Google Apps、Box、ServiceNow、Dropbox 也支持。

除了免费版本,微软还提供了三个收费版本(注:此三个版本在 Azure China 没有提供,只能在 Azure Global 等其他环境里提供):

- Azure Active Directory Basic;
- Azure Active Directory Premium P1;
- Azure Active Directory Premium P2。

它们都是基于免费版本,提供了更多的企业级收费服务,如 Spanning Self Service、增强的监测、安全报告、多因子登录、手持设备登录等。

3. Azure Active Directory Basic 级别收费服务

它是为那些云优先的工作者设计的,这个版本提供了云中心的应用访问和自我服务的 ID 管理解决方案。用户将获得生产力的增强和费用的减少,如组的访问控制、自我服务的密码重置,还有 Azure Active Directory 的代理(使用 Azure Active Directory 用于部署 Web 应用),所有这些都在一个 99.9% 上线时间的企业级服务级别协议 SLA(Service Level Agreement)中。

4. Azure Active Directory Premium P1 级别收费服务

它是为那些有更多 ID 管理和访问控制需求的企业级客户所设计的。Azure Active Directory Premium P1 提供了功能丰富的企业级 ID 管理能力和允许混合的用户无缝地访问企业内云资源的能力。此版本的 Azure Active Directory 提供了所有信息工作者和 ID 管理者需要的功能、在混合的环境(企业内部网络和 Azure 云结合起来)跨应用的访问、自我服务的 ID 认证和访问控制管理,以及 ID 在 Azure 云中的保护和安全机制。此级别的收费服务支持高级的管理和代理权限给另外的账号来管理,以及动态组和自我服务的组管理。它包括微软 ID 管理服务(一个 On-PremiseID 和访问管理套件),并提供云 Write-Back 能力,使 On-Premise 用户能自我服务地进行密码重置。

5. Azure Active Directory Premium P2 级别收费服务

它为所有用户和管理员提供了高级保护措施。这个服务提供包括所有 P1 中提供的服务，在 P1 基础上增加了 ID 保护、ID 权限管理。Azure Active Directory ID 保护对上百万个信号产生影响，提供基于风险的条件访问（risk-basedconditional access）你的应用和关键的公司数据。微软用 Azure Active Directory Privileged Identity Management 帮助用户管理和保护有权限的账号，所以你可以发现、限制和监控管理员们和他们的资源访问权限，并且当需要的时候提供即时（just-in-time）的访问权限。

注：有一些版本的服务作为用多少支付多少的模式（pay as you go）。

Active Directory B2C 是一个面对消费者应用的 ID 和访问管理解决方案。除了可以给自己的员工建立账号外，还可以给消费者建立账号。

Azure Multi-Factor Authentication Azure 多因子登录能在较强的登录安全情况下使用，其收费方式可以基于每个用户或者每次登录。

5.2　开始使用 Azure AD

要使用 Azure Active Directory Premium，需要有订阅 Subscription。如下几种订阅都可以：

- Azure 或者 Office 365 的订阅；
- 企业级移动和安全 License 计划；
- 微软 Volume license 计划。

需要登录 Azure，来确认可以使用 Azure Active Directory Premium 级别的服务。

然后需要加自定义的域名。默认有一个域名：domainname.onmicrosoft.com，而你的企业有另外一个域名，则可以利用已有的域名建一个子域名，然后将这个子域名绑到 Azure AD 上。绑域名的操作可以在 Azure Portal 中选择菜单 Azure Active Directory，然后选择菜单 Custom domain names，再选择 Add custom domain，如图 5-1 所示。

接下来输入准备好的域名。这样就给 Azure AD 加上了自己的域名，那么用户就可以用熟悉的域名访问你的服务。输入自定义的域名，如图 5-2 所示。

最后需要确保绑定的域名是 Verified 状态。这表示绑定成功。

现在需要定制你的企业的登录页面，找到 Customized sign-in page，编辑定制的页面内容，然后定制好登录页面的样式、风格等。

下一步可以建立用户账号，有两个方法。一种方法是如果你的用户账号都是从已有的 Windows Active Directory 目录中获得的，则可以进行一个导入操作，将所有 Windows Active Directory 的用户账号都导入到 Azure AD 中。这种方法兼容性比

图 5-1　给 Azure AD 加上了自己的域名

较好。另外一种方法是在 Azure Portal 中一个一个地建立用户账号。这个是在之前用户账号的情况下采用。

下面要保证使用了正确的 License。如果打算使用 Azure AD Premium，则需要在此界面购买服务，如图 5-3 所示。

单击 Try/Buy 即可开始尝试或者购买服务。

图 5-2　输入自定义的域名

图 5-3　购买 Azure AD Premium 的服务

1. 分配 License

虽然获得订阅可以获得那些收费服务，但是还需要给每个用户分配 License，这样获得分配 License 的用户才能使用那些收费服务。License 分配必须得是用户和

Azure 收费服务(如 Azure AD Premium、Basic、Enterprise Mobility＋Security)的映射,可以用收费模式管理这些用户。你可以按组类给用户设计制定规则,例如:
- 所有用户自动获得 License;
- 那些有特定职级的人获得 License。

或者将这些决定代理给一些经理,然后由这些经理决定哪些用户需要获得收费的 Azure AD 服务(使用自我服务的组)。

2. 配置自我服务的密码重置

自我服务的密码重置允许管理员设置一个功能,允许用户自行解锁账号或者重置密码。系统会提供报告,可以查看用户使用记录,可以查看到用户的不正确使用和滥用的情况。在 Azure Portal 的 Azure Active Directory 中选择 Password reset,确保正确的用户和组分配了此权限。

3. 自行密码重置

首先需要建立一个租户(tenant)。在已有的租户中,选择 Azure Portal,然后选择 Azure Active Directory,再选择 Password reset。

在 Properties 页面,在 Self Service Password Reset Enabled 之下,选择:
- None:表示没有人可以用自行密码重置。
- Selected:表示选择了的用户可以用自行密码重置。建议建立一个专门的组来组织这些用户。
- All:表示所有用户都可以自行密码重置。

4. Authentication Methods 登录方法页面

重置密码所需的方法数(Numberof methods required to reset),通常只有一个,最大两个。

用户可用的登录方法如下:
- 电子邮件:发送一个动态码到预先配置的电子邮件地址,然后输入这个动态码就可以了。
- 移动电话:发送一个短信息,将动态码发给用户或者给此电话一个拨号,用语音传送一个动态码,用户得到动态码之后即可输入动态码。这样即可验证用户。
- 办公室电话:用预先存的用户办公室电话,系统打一个语音电话,通过语音传递一个动态码,用户得到动态码后就可输入验证。

5. 安全问题

这需要进行预先的设置。

① 选择每次注册需要多少个安全问题。然后就有多少个安全问题的用户重置用户密码。

② 选择安全问题。Azure 系统已经有一些安全问题,可以从中选择几个;如果

不满意，还可以选择 Custom 的选项，然后自己输入一些安全问题。这些安全问题选择好之后，用户注册的时候，就需要回答这些安全问题。这些安全问题的回答对于用户个人来说比较隐私，一般只有用户自己才知道答案，如图 5-4 所示。

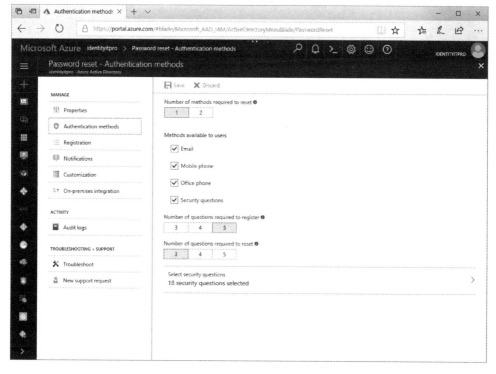

图 5-4　选择用户可以用哪些方式重置密码

这样用户就有多种选项来表明他/她是谁。

一旦系统可以确定用户身份，就可以让用户进行属于用户个人的一些操作。比如密码重置、解锁账号。这样用户自行密码重置的设置就设好了。

6. 在 Azure Active Directory 中加入公司的私密信息

一个租户的管理员可以给 Azure Active Directory 加入公司的私密信息。强烈建议将公司的私密政策和泄密后的联系人或者联系工作组都在 Azure Active Directory 中公布，这样企业的员工和外部客户都可以阅读到这些私密政策和联系人信息。具体操作如下：

在 Azure Portal 中，最左边菜单选择 Azure Active Directory，然后选择 Properties，如图 5-5 所示。

输入 Name，一般输入公司名字。

输入技术支持的邮件地址，这是一个关于 Azure AD 及账号的技术支持邮件地址。

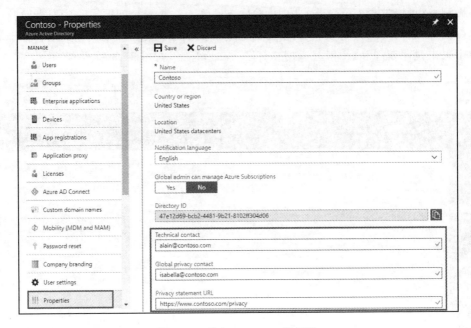

图 5-5　输入 Azure AD 的属性

输入全球私密联系邮箱,这个邮箱可以咨询个人数据私密性情况。

公司的私密政策的链接如图 5-5 中所示。通过此链接,员工和外部客户可以了解公司的私密政策。

7. 创建用户组

在 Azure Portal 中,选择 All services 菜单,然后搜索 Users and groups,如图 5-6 所示。

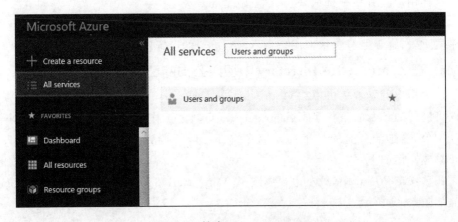

图 5-6　搜索 Users and groups

点击下面结果中出来的 Users and groups,出现对应的 Users and groups 页面。

选择 All groups 菜单，如图 5-7 所示。

图 5-7　显示 All groups 列表

右边出现了所有已创建的 groups 的列表。

单击 Add 按钮就去创建新的 groups，新建 groups 的界面出现了，如图 5-8 所示。

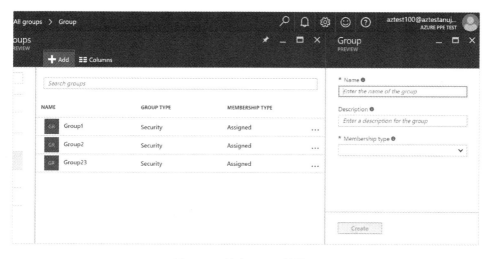

图 5-8　新建 groups 的界面

输入 Group Name，然后选择成员类型 Membership type：Assigned，即分配的。然后 Members 选项就出来了，点击 Members 就可以往此 Group 中加入成员，如图 5-9 所示。

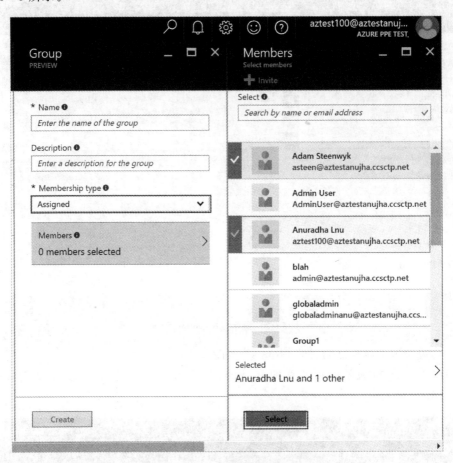

图 5-9　往组中加入成员

右边列表会把已建立好的用户和 Group 都列出来，因为 Group 之下，还可以有子 Group，选择好之后，单击 Select 按钮，此 Group 中就有成员了。当已经加入了成员后，就单击 Create 按钮，此 Group 就创建好了，如图 5-10 所示。

8. 管理 Group Members

除了在创建 Group 的时候可以加入 Member 外，在之后也可以管理 Member。

进入 Azure Portal，选择 All Services→Users and Groups→All groups→group→Members，然后可以点击 Add Members，如图 5-11 所示。

在列出的 Members 中选择一个。

删除与之类似，如图 5-12 所示。

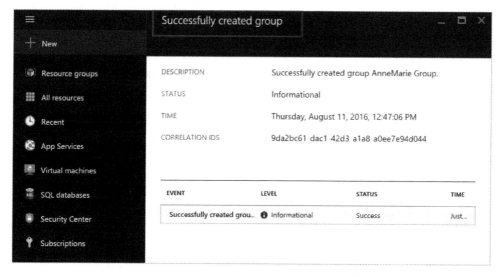

图 5-10　成功地创建了组

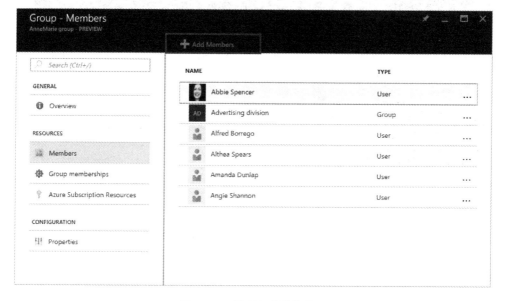

图 5-11　管理组成员的界面

查看 Group 中 Members 列表时,选中一个 Member,然后右边出现一个菜单,单击 Remove,再确认 Yes,即可删除此用户。

9. 注册 App

一个 App 要能跟 Azure AD 集成起来需要向 Azure Portal 注册。注册了以后就是备案了。备了案的才可以跟 Azure AD 互动,不然会被拒绝。

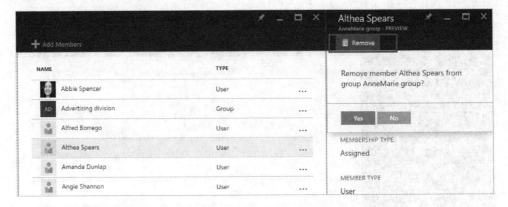

图 5-12 删除一个组成员

进入 Azure Portal,单击菜单项 Azure Active Directory,再找到 App registration 的菜单项,如图 5-13 所示。

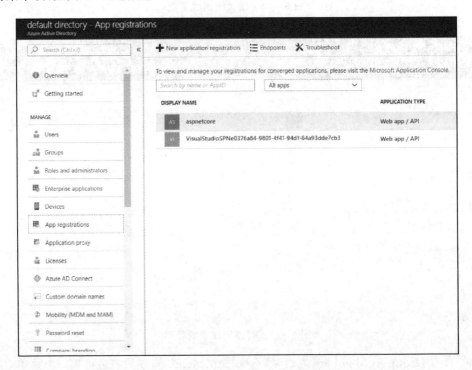

图 5-13 进行 App registration 注册

选择 New application registration 去新加一个 App registration 项(图 5-13 中的 aspnetcore 项就是加上去的),出现如图 5-14 所示的界面。

在 Name 下输入你所想的 Application 的名字。

图 5-14 新建一个 App registration

Application Type 只有 Web App/API 和 Native 两个选项。如果是在开发本地化的应用,可以选 Native;其他就选择 Web App/API。

在 Sign-on URL 下输入应用的登录 URL,假设域名是 contoso.com,那么登录的 URL 就是 contoso.com/login,这里用了一个微软官方的样例程序,所以暂时输入此样例程序的 URL,即 https://localhost:44353/,注意它是利用 IIS Express 启动的。然后单击 Create 按钮创建此 Application registration 项,如图 5-15 所示。

创建好的 Application 就出现在列表中。

然后要检查其 Reply URLs。单击 Settings,再单击 Reply URLs,如图 5-16 所示。

对于此样例程序,图中设置了两个 Reply URLs:

https://localhost:44353/signin-oidc;

https://localhost:44353/。

这两个都是可以的。

实际上后面这个是默认的。这个样例程序用了 OpenID connect 的协议。所以其 Reply URLs 也有一定的特殊性。后面讲到样例的时候会用到这些。

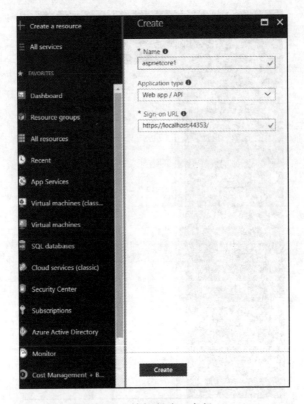

图 5-15 单击 Create 按钮创建一个新 App registration

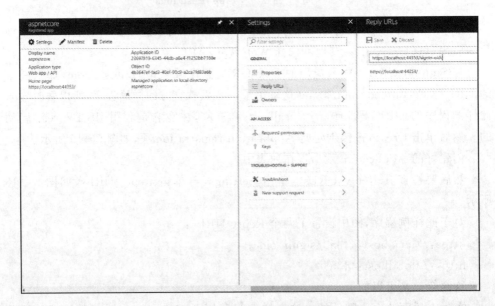

图 5-16 设置 Reply URLs

5.3　Azure AD 流程

这里有一些概念应明确：

① 目录(Directory)：一个 Active Directory 的实例。其内部存储了一个目录树，体现了组织的层级结构，有用户、组等。

② 租户(Tenant)：租户是针对多个机构用户的应用程序来说的。机构用户本身就是一个组织，它有一个组织结构，对应着目录。这个机构用户在此应用程序中是一个租户，租用了此应用程序的一定的存储空间，利用其计算资源、软件等，所以是其中的一个租户。一般一个租户对应一个目录。

③ 用户及其属性：用户是指某个租户中的单个用户，即某公司中的某一个特定的用户，如 contoso 公司的 Alice 用户。她有一个 Alice.Marryland@contoso.com 的邮箱，还有个人 ID、电话等。

④ 声明(Claim)：声明就是对某个个体的某项属性的一个确定的陈述。如：Alice 的邮箱是 Alice.Marryland@contoso.com，即是一个陈述。

安全令牌(Security Token)：安全令牌是声明的集合，即下面会说到的 JWT-Token。它内部有若干属性，还有 header，以及对属性的数字签名。

Azure AD 支持几种验证用户的方式。这里通过图 5-17 来表示。

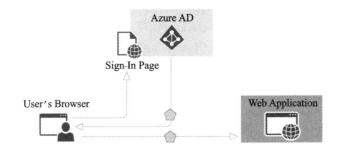

图 5-17　Azure AD 验证用户的方式

这个场景是用户使用一个浏览器(Browser)，需要向右侧的 Web Application 来验证其合法性。此 Web Application 发现此用户还没有获得验证，预先已设置了需要使用 Azure AD，因此将此用户重定向到 Azure AD 的服务器，用户在 Sign-in 页面里输入合法的用户名和密码。Azure AD 验证了用户的合法性之后，就给用户一个 Token，一般这个 Token 是一个 JWT Token，这个 Token 由用户浏览器存储，一般是存储在 Cookie、Session Storage、Local Storage 等地方。在用户获得 Azure AD 的验证之后，用户浏览器之后的请求(Request)都会有在请求中发送获得的 Token，如果存储在 Cookie 中，则每次发给服务器的请求都会带上这个 Cookie；如果是存储在别的地方，则浏览器发给 Web Application 的请求中都会带上这个 Token，比如在请求

Header 中,或者在页面 Form 的内容中,这样后续的请求都可被 Web Application 服务器所承认。图 5-17 中的五边形指的是一个 Token。

在这个场景下,Azure AD 作为 ID 提供者,负责验证用户的 ID 是否合法。而 Web Application 是将验证用户合法身份的工作交给了 Azure AD。Azure AD 存储了用户的用户名和密码,以及重置密码所需的安全问题回答等。只有合法的用户才知道这些。Web Application 在 Azure AD 中进行了注册,所以 Azure AD 可以接受此 Web Application 的请求,否则会拒绝来自 Web Application 的请求。Web Application 和 Azure AD 验证用户所用的协议包括 OAuth 2.0、OpenID Connect、WS-Federation、SAML 2.0 等。微软提供了一系列的库,包括浏览器中的.js 库、.ts 库、.net dll、Java 的 Jar 等,都是为了帮助用户进行这些协议的通信,以验证用户的合法身份。如 JWT Token 中包括一个此次登录的有效时间,这些库读到了 JWT Token 都会将 JWT Token 解析,了解其中包括的内容,验证 JWT Token 的合法性,检查用户的登录有效时间是否已到,是否需要重新登录等。一般一次登录的有效时间是 1 小时。过了 1 小时需要再次获得新的 Token。再次获得 Token 的时候,如果是设置长时间登录,则可从 Azure AD 服务器自动获得一个 Token;如果不是设置的长时间登录,则需要再输入用户名和密码才能获得新的 Token。

JWT Token 是一个自带比较多内容的 Token,详细内容见后面的章节。

任何想将登录验证工作交给 Azure AD 的应用,都必须向 Azure AD 的一个目录注册。Azure AD 需要知道此应用的几个信息:

① Application ID URL。这个 URL 可以唯一标识此应用程序。这个 Application ID URL 会被登记到 Token 中,当 Azure AD 收到一个 Token 时,Token 中相关的值标识了此 Token 的验证目标是哪个 Application ID。通过比较 Token 中的 Application ID 和 Azure AD 中记录的 Application ID,则能判断 Token 是否为此 Application 的 Token。

② Reply URL 或者 Redirect URL。这个 URL 是 Azure AD 用来传递已验证信息的 URL,当用户输入了合法的用户名和密码后,Azure AD 就向此 URL 通知用户已经经过验证,然后应用程序的此 URL 就根据 Azure AD 返回的信息来给予用户相应的身份并访问相应的资源。

③ Application ID。这个 ID 是 Azure AD 产生的一个 ID,用来唯一标识此应用。用来验证用户的时候,发出的请求中都带有这个 Application ID,Azure AD 接收到此 Application ID,就可以得到对应此应用程序的其他信息。

Key 是当调用应用程序 API 时与 Application ID 一起发给 Azure AD 的。

Azure AD 处理登录请求,与应用程序交换 Token 信息。Azure AD 必须确保应用程序有足够的权限访问此目录,同时有权限访问其他应用程序。

与 Azure AD 集成的应用程序主要是有两类:

① 单租户的应用程序。此应用程序只能有一个租户,对应一个目录树。目录树

可以有多层结构。此目录树属于一个公司组织。此目录由此公司的开发者注册。

② 多租户的应用程序。多租户的应用程序将被多个公司组织同时使用。这是一个典型的软件即服务(Software as a service)的应用程序。它有多个目录树,每个租户对应一个目录树。每个租户用租户的标识来与其他租户区别开。比如登录用户名中有后缀名,后缀名是公司名.com,通过这些区别,多租户互相之间就可以区别开。

在单租户的程序中,Azure AD 服务器有一个公共的端点:https://login.microsoftonline.com/common,这个端点适合所有的 Azure 目录。而在多租户的情况下,其端点是这样的:https://login.microsoftonline.com/contoso.onmicrosoft.com,此端点 URL 中有一个 contoso.onmicrosoft.com,这就是租户的 URL,代表的是 contoso 这个组织,当其用户向此 URL 发送登录请求时,Azure AD 检测到 URL 中的这个部分,就知道是租户 contoso 的用户。别的租户的这个部分是不同的。

1. 应用程序的类型和场景

应用程序可以是如下类型:

Web 浏览器和 Web Application,客户端即浏览器。Web Application 是由 Azure AD 保护的。用户需要登录。

① 单页应用:应用程序只有一页,其余的内容都是换页面的某个部分。此单页应用也是由 Azure AD 保护。

② Native Application 本地应用和 Web API:Native 应用可以是手机的本地应用,通过 RESTFul API 调用后端的 Web API,而此 Web API 由 Azure AD 保护。

③ Web Application 和 Web API:用户访问的是 Web Application,Web Application 后端访问的是 Web API。而 Web API 是由 Azure AD 保护,Web Application 可以由 Azure AD 保护,也可以不由 Azure AD 保护。这取决于具体应用。

④ Daemon Application、Server Application 和 Web API:此为后台监控程序或者服务器端的服务程序,调用一些 Web API,而此 Web API 由 Azure AD 来保护。

详见图 5-18。

通过上面的叙述,可以看出 Azure AD 适合保护应用程序系统中的某一部分,这部分要含有 http 或者 https 协议的访问接口。而那些运行在手机端的 Native 应用以及运行在服务器上的监控程序等,并不适合直接由 Azure AD 来保护。

2. Web 浏览器和 Web Application 的验证过程

此场景支持 WS-Federation、SAML 2.0、Open ID Connect 协议,如图 5-19 所示。

此流程如下:

① 用户浏览 Web Application。

② Web Application 检测到用户没有登录,向用户发送重定向的回复。此回复

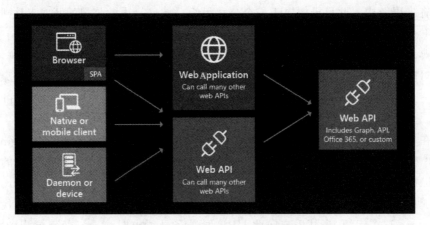

图 5-18　Azure AD 适合保护的应用

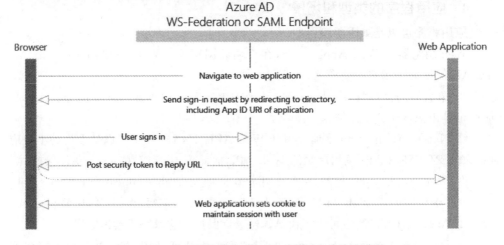

图 5-19　Web 浏览器和 Web Application 的验证过程

中包括 Azure AD 目录地址、App ID URL 等。

③ 用户浏览器接收到重定向回复,就自动重定向到 Azure AD 服务器。

④ Azure AD 根据目录名、App ID URL 等,显示相应的 Sign in 页面。

⑤ 用户输入合法的用户名和密码,Azure AD 通过,就向预先配置的 Reply URL 发送 Post Security Token(登录后的 Token)。

⑥ 应用程序的 Reply URL 接收到登录后的 Token,就设置 Cookie 以标识用户已登录。用户后续的请求中都有此 Cookie。

注:此处设置 Cookie 只是其中一种方法,还有可以将登录后 Token 存储到 Session Storage 或者 Local Storage 的方法。

此中场景不包括 OAuth 2.0 协议。

3. 单页应用的 Azure AD 验证流程

此场景下支持 OAuth 2.0、SAML 2.0、OpenID Connect、WS-Federation，如图 5-20 所示。

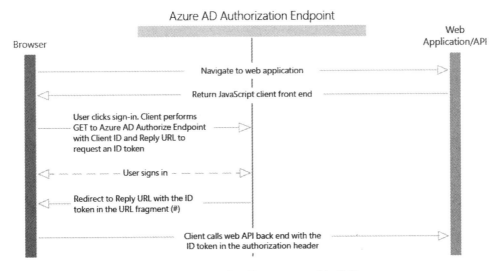

图 5-20 单页应用的 Azure AD 验证流程

单页应用流程如下：

① 用户访问此 Web Application。

② Web Application 向浏览器返回带 Java Script 的前端代码。

③ 用户点击 Sign-in(登录)，客户端执行 GET 请求到 Azure AD 的端点，请求中有 Client ID 和 Reply URL，去得到一个 ID Token。

④ 用户输入用户名和密码去 Sign-in。

⑤ 重定向用户到 Reply URL，并带着 ID Token。

⑥ 后续的请求，客户端调用 Web API 都带有这个 ID Token 在 http header 中。

ID Token 即可验证用户已经登录。客户端 Java Script 收到此 ID Token，可以验证此 Token 的有效性。比如首先验证数字签名是否正确，然后看此 ID Token 是否过期，看正确的 Azure 目录是否发布。

这个客户端的 Java Script 程序包括一个 Adaj.js 的库，此库由微软提供，此库可以帮助用户处理 ID Token 等，并接收 Token，存储 Token，验证 Token 的合法性；除了处理 ID Token 外，还可以处理 Access Token 等，还支持跨域处理(Cross-Origin Resource Sharing CORS)。

4. Native Application 本地化应用和 Web API

本地化应用调用 Web API 并获得 Access Token，获得 Access Token 后就去用 Access Token 来访问资源，如图 5-21 所示。

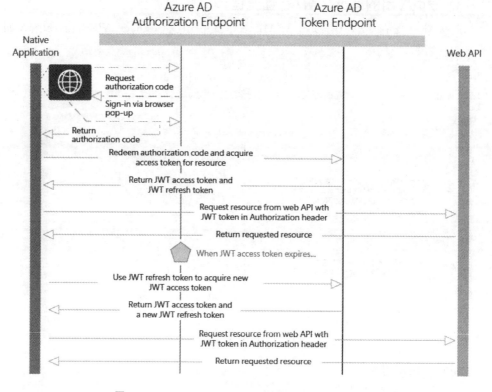

图 5-21 Native Application 本地化应用和 Web API

本地化应用获得 Token 的流程如下：

① 用户使用本地化应用，本地化应用向 Azure AD 授权（authorization）端点请求授权码，然后弹出浏览器登录窗口。

② Azure AD 授权端点返回授权码。

③ 根据授权码向 Azure AD Token 端点申请 Access Token。

④ Azure AD Token 端点返回 JWT Access Token 和 JWT Refresh Token。

⑤ 在获得 Access Token 之后，向 Web API 发送请求，申请所需的资源，并带上 Access Token 作为 http authorization header。

⑥ Web API 返回所请求的资源。

⑦ 当 JWT Access Token 过期时，本地化应用就用 JWT Refresh Token 去向 Azure AD Token 端点申请一个新的 Access Token。

⑧ Azure AD Token 端点生成一个新的 Access Token 和一个新的 Refresh Token 返回给本地化应用。

⑨ 在获得刷新的 Access Token 之后，本地化应用就向 Web API 继续申请资源。

⑩ Web API 验证了新 Access Token 后，就返回给本地化应用所申请的资源。

5. Web Application 和 Web API 的验证流程

通常 Web Application 和 Web API 都有 Azure AD 保护。Web API 的全部和 Web Application 的一部分路径需要用户登录后才能访问。此场景下能支持 OAuth 2.0、SAML 2.0、Open ID connect、WS-Federation 等协议。此场景下需要的是 ID Token，而不是 Access Token，如图 5-22 所示。

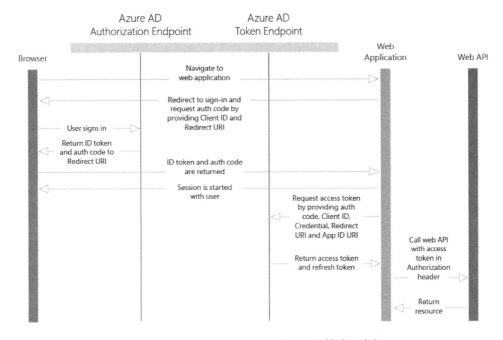

图 5-22 Web Application 和 Web API 的验证流程

① 用户浏览器向 Web Application 发出请求。

② Web Application 将用户重定向到 Sign-in 页面，并提供 Client ID 和重定向 URL 请求 Auth 码。

③ 用户去 Sign-in。

④ Azure AD Authorization 端点会发送 ID Token 和 Auth Code 并重定向到 Redirect URL。

⑤ 用户浏览器将返回的 ID Token 和 Auth Code 发给 Web Application。

⑥ Web Application 验证用户 ID Token，合法后就创建用户的 Session。

⑦ Web Application 向 Azure AD Token 端点申请 Access Token。

⑧ Azure AD Token 端点返回 Access Token 和 Refresh Token，之后，Web Application 用 Access Token 向 Web API 申请资源，方式是在请求 Authorization Header 时附上 Access Token。

⑨ Web API 返回所申请的资源。

从上可以看出，Token 主要有 ID Token、Access Token、Refresh Token。ID Token 是用于验证用户身份的，一旦 ID Token 过期了，用户就需要再获得一个新的 ID Token。要么是再输入一次密码，要么是保持 Session 久一点，Azure AD 会自动给用户再分配一个新的 ID Token。Access Token 是用于用户访问某一资源的。Refresh Token 是用于当 Access Token 过期后，再申请一个 Access Token 的。一般 Refresh Token 用过以后，就会产生一对新的 Access Token 和 Refresh Token。

5.4 JWT Token

JWT 是 JSON Web Token 的缩写。JWT 是一种紧凑且自包含的、用于在多方传递 JSON 对象的技术。传递的数据可以使用数字签名增加其安全性。可以用 HMAC 加密算法或 RSA 公钥/私钥加密的方式。其特点是：

① 紧凑：数据小，可以通过 URL、POST 参数、请求头发送，且数据小代表传输速度快。

② 自包含：使用 Payload 数据块记录用户必要且不隐私的数据，可以有效地减少数据库的访问次数，提高代码的性能。

JWT 一般用于处理用户身份验证或数据信息交换。

① 用户身份验证：一旦用户登录，每个后续请求都将包含 JWT，允许用户访问该令牌允许的路由、服务和资源。单点登录是当今广泛使用 JWT 的一项功能，因为它的开销很小，并且能够轻松地跨不同域使用。

② 数据信息交换：JWT 是一种非常方便的多方传递数据的载体，因为其可以使用数据签名来保证数据的有效性和安全性。

1. JWT 数据结构

JWT 的数据结构主要是三大块。由字符点"."来分隔三部分数据。

- Header(头信息)；
- Payload(内容载荷)；
- Signature(对内容载荷的数字签名)。

(1) Header 部分包括的内容

① alg 是加密算法定义内容，如 HMAC SHA256 或 RSA。

② typ 是 Token 类型，这里固定为 JWT。

(2) Payload 部分包括的内容

Payload 数据块一般用于记录实体(通常为用户信息)或其他数据，主要分为三个部分，分别是：已注册信息(registered claims)、公开数据(public claims)、私有数据(private claims)。

Payload 中常用的信息有：iss(发行者)、exp(到期时间)、sub(主题)、aud(受众)等。前面列举的都是已注册信息。

公开数据部分一般都会在 JWT 注册表中增加定义,以避免和已注册信息冲突。
公开数据和私有数据可以由程序员任意定义。

注意:即使 JWT 有签名加密机制,但是 Payload 内容都有明文记录,除非记录的是加密数据,否则不排除泄露隐私数据的可能。不推荐在 Payload 中记录任何敏感的数据。

(3) Signature 对内容载荷的签名

对内容载荷的数字签名信息是基于密码学的数字签名得到的,用以确保内容载荷没有被篡改。这是一个由 Azure AD 系统提供的信息,是验证服务器传递的数据是否有效安全的标准。在生成 JWT 最终数据之前,先使用 Header 中定义的加密算法,将 Header 和 Payload 进行加密,并使用点进行连接,如加密后的 Head、加密后的 Payload;再使用相同的加密算法,对加密后的数据和签名信息进行加密,得到最终结果。

2. JWT 执行流程

(1) 基于 JWT 机制的单点登录

其实现详见后面的代码。

注意:使用 JWT 实现单点登录时,需要注意 Token 的时效性。Token 是保存在客户端的令牌数据,如果永久有效,则有被劫持的可能。Token 在设计的时候,可以考虑一次性有效或一段时间内有效。如果设置有效时长,则需要考虑是否需要刷新 Token 有效期的问题。

(2) Token 保存位置

使用 JWT 技术生成的 Token,客户端在保存的时候可以考虑 Cookie 或 Local Storage。Cookie 保存方式,可以实现跨域传递数据。Local Storage 是域私有的本地存储,无法实现跨域。

(3) Webstorage

Web Storage 可保存的数据容量为 5 MB,且只能存储字符串数据。

Web Storage 分为 Local Storage 和 Session Storage。

Local Storage 的生命周期是永久的,关闭页面或浏览器之后,Local Storage 中的数据也不会消失。Local Storage 除非主动删除数据,否则数据永远不会消失。

Session Storage 是会话相关的本地存储单元,生命周期是仅在当前会话下有效。Session Storage 引入了一个"浏览器窗口"的概念,Session Storage 是在同源的窗口中始终存在的数据。只要这个浏览器窗口没有关闭,即使刷新页面或者进入同源另一个页面,数据也依然存在。但是 Session Storage 在关闭了浏览器窗口后就会被销毁。同时,独立地打开同一个窗口同一个页面,Session Storage 也是不一样的。

样例 JWT Token 如下:(这个样例 Token 从 https://docs.microsoft.com/en-us/azure/active-directory/develop/active-directory-token-and-claims 页面拷贝而来)

eyJ0eXAiOiJKV1QiLCJhbGciOiJub25lIn0.eyJhdWQiOiIyZDRkMTFhMi1mODE0LTQ2YTctODkwYS0yNz-
RhNzJhNzMwOWUiLCJpc3MiOiJodHRwczovL3N0cy53aW5kb3dzLm5ldC83ZmU4MTQ0Ny1kYTU3LTQzODUtYmVjYi-
i02ZGU1N2YyMTQ3N2UvIiwiaWF0IjoxMzg4NDQwODYzLCJuYmYiOjEzODg0NDA4NjMsImV4cCI6MTM4ODQ0NDc-
MywidmVyIjoiMS4wIiwidGlkIjoiN2Z1ODE0NDctZGE1Ny00Mzg1LWJlY2ItNmRlNTdmMjE0NzdlIiwib2lkIjo-
iNjgzODlhZTItNjJmYS00YjE4LTkxZmUtNTNkZDEwOWQ3NGY1IiwidXBuIjoiZnJhbmttQGNvbnRvc28uY29-
iwidW5pcXVlX25hbWUiOiJmcmFua21AY29udG9zby5jb20iLCJzdWIiOiJKV3ZZZENQaGhTUGhCVGMVpzZjd5WVV4-
U2hVd3RVbTV5elBtdl8tangzZkhZIiwiZmFtaWx5X25hbWUiOiJNaWxsZXIiLCJnaXZlbl9uYW1lIjoiRnJh-bm-
sifQ.

初一看它是用Base64编码的。打开浏览器，访问jwt.io。将此样例Token粘贴到jwt.io中解码，如图5-23所示。

图5-23 JWT网站解析出来的Token信息

它的Header如下：

{
"typ": "JWT",
"alg": "none"
}

alg为none，说明它没有用算法做数字签名。如果有，则是数字签名过的Token。而其Payload载荷部分如下：

{
"aud": "2d4d11a2-f814-46a7-890a-274a72a7309e",
"iss": "https://sts.windows.net/7fe81447-da57-4385-becb-6de57f21477e/",
"iat": 1388440863,
"nbf": 1388440863,
"exp": 1388444763,
"ver": "1.0",

```
"tid": "7fe81447 - da57 - 4385 - becb - 6de57f21477e",
"oid": "68389ae2 - 62fa - 4b18 - 91fe - 53dd109d74f5",
"upn": "frankm@contoso.com",
"unique_name": "frankm@contoso.com",
"sub": "JWvYdCWPhhlpS1Zsf7yYUxShUwtUm5yzPmw_ - jX3fHY",
"family_name": "Miller",
"given_name": "Frank"
}
```

其中比较重要的是 exp,即过期时间。"exp": 1388444763,这个数是一个日期。可以用浏览器的调试窗口,输入 new Date(1388444763),然后看一下此数是什么日期时间,也就是此 Token 什么时候过期。通常的情况下,ID Token 过期时间是一个小时,即从发布这个 Token 的时间往后延一个小时。

5.5 使用 Azure AD 样例

样例工程从网址 https://azure.microsoft.com/en-us/resources/samples/active-directory-dotnet-webapp-openidconnect-aspnetcore-v2/处下载。

打开以后如图 5 - 24 所示。

图 5 - 24   Open ID Connect 样例程序打开后的样子

首先来看 config.json 文件:

```
{
    "AzureAd": {
        "ClientId": "23697819 - 6345 - 44db - a6e4 - f1252bb7160e",
```

```
        "AadInstance": "https://login.microsoftonline.com/contoso.onmicrosoft.com",  // This is the public instance of Azure AD
        "PostLogoutRedirectUri": "https://localhost:44353/"
    }
}
```

这个 Client Id 是从对应的 App Registration 那里获得的,如图 5-25 所示。

图 5-25　在 App Registration 中的 Application ID

注册 Application 之后,会有一个 Application ID,见图 5-24,点击 Click to copy 就拷贝到剪贴板了。然后就放到这个 Client Id 中来。

Aad Instance 默认的值是 https://login.microsoftonline.com/{0}{1},把它替换成 https://login.microsoftonline.com/contoso.onmicrosoft.com,其中 contoso.onmicrosoft.com 是属于这个特定的目录的域名,如前文所述,可以加入一个自定义的域名,如 <公司英文名>.com 这样的格式,那么就可以用这个自定义的域名来代替 contoso.onmicrosoft.com 部分。对于 Azure AD 来说,这个自定义的域名或者 contoso.onmicrosoft.com 都是可以认识的。因为这两个都是你这个组织的域名,只不过自定义的域名会更好读。

PostLogoutRedirectUri 这个项就保持不变。当然了,如果是正式发布一个应用,这个参数也是要根据生产环境的域名和 URL 来改的。

再来看 Controller/AccountController.cs 文件。

```
using System;
using System.Collections.Generic;
using System.Linq;
using System.Threading.Tasks;
using Microsoft.AspNetCore.Authentication.Cookies;
using Microsoft.AspNetCore.Authentication.OpenIdConnect;
using Microsoft.AspNetCore.Http.Authentication;
using Microsoft.AspNetCore.Mvc;


// For more information on enabling MVC for empty projects, visit http://go.microsoft.
```

//com/fwlink/? LinkID = 397860

```csharp
namespace WebApp_OpenIDConnect_DotNet.Controllers
{
    public class AccountController : Controller
    {
        // GET: /Account/Login
        [HttpGet]
        public async Task Login()
        {
            if (HttpContext.User == null || !HttpContext.User.Identity.IsAuthenticated)
                await HttpContext.Authentication.ChallengeAsync(OpenIdConnectDefaults.AuthenticationScheme,
                    new AuthenticationProperties { RedirectUri = "/" });
        }

        // GET: /Account/LogOff
        [HttpGet]
        public async Task LogOff()
        {
            if (HttpContext.User.Identity.IsAuthenticated)
            {
                await HttpContext.Authentication.SignOutAsync(OpenIdConnectDefaults.AuthenticationScheme);
                await HttpContext.Authentication.SignOutAsync(CookieAuthenticationDefaults.AuthenticationScheme);
            }
        }
    }
}
```

它有两个方法：一个是 login，一个是 logout。

它利用了 OpenIdConnectDefaults.AuthenticationScheme，用的是 Open ID Connect 的协议。

HttpContext.User.Identity.IsAuthenticated 可以告诉我们用户是否登录。所以在需要用户登录的地方可以用此代码进行检测。

根据需要，也可以考虑换成别的协议。这得看实际的情况来定。

其应用 CookieAuthenticationDefaults.AuthenticationScheme，是用 ASP.NET Core 2.1 的特性存储获得的 Azure AD 的 Token 到一个 Cookie 中。这个 Signout

过程就是将此 cookie 清除。

再看 Program.cs 文件：

```
using System;
using System.Collections.Generic;
using System.IO;
using System.Linq;
using System.Threading.Tasks;
using Microsoft.AspNetCore.Hosting;

namespace WebApp_OpenIDConnect_DotNet
{
    public class Program
    {
        // Entry point for the application.
        public static void Main(string[] args)
        {
            var host = new WebHostBuilder()
                .UseKestrel()
                .UseContentRoot(Directory.GetCurrentDirectory())
                .UseIISIntegration()
                .UseStartup<Startup>()
                .Build();

            host.Run();
        }
    }
}
```

此文件用了一句 new WebHostBuilder().UseKestrel()。

此 Kestrel 服务器是一个开发用的服务器。在生产环境中不需要这句，可以用 IIS 或者 Azure 的云环境部署应用。

UseContentRoot(Directory.GetCurrentDirectory())指定当前目录为应用的访问目录。

UseIISIntegration()在应用运行于 ASP.NET Core module 的情况下配置应用的基础路径和端口，此方法在 IIS 和 Azure App Service 服务环境中都可以运行。

UseStartup<Startup>()指定 Startup 类为 WebHost 的起始配置类。

Startup 类在 Startup.cs 中。我们来看 Startup.cs 文件。

```
using System;
using System.Collections.Generic;
using System.Linq;
```

```csharp
using System.Threading.Tasks;
using Microsoft.AspNetCore.Authentication;
using Microsoft.AspNetCore.Authentication.Cookies;
using Microsoft.AspNetCore.Authentication.OpenIdConnect;
using Microsoft.AspNetCore.Builder;
using Microsoft.AspNetCore.Hosting;
using Microsoft.Extensions.Configuration;
using Microsoft.Extensions.DependencyInjection;
using Microsoft.Extensions.Logging;
using Microsoft.IdentityModel.Protocols.OpenIdConnect;
using Microsoft.IdentityModel.Tokens;
using System.Globalization;

namespace WebApp_OpenIDConnect_DotNet
{
    public class Startup
    {
        public Startup(IHostingEnvironment env)
        {
            // Set up configuration sources.
            Configuration = new ConfigurationBuilder()
                .SetBasePath(env.ContentRootPath)
                .AddJsonFile("config.json")
                .AddJsonFile("appsettings.json")
                .Build();
        }

        public IConfigurationRoot Configuration { get; set; }

        // This method gets called by the runtime. Use this method to add services to the
        // container.
        public void ConfigureServices(IServiceCollection services)
        {
            // Add MVC services to the services container.
            services.AddMvc();

            // Add Authentication services.
            services.AddAuthentication(sharedOptions =>
sharedOptions.SignInScheme = CookieAuthenticationDefaults.AuthenticationScheme);
        }

        // This method gets called by the runtime. Use this method to configure the HTTP
```

```csharp
// request pipeline.
public void Configure(IApplicationBuilder app, IHostingEnvironment env, ILoggerFactory loggerFactory)
{
    // Add the console logger.
    loggerFactory.AddConsole(Configuration.GetSection("Logging"));

    // Configure error handling middleware.
    app.UseExceptionHandler("/Home/Error");

    // Add static files to the request pipeline.
    app.UseStaticFiles();

    // Configure the OWIN pipeline to use cookie auth.
    app.UseCookieAuthentication(new CookieAuthenticationOptions());

    // Configure the OWIN pipeline to use OpenID Connect auth.
    app.UseOpenIdConnectAuthentication(new OpenIdConnectOptions
    {
        ClientId = Configuration["AzureAD:ClientId"],
        Authority = string.Format(CultureInfo.InvariantCulture, Configuration["AzureAd:AadInstance"], "common", "/v2.0"),
        ResponseType = OpenIdConnectResponseType.IdToken,
        PostLogoutRedirectUri = Configuration["AzureAd:PostLogoutRedirectUri"],
        Events = new OpenIdConnectEvents
        {
            OnRemoteFailure = RemoteFailure,
            OnTokenValidated = TokenValidated
        },
        TokenValidationParameters = new TokenValidationParameters
        {
            // Instead of using the default validation (validating against
            // a single issuer value, as we do in line of business apps),
            // we inject our own multitenant validation logic
            ValidateIssuer = false,

            NameClaimType = "name"
        }
    });

    // Configure MVC routes
```

```csharp
        app.UseMvc(routes =>
        {
            routes.MapRoute(
                name: "default",
                template: "{controller = Home}/{action = Index}/{id?}");
        });
    }

    private Task TokenValidated(TokenValidatedContext context)
    {
        /* --------------------
        // Replace this with your logic to validate the issuer/tenant
        --------------------
        // Retriever caller data from the incoming principal
        string issuer = context.SecurityToken.Issuer;
        string subject = context.SecurityToken.Subject;
        string tenantID = context.Ticket.Principal.FindFirst("http://schemas.microsoft.com/identity/claims/tenantid").Value;

        // Build a dictionary of approved tenants
        IEnumerable <string> approvedTenantIds = new List <string>
        {
            " <Your tenantID> ",
            "9188040d-6c67-4c5b-b112-36a304b66dad" // MSA Tenant
        };

        if (!approvedTenantIds.Contains(tenantID))
            throw new SecurityTokenValidationException();
        -------------------- */

        return Task.FromResult(0);
    }

    // Handle sign-in errors differently than generic errors.
    private Task RemoteFailure(FailureContext context)
    {
        context.HandleResponse();
        context.Response.Redirect("/Home/Error? message = " + context.Failure.Message);
        return Task.FromResult(0);
    }
}
```

}

这里重要的一句是：

```
// Add Authentication services.
services.AddAuthentication(sharedOptions =>
sharedOptions.SignInScheme = CookieAuthenticationDefaults.AuthenticationScheme);
```

这句是加入验证用户账号的机制。采用 Cookie 存储登录 Token 的机制。

```
public Startup(IHostingEnvironment env)
{
    // Set up configuration sources.
    Configuration = new ConfigurationBuilder()
        .SetBasePath(env.ContentRootPath)
        .AddJsonFile("config.json")
        .AddJsonFile("appsettings.json")
        .Build();
}
```

此方法是进行 WebHost 的配置项设置。

.SetBasePath(env.ContentRootPath)用来设置应用的基础目录。它用的是环境变量 ContentRootPath。

随后这两句：

.AddJsonFile("config.json")

.AddJsonFile("appsettings.json")

是读取两个 json 文件作为配置文件。

Config.json 文件已经交代过。

Appsettings.json 文件内容如下：

```
{
    "Logging": {
        "IncludeScopes": false,
        "LogLevel": {
            "Default": "Debug",
            "System": "Information",
            "Microsoft": "Information"
        }
    }
}
```

这个文件对 Logging 进行了设置。Loglevel 进行了多个级别的设置。

第6章

Service Bus

6.1 基础

Service Bus 简单地说是一个消息总线(Message Bus)，主要用于消息处理。架构设计上曾经有 ESB。ESB 的全称为 Enterprise Service Bus，即企业服务总线。它是传统中间件技术与 XML、Web 服务等技术结合的产物。ESB 提供了网络中最基本的连接中枢，是构筑企业神经系统的必要元素。ESB 从面向服务体系架构(Service-Oriented Architecture，SOA)发展而来，是传统中间件技术与 XML、Web 服务等技术结合的产物。

ESB 提供了网络中最基本的连接中枢，是构筑企业神经系统的必要元素。ESB 采用了"总线"这样一种模式来管理和简化应用之间的集成拓扑结构，以广为接受的开放标准为基础来支持应用之间在消息、事件和服务级别上动态的互联互通，是一种在松散耦合的服务和应用之间标准的集成方式。它可以作用于：

① 面向服务的架构：各服务间互相调用，传递数据。

② 面向消息的架构：消息是驱动整个系统的核心元素，包含消息队列、先进先出处理、子系统间传递消息和处理消息。

③ 事件驱动的架构：事件源、事件侦听、事件响应、事件消息传递、事件消息处理。

ESB 的出现改变了传统的 SOA 软件架构，可以提供比传统中间件产品更为低廉的解决方案；同时，ESB 减少了子系统之间的耦合，提高了内聚。只要准寻接口的消息标准，上下游就可以处理消息。允许不同平台、不同编程语言的应用之间的技术差异，让不同的应用服务器协调运作，实现了不同服务之间的通信与整合。ESB 是传统中间件技术与 XML、Web 服务等技术相互结合的产物，用于实现企业对不同消息和信息的准确、高效和安全传递，让不同的应用服务协调运作，实现不同服务之间的通信与整合。从功能上看，ESB 提供了事件驱动和文档导向的处理模式，以及分布式的运行管理机制，它支持基于内容的路由和过滤，具备了复杂数据的传输能力，并可以提供一系列的标准接口。

1. 企业服务总线 ESB 的功能

(1) 服务的 Meta Data 管理

在总线范畴内对服务的注册命名及寻址进行管理。

(2) 集成服务

提供基础的服务和定制的服务；支持集成服务模式；支持服务分解、服务调度和路由、服务封装、服务组合。

(3) 公用服务

提供内置的各种公用服务，例如认证服务、日志服务等。

(4) 传输服务

确保通过企业总线互连的业务流程间消息的正确交付，还包括基于内容的路由功能。

(5) 中 介

提供位置透明的路由和定位服务；提供多种消息传递形式；支持广泛使用的传输协议。

(6) 服务协议转换

通过把不同的通信协议转换成标准的报文，屏蔽异构系统的底层技术差异。

(7) 服务监控

提供服务等级管理及流量管理。提供多角度的服务实时监控、报警与交易分析报表。

(8) 安全体系

提供多种安全机制并支持和第三方安全系统的有效集成，提供有效的安全监控机制。

2. ESB 和 SOA 的区别

SOA(Service-Oriented Architecture，面向服务的体系结构)是一个组件模型，它将应用程序的不同功能单元(称为服务)通过这些服务之间定义的良好的接口和契约联系起来。接口是采用中立的方式进行定义的，它应该独立于实现服务的硬件平台、操作系统和编程语言。这使得构建在各种各样的系统中的服务可以使用一种统一和通用的方式进行交互。

ESB(Enterprise Service Bus，企业服务总线)是一种开放的、基于标准的分布式同步或异步信息传递中间件。通过 XML、Web Service 接口以及标准化基于规则的路由选择文档等支持，ESB 为企业应用程序提供安全互用性。

二者的区别如下：

SOA 主要使用服务来作为接口。其对外接口是一致的。调用方和被调用方都遵循统一的接口。

ESB 主要使用消息来交互。系统各组件之间用消息来驱动，交换的是消息。消息是 xml 格式的或者其他格式的。

在云计算时代之前,有多家公司提供了不同的产品:Oracle Service Bus、IBM Web Sphere ESB,微软有 Microsoft Biz Talk Server 系列。还有一些开源产品,如 Jboss ESB、Mule ESB、Service Mix/FUSE ESB、Synapse/WSO2 ESB 等。进入云计算时代之后,微软在 ESB 这部分的产品就是 Azure 平台上的 Service Bus 服务。

Azure Service Bus 主要有 Queue、Topic、Relays 等服务。

Queue 就是消息队列。消息产生端可以是任何应用、Web 应用、移动应用,或者一个 Service,如 RESTFul 的服务。给这个 Queue 发送消息,这些消息在被取出之前就存储在 Queue 中。在过期之前,侦听消息的应用就会将消息取出并处理。这样的模式,就将消息发送者和消息接收者解耦,它们之间就不必有比较强的依赖。它们之间的依赖就变成了消息格式的定义。消息格式是很重要的,相当于一个合同。消息队列的流程如图 6-1 所示。

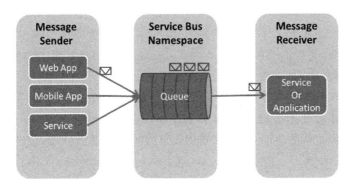

图 6-1 一个消息队列的发送-接收图

一般地,Queue 中的消息是符合先进先出的规律的,但是在有些选项下,Queue 中的消息不一定是先进先出,即 Enable session 为否的时候,则不能保证消息的先进先出,而有可能部分消息是先进先出,另外一些则可能是后进先出。

发送消息的过程很简单:发送方将消息发送至 Service Bus 队列,接收方在随后的某个时间内接收该消息。一个队列只能有一位接收方,如图 6-1 所示;或者多个应用程序可从同一队列中读取消息。在后一种情况下,每条消息通常仅由一位接收方读取,队列不提供多播服务。如果是多线程的情况,则可以有多个线程作为客户端来访问这些消息。

每条消息由两部分组成:消息正文和一个键/值对,每个消息都可以定义一些自定义的属性,使用的方式取决于应用程序尝试执行的操作。例如,发送近期销售消息的应用程序可能包含自定义属性,根据这些自定义属性,应用可以进行相应的数据处理。消息正文可能包含已签署的销售合同的扫描图像,如果不包含该合同,则只需留空。

接收方可采用两种不同方式从 Service Bus 队列中读取消息。第一种方式称作

Receive And Delete，即从队列中接收消息并立即将其删除。此操作很简单，但如果接收者在完成处理消息之前崩溃，则该消息将丢失。因为消息已从队列中移除，所以接收方无法访问该消息。

另外一种方式 Peek/Lock 可以帮助解决这个问题。与 Receive And Delete 一样，Peek Lock 可从队列中移除消息。不过，它不会马上删除该消息；相反，它会锁定该消息，使其对其他接收方不可见，然后等待以下三个事件之一：

① 如果接收方成功处理了该消息，将调用 complete，并且队列将删除该消息。

② 如果接收方判定它无法成功处理该消息，将调用 Abandon。队列即会解除对该消息的锁定，使其可供其他接收方使用。

③ 如果接收方在可配置时间段（默认为 60 s）内没有调用这两个命令，则队列将假定接收方失败。在这种情况下，队列的行为就像接收方已调用"放弃"一样，即使消息可供其他接收方使用。

请注意，可能发生的情况：同一条消息可能被发送两次，即或许将其发送给两个不同的接收方。使用 Service Bus 队列的应用程序必须为此做好准备。为了更轻松地进行重复检测，每条消息都拥有一个唯一的 Message ID 属性，无论从队列中读取消息多少次，该属性默认情况下都始终保持不变。

队列在很多情况下都非常有用。即使两个应用程序没有同时运行，队列也可使这两个应用程序之间相互通信，这对于批处理和移动应用程序尤为方便。当所发送的消息传播给多个接收方时，具有这些接收方的队列还提供自动负载平衡。

3. Topic 主题

Topic 的流程如图 6-2 所示。

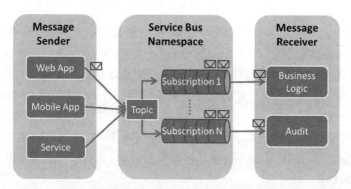

图 6-2 Topic 主题的发送-接收图

Topic 主题实现的是按意愿将多个客户端多播。就像广播一样，节目好，听众还想听，就会锁定该频率，相当于订阅了你的广播。电视节目也类似，频道多，观众可以选择自己喜欢的频道来看。这个 Topic 相当于电视的频道，或者电台的频率，受众可以接收其信息。可以有很多受众，所有受众接收的都是一样的内容。Topic 接收消息,消息会给每一个 Subscription 订阅一份。主题在很多方面与队列类似。发送方将消息提交至主题的方式与将消息提交至队列的方式相同,这些消息与使用队列的消息看起来一样。最大的区别是主题让每个接收应用程序通过定义筛选器(Filter)创建其自己的订阅。然后,订户将只能看到与该筛选器匹配的消息。与队列一样,某主题的订户可使用 Receive And Delete 或 Peek Lock 读取消息。不过与队列不同的是,发送至主题的单个消息可由多个订户接收。此方法通常称作发布和订阅,在当多个应用程序对相同消息感兴趣时非常有用。通过定义适当的筛选器,每位订户可以只访问需要查看的消息流部分。

4. Relay 中继

队列和主题均通过代理提供单向异步通信。流量只按一个方向流动,发送方和接收方之间没有直接连接。但如果不希望这样,该怎么办?假设应用程序需要同时发送和接收消息,或者可能希望应用程序之间进行直接链接,而不需要在某个位置来存储两者之间的消息。为解决此类问题,Service Bus 提供了中继,如图 6-3 所示。

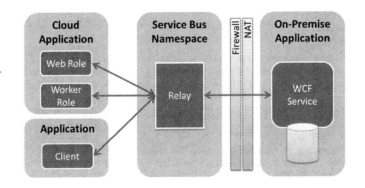

图 6-3 Relay 中继的发送和接收图

Azure 中继可用于在现有企业环境中托管 WCF 服务。然后,可以将侦听传入会话和请求这些 WCF 服务的任务委托给在 Azure 内运行的中继服务。这使你能够向 Azure 中运行的应用程序代码或者向移动工作者或 Extranet 合作伙伴环境公开这些服务。中继允许精确、安全地控制谁可以访问这些服务。它提供了一种强大且安全的方式,从现有企业解决方案公开应用程序功能和数据并从云中利用这些功能和数据。

本书探讨如何使用 Azure 中继创建 WCF Web 服务,并使用可实现双方之间安全对话的 TCP 通道绑定公开该服务。

若要开始在 Azure 中使用服务总线消息实体,必须先使用在 Azure 中唯一的名称创建一个命名空间。命名空间提供了用于对应用程序中的 Service Bus 资源进行寻址的范围容器。

6.2 队列(Queue)入门

首先要创建一个 Service Bus。如图 6-4 所示，在左边菜单里选择 Service Bus，然后单击 Add 按钮，出现如图 6-5 所示界面。

图 6-4 单击 Add 按钮新建 Service Bus

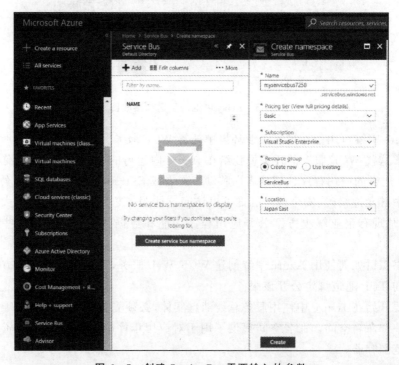

图 6-5 创建 Service Bus 需要输入的参数

在这里创建一个 Service Bus Namespace。输入一个 Name，如果有提示说名字重复，就得改一下名字，直到名字跟别人的 Service Bus 名字不重复为止。然后选择 Pricing Tier，一般生产环境需要选择 Premium 级别。本处是演示教学目的，就可以选择 Basic 级别的 Pricing Tier。Premium 级别是相当完善的级别，有消息的异地备份，可以保证大量的消息存储、输入，有完善的事务机制。异地部署的服务可以无缝地接管主实例。万一发生了一些事件导致主消息服务器无法提供服务，异地实例就会无缝地接管主实例。这个级别相当于有一个消息集群服务器。

确定好之后单击 Create 按钮，创建过程就开始了，如图 6-6 所示。

图 6-6　Service Bus 正在被创建

创建好之后，此 Service Bus Namespace 就出现在列表中，如图 6-7 所示。

图 6-7　新建的 Service Bus 被显示在列表中

单击此 Service Bus Namespace，进入其中，如图 6-8 所示。中间有菜单，这里介绍几个比较重要的菜单。

图 6-8　查看 Service Bus 的详情

1. Overview 菜单

如图 6-8 所示，这里显示此 Service Bus Namespace 的大概情况。

这里有＋Queue 的按钮，用来创建一个 Queue(队列)，如图 6-9 所示。

图 6-9　新建队列 Queue 的按钮＋Queue

输入 Queue 的名字，然后选择 Max Queue Size(最大 Queue 的大小)，它的范围是 1～5 GB，请根据需要来选择。此处用的是 Basic 级别。Premium 的级别可以建立最大 80 GB 的存储。

选择 Message time to live(即消息存活时间)，默认是 14 天。可以根据实际需要来设置。

设置 Lock duration，默认是 30 秒。可以根据实际情况进行调整。

下面有几个选项：

① Enable duplicate detection，允许检测是否有重复项。如果勾选了，就会有一个消息历史记录，按照设定的时间段保存消息历史，一段时间内的消息如果出现重复，系统就会告警。不接收重复的消息数据。

② Enable dead lettering on message expiration，允许消息过期后变成死信，每个消息都存在一个消息过期时间，如果在设定的时间内未能将消息提交到此消息该去的客户端，则此消息就变成了过期消息。如果没有这个设置，则过期消息需要删除。有了这个设置之后，就会让过期消息进入死信队列，在死信队列中是没有过期时间的。消息可以在死信队列中停留任意长时间。

③ Enable sessions，允许使用 Session 会话机制。如果使用 Session 会话机制，则能确保消息队列是先进先出；如果没有使用 Session 机制，就不能保证消息的先进先出，有可能有的消息晚进而先出。

④ Enable partitioning，允许分区。消息分区后，就会有多个消息代理和消息存储分区。这样多个消息代理互为备份，同时消息在不同分区中存储，每个分区有对应的一个消息代理管理，用户使用的时候完全不知道分区的存在。消息都是由这些消息代理提供出来。

单击 Create 按钮，即创建此 Queue。

2. Shared access policies 菜单

这个 Shared access policies 用于给别的公司访问 Service Bus 授权。由 Azure 账号来创建授权，如图 6-10 所示。

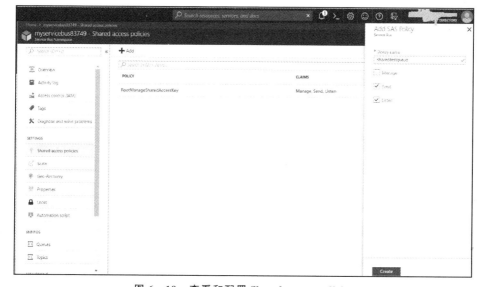

图 6-10　查看和配置 Shared access policies

Manage 是允许管理权限。允许创建新资源,一般这个权限不会给别的公司,而是根据实际需要来分配 Send 或者 Listen 权限。有可能只给一个,如图 6-11 所示。

图 6-11 新建 Shared access policies

创建了之后,就多了一个 Shared access policies。这个新的 Policies 也有 Connection String,客户端如果用了这个链接串,则只能使用预先配置的权限,比如只能 Listen 或者只能 Send 等。

3. Scale 菜单

用于配置 Service Bus 的伸缩级别。Scale 的伸缩级别与 Service Bus Namespace 的级别是一样的。Service Bus Namespace 的级别是 Standard,这些级别的伸缩安排都是微软设置好的。当消息发送和接收压力比较大的时候,会进行自动的伸缩,让多台服务器来响应客户端的请求。级别越高的,可伸缩性就越好。

4. Geo Recovery 菜单

这是 Premium 级别的服务才有的功能,当发生了一些局部的自然灾害的时候,可异地灾备恢复,可以帮助用户迅速恢复业务的运转。

如果是 Standard 级别以上的 Service Bus Namespace,则+Topic 按钮也是可以点击的。而 Basic 级别的不能点击。+Topic 按钮可以用来创建一个 Topic(话题)。这为消息订阅机制提供了可能。在一个 Topic 之下,可以有多个订阅者,这个 Topic 的消息由此 Topic 的所有订阅者所阅读和处理。除非有特别的设定,某些符合条件的消息只给某一部分订阅者。

单击+Topic 按钮,如图 6-12 所示。

输入 Topic name,其余选择与创建 Queue 时差不多,就不重复了。

单击 Create 按钮,等一会儿此 Topic 创建成功,如图 6-13 所示。

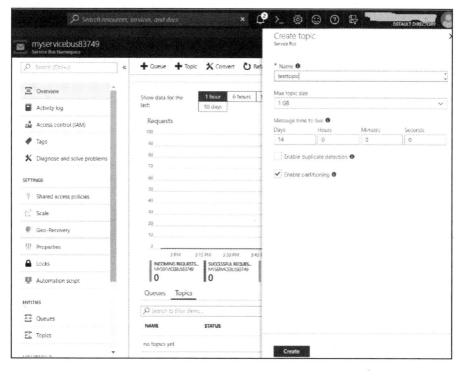

图 6-12　创建 Topic

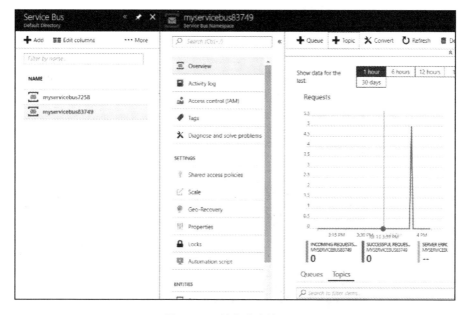

图 6-13　创建成功的 Topic

新创建的 Topic 就出现在列表中。单击此 Topic，可以进入其界面，如图 6-14 所示。

图 6-14　查看新建的 Topic

可以在此界面中单击＋Subscription 按钮来创建 Subscription，如图 6-15 所示。

图 6-15　创建订阅

下面要用工具 Service Bus Explorer。https://code.msdn.microsoft.com/windowsapps/service-bus-explorer-f2abca5a 是微软官方开发的工具，如图 6-16 所示。

图 6-16 Service Bus Explorer

这是其界面，需要建立链接，File→Connect，出现如图 6-17 所示界面。

图 6-17 建立连接窗口

需要输入链接串来链接，要找链接串，可以点击 Shared Access Policies，如图 6-18 所示。

默认创建 Service Bus namespace 时，在 Shared Access Policies 中都会有一个 Root Manage Shared Access Key，这是用于给 Azure 账号拥有者访问 Server Bus 用的。点击 Primary Connection String 右边的拷贝按钮，将链接串拷贝到剪贴板。然后回到 Service Bus Explorer，将剪贴板的链接串粘贴到窗口中，如图 6-19 所示。

此 Service Bus Namespace 中的 Queue、Topic 等资源都列出来了，如图 6-20 所示。

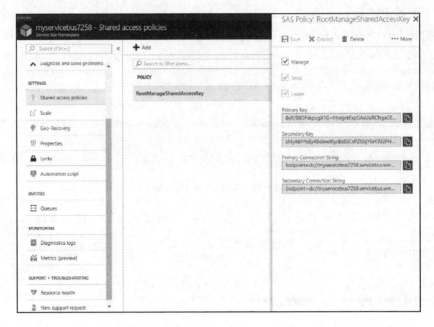

图 6-18　从 Azure Portal 中找到 Shared Access Policy，然后找到链接串

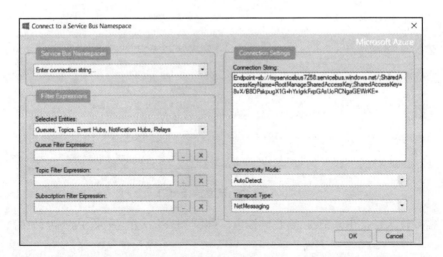

图 6-19　用链接串来链接 Service Bus

5. Create Queue 菜单

如图 6-21 所示，右键弹出菜单。

在这里可以创建 Queue。

右键选择了之后，右边就会要求用户输入 Relative URI，还有其他一些选项，比如 Duplication Detection History Window。如果没有输入任何信息，就是不需要重复项检测；输入了一些信息，则是需要进行重复项检测。Default Message time to live,默认是 14 天。Queue Settings 中的一些选项可根据实际需要来进行设置。这些项都设置好之后,就可创建 Queue 了。

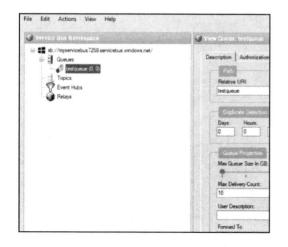

图 6-20 列出的 Service Bus 资源

图 6-21 右键弹出菜单

如果需要删除一个 Queue,则右键点击到此账号对应的 Queue 之后,出现上下文菜单,选择 Delete Queue,确认后,Queue 就被删除,如图 6-22 所示。

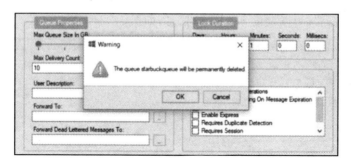

图 6-22 确认删除 Queue 的对话框

Queue 上一个比较常用的菜单是 Send Messages 菜单。我们可以发送消息到 Queue 当中,可以用来调试一些程序。Send Messages 菜单出来的窗口如图 6-23 所示。

左边 Message Text 中可以输入信息,可以是任何文本信息。这个信息的最大长度受所申请的 Service Bus Namespace 的限制,一般的内容都可以接受。

Message Properties 可以加入一些自定义的属性。这样,可以根据自定义的属性来做一些不同的处理。

此例中发送的是一个 json 信息。除了 json 信息格式,xml、html 或者其他格式都可以,格式没有限制。

然后有 File 页签,可以上传一个文件附加到这个消息之上。这样消息处理程序

图 6-23 Send Messages 对话框

就不仅可以读取消息,还可以读取并处理附件文件。可以通过 Select Files 菜单选择一个或者多个文件。这些文件可以是任何类型,如二进制、json、普通文本或者 xml 之类。

如果还有需要,可以设置 Sender 的信息,如图 6-24 所示。

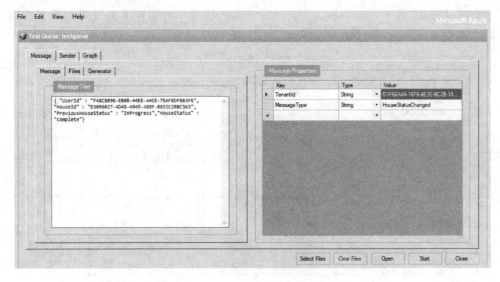

图 6-24 设置 Sender 的属性

可以把 MessageId、To、Reply To、Context Type 等信息填好,一般 Context Type 是比较常用的字段。Body Type 可以选择 Stream、String、WCF 等。Stream 代表着消息 Body 是一个流。Body Type 为 String 就比较简单,即一个字符串。WCF 说明 Body 类型是一个 WCF 服务。

以上都填写好了以后，可以单击 Start 按钮发送到此 Queue 当中。然后可以看到 Log 窗口有类似这样的信息：

<14:52:18> Sender[0]:   Message sent. MessageId = [b3b61138 - 67f9 - 4b38 - b38f - 2ba08e2debbc] SessionId = [NULL] Label = [Service Bus Explorer] Size = [336]

<14:52:18> Sender[0]:

— Message Count = [1] Messages Sent/Sec = [0.593119810201661] Total Elapsed Time (ms) = [1686]

— Average Send Time (ms) = [1686] Minimum Send Time (ms) = [1686] Maximum Send Time (ms) = [1686]

右键单击对应的 Queue，选择 Refresh Queue 菜单。可以看到此 Queue 中加入了一个消息（其显示数字 1,0），表示有新消息，如图 6 - 25 所示。

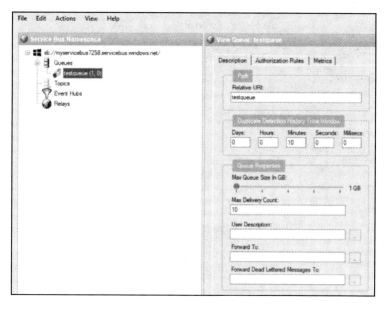

图 6 - 25　接收到消息的 Queue

右键单击此 Queue，可以选择 Receive Messages，如图 6 - 26 所示。
然后右边出现如图 6 - 27 所示界面。
Receive Mode 中有两个选项，Peek 代表只读取消息，而不移除消息；Receive and Delete 就会标记消息已经接收到，并删除此消息。如果希望此操作不删除消息，则选择 Peek 选项。如果此处需要删除消息，不让其他程序读到消息，则可以选择 Receive and Delete。其他的选项包括，Message Count，要么是 All，即全部；要么是 Top，这个 Top 的大小需要输入，默认是 10 个，即只取最上面的 10 个消息。

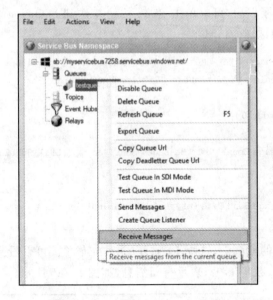

图 6-26 Receive Messages 菜单

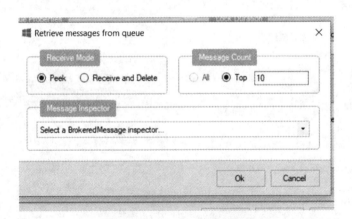

图 6-27 Receive Messages 对话框

接收的消息如图 6-28 所示。

这些消息内容即刚刚输入的消息。同时,也可看到此消息的自定义 Properties。所有这些东西用户都可以用一个程序来 Listen(侦听)某一个 Queue,然后对进入此 Queue 的消息进行响应。

Basic 的 Service Bus Namespace 不能建立 Topic。所以在 Service Bus Explorer 中其 Topic 就是灰色的,意思是不能进行任何操作。只有 Standard 以上的级别可以有 Topic。

连接一个 Standard 的 Service Bus Namespace,如图 6-29 所示。

可以看到 Topic 是可以操作的。其下有一些 Topic。

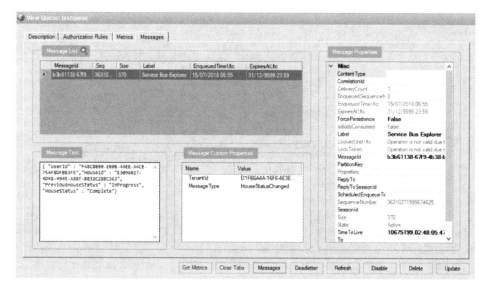

图 6-28　接收的消息

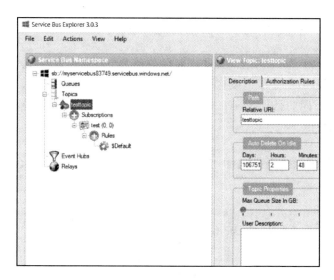

图 6-29　Topic 在 Service Bus Explorer 中的显示

用类似的方法,可以建立或者删除 Topic,以及建立或者删除 Subscription(订阅)。Topic(主题)-订阅模式是一对多的。一个主题可以有多个订阅,如图 6-30 所示。

一个主题可以有很多订阅,这样允许很多客户端订阅某一个主题,这些订阅者都能收到同样的消息内容。每个订阅者都可以分别按自己的方式来处理这些订阅的消息。每个订阅者都可以不必了解别的订阅者是如何工作的,只需要按预定的方式将自己这份消息处理好就可以了。

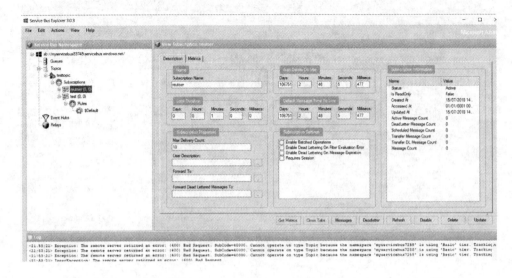

图 6-30　一个 Topic 可以有多个订阅

一个消息进来，有多少个订阅者，系统就会复制多少份，给每个订阅者一份此消息的拷贝。

如果右键单击 Topic 节点，就可以选择菜单 Send Messages，这个菜单可以帮助用户进行程序的调试，手工地发送一个样例消息，这样可以驱动程序处理消息。

与 Queue 不同的是，Receive Messages 菜单在每个订阅上都有此菜单项。其操作与前文所述类似。

6.3　用 Visual Studio 开发 Service Bus Queue

可以打开 https://azure.microsoft.com/en-us/resources/samples/service-bus-dotnet-manage-queue-with-basic-features/，下载其 Example 代码。

解压代码到一个目录，如 E:\example\service-bus-dotnet-manage-queue-with-basic-features-master，然后打开 ServiceBusQueueBasic.csproj，如图 6-31 所示。

打开 Readme.md 文件。它要求我们访问页面：https://github.com/Azure/azure-sdk-for-net/blob/Fluent/AUTH.md 来获得一个文件 Auth.md。访问此页面，找到黑框中部分，如图 6-32 所示。

将其复制到剪贴板，在 Solution 中创建一个文件叫 Auth.md，然后把这个剪贴板中的内容粘贴到 Auth.md 中，如图 6-33 所示。

注意，这个文件中用的微软官方样例提供的账号，可以用来运行样例程序，如果用户自己开发应用，则需要将这些信息改成相应用户的登录信息。

Service Bus

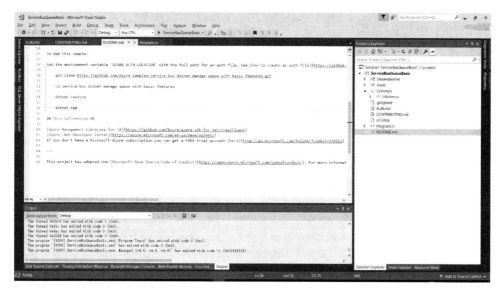

图 6-31　Service Bus 样例程序打开后工程的样子

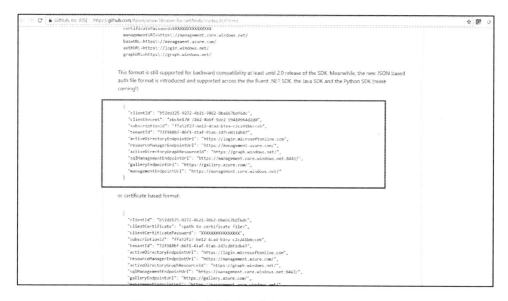

图 6-32　找到我们需要的 Auth.md 文件的内容

下面来看看 Program.cs 文件：

// Copyright (c) Microsoft Corporation. All rights reserved.
// Licensed under the MIT License. See License.txt in the project root for license infor-
// mation.

using Microsoft.Azure.Management.Fluent;

图 6-33 将内容复制到此文件中

```csharp
using Microsoft.Azure.Management.ResourceManager.Fluent;
using Microsoft.Azure.Management.ResourceManager.Fluent.Core;
using Microsoft.Azure.Management.Samples.Common;
using Microsoft.Azure.Management.ServiceBus.Fluent;
using Microsoft.Azure.Management.ServiceBus.Fluent.Models;
using System;
using System.Linq;
using System.Text;

namespace ServiceBusQueueBasic
{
    public class Program
    {
        /**
         * Azure Service Bus basic scenario sample.
         *  - Create namespace with a queue.
         *  - Add another queue in same namespace.
         *  - Update Queue.
         *  - Update namespace
         *  - List namespaces
         *  - List queues
         *  - Get default authorization rule.
         *  - Regenerate the keys in the authorization rule.
         *  - Get the keys from authorization rule to connect to queue.
         *  - Send a "Hello" message to queue using Data plan sdk for Service Bus.
         *  - Delete queue
         *  - Delete namespace
         */
        public static void RunSample(IAzure azure)
        {
```

```csharp
//产生几个随机的资源名字。SdkContext 类是微软 AzureSDK 中的类,帮助用户产生随机名字
var rgName = SdkContext.RandomResourceName("rgSB01_", 24);
var namespaceName = SdkContext.RandomResourceName("namespace", 20);
var queue1Name = SdkContext.RandomResourceName("queue1_", 24);
var queue2Name = SdkContext.RandomResourceName("queue2_", 24);
try
{
    // ============================================================
    // Create a namespace. 产生的是 Premium 的 Service Bus Namespace

    Utilities.Log("Creating name space " + namespaceName + " in resource group" + rgName + "...");

    var serviceBusNamespace = azure.ServiceBusNamespaces
                .Define(namespaceName)
                .WithRegion(Region.USWest)
                .WithNewResourceGroup(rgName)
                .WithSku(NamespaceSku.PremiumCapacity1)
                .WithNewQueue(queue1Name, 1024)
                .Create();

            Utilities.Log("Created service bus" + serviceBusNamespace.Name);
            Utilities.Print(serviceBusNamespace);

            var firstQueue = serviceBusNamespace.Queues.GetByName(queue1Name);
            Utilities.Print(firstQueue);

    // ============================================================
    // Create a second queue in same namespace 在同一个 namespace 中创建第二个 queue

            Utilities.Log("Creating second queue " + queue2Name + " in namespace " + namespaceName + "...");

                var secondQueue = serviceBusNamespace.Queues.Define(queue2Name)
                    .WithExpiredMessageMovedToDeadLetterQueue()
                    .WithSizeInMB(2048)
                    .WithMessageLockDurationInSeconds(20)
                    .Create();

            Utilities.Log("Created second queue in namespace");

            Utilities.Print(secondQueue);
```

```
//=========================================================
// Get and update second queue. 获取并更新第二个queue,更新其存储空间大小

            secondQueue = serviceBusNamespace.Queues.GetByName(queue2Name);
            secondQueue = secondQueue.Update().WithSizeInMB(3072).Apply();

            Utilities.Log("Updated second queue to change its size in MB");

            Utilities.Print(secondQueue);

            //=============================================
            // Update namespace
            Utilities.Log("Updating sku of namespace " + serviceBusNamespace.
Name + "...");

            serviceBusNamespace = serviceBusNamespace
                    .Update()
                    .WithSku(NamespaceSku.PremiumCapacity1)
                    .Apply();
            Utilities.Log("Updated sku of namespace " + serviceBusNamespace.
Name);

            //=============================================
            // List namespaces 列出所有 Service Bus 名称空间

            Utilities.Log("List of namespaces in resource group " + rgName
+ "...");

            foreach (var serviceBusNamespace1 in azure.ServiceBusNamespaces.
ListByResourceGroup(rgName))
            {
                Utilities.Print(serviceBusNamespace1);
            }

            //=============================================
            // List queues in namespaces 列出 ServiceBus 名称空间中的 queue 列表

            var queues = serviceBusNamespace.Queues.List();
            Utilities.Log("Number of queues in namespace :" + queues.Count());

            foreach (var queue in queues)
```

Service Bus

```csharp
            {
                Utilities.Print(queue);
            }

            //==========================================
            // Get connection string for default authorization rule of namespace
            // 获得默认名称空间的默认访问 Rule 中的链接串

            var namespaceAuthorizationRules = serviceBusNamespace.AuthorizationRules.List();
            Utilities.Log("Number of authorization rule for namespace :" + namespaceAuthorizationRules.Count());

            foreach (var namespaceAuthorizationRule in namespaceAuthorizationRules)
            {
                Utilities.Print(namespaceAuthorizationRule);
            }

            Utilities.Log("Getting keys for authorization rule ...");

            var keys = namespaceAuthorizationRules.FirstOrDefault().GetKeys();
            Utilities.Print(keys);
            Utilities.Log("Regenerating secondary key for authorization rule ...");
            keys = namespaceAuthorizationRules.FirstOrDefault().RegenerateKey(Policykey.SecondaryKey);
            Utilities.Print(keys);

            //==========================================
            // Send a message to queue. 发送一个消息到 queue
            Utilities.SendMessageToQueue(keys.PrimaryConnectionString, queue1Name, "Hello");
            //==========================================
            // Delete a queue and namespace 删除 queue 和对应的 ServiceBus 名称空间
            Utilities.Log("Deleting queue " + queue1Name + " in namespace " + namespaceName + "...");
            serviceBusNamespace.Queues.DeleteByName(queue1Name);
            Utilities.Log("Deleted queue " + queue1Name + "...");

            Utilities.Log("Deleting namespace " + namespaceName + "...");
            // This will delete the namespace and queue within it. 删除名称空间和
```

```csharp
            // 其中的 queue
            try
            {
                azure.ServiceBusNamespaces.DeleteById(serviceBusNamespace.Id);
            }
            catch (Exception)
            {
            }
            Utilities.Log("Deleted namespace " + namespaceName + "...");
        }
        finally
        {
            try
            {
                Utilities.Log("Deleting Resource Group: " + rgName);
                azure.ResourceGroups.BeginDeleteByName(rgName);
                Utilities.Log("Deleted Resource Group: " + rgName);
            }
            catch (NullReferenceException)
            {
                Utilities.Log("Did not create any resources in Azure. No clean up is necessary");
            }
            catch (Exception g)
            {
                Utilities.Log(g);
            }
        }
    }

    public static void Main(string[] args)
    {
        try
        {
            //=====================================================
            // Authenticate 用 Auth.md 文件中的登录信息来登录,需要设置一个环境变量 AZURE_AUTH_
            // LOCATION,然后重启 VisualStudio,以便 Visual Studio 可以读取到新设置的环境变量
            var credentials = SdkContext.AzureCredentialsFactory.FromFile(Environment.GetEnvironmentVariable("AZURE_AUTH_LOCATION"));

            var azure = Azure
                .Configure()
                .WithLogLevel(HttpLoggingDelegatingHandler.Level.Basic)
```

```csharp
                    .Authenticate(credentials)
                    .WithDefaultSubscription();

                // Print selected subscription 打印这个订阅的详细信息
                Utilities.Log("Selected subscription: " + azure.SubscriptionId);

                RunSample(azure);
            }
            catch (Exception e)
            {
                Utilities.Log(e.ToString());
            }
        }
    }
}
```

可以看到此 cs 文件应用了 Utilities 类。

Utilities.cs 文件部分源代码(源文件很长,有 2 000 多行)如下:

这是发送消息到 Topic 的代码:

```csharp
public static void SendMessageToTopic (string connectionString, string topicName, string message)
        {
            if (!IsRunningMocked)
            {
                try
                {
                    var topicClient = new TopicClient(connectionString, topicName);
                    topicClient.SendAsync(new Message(Encoding.UTF8.GetBytes(message))).Wait();
                    topicClient.Close();
                }
                catch (Exception)
                {
                }
            }
        }
```

这是发送消息到 Queue 的代码:

```csharp
publicstaticvoid SendMessageToQueue(string connectionString, string queueName, string message)
        {
```

```csharp
            if (!IsRunningMocked)
            {
                try
                {
                    var queueClient = new QueueClient(connectionString, queueName, ReceiveMode.PeekLock);
                    queueClient.SendAsync(new Message(Encoding.UTF8.GetBytes(message))).Wait();
                    queueClient.Close();
                }
                catch (Exception)
                {
                }
            }
        }
```

这个方法将打印所有 queue 的属性：

```csharp
        public static void Print(IQueue queue)
        {
            StringBuilder builder = new StringBuilder()
                    .Append("Service bus Queue: ").Append(queue.Id)
                    .Append("\n\tName: ").Append(queue.Name)
                    .Append("\n\tResourceGroupName: ").Append(queue.ResourceGroupName)
                    .Append("\n\tCreatedAt: ").Append(queue.CreatedAt)
                    .Append("\n\tUpdatedAt: ").Append(queue.UpdatedAt)
                    .Append("\n\tAccessedAt: ").Append(queue.AccessedAt)
                    .Append("\n\tActiveMessageCount: ").Append(queue.ActiveMessageCount)
                    .Append("\n\tCurrentSizeInBytes: ").Append(queue.CurrentSizeInBytes)
                    .Append("\n\tDeadLetterMessageCount: ").Append(queue.DeadLetterMessageCount)
                    .Append("\n\tDefaultMessageTtlDuration: ").Append(queue.DefaultMessageTtlDuration)
                    .Append("\n\tDuplicateMessageDetectionHistoryDuration: ").Append(queue.DuplicateMessageDetectionHistoryDuration)
                    .Append("\n\tIsBatchedOperationsEnabled: ").Append(queue.IsBatchedOperationsEnabled)
                    .Append("\n\tIsDeadLetteringEnabledForExpiredMessages: ").Append(queue.IsDeadLetteringEnabledForExpiredMessages)
                    .Append("\n\tIsDuplicateDetectionEnabled: ").Append(queue.IsDuplicateDetectionEnabled)
                    .Append("\n\tIsExpressEnabled: ").Append(queue.IsExpressEnabled)
                    .Append("\n\tIsPartitioningEnabled: ").Append(queue.IsPartitioningEnabled)
```

```csharp
                    .Append("\n\tIsSessionEnabled: ").Append(queue.IsSessionEnabled)
                    .Append("\n\tDeleteOnIdleDurationInMinutes: ").Append(queue.DeleteOnIdleDurationInMinutes)
                    .Append("\n\tMaxDeliveryCountBeforeDeadLetteringMessage: ").Append(queue.MaxDeliveryCountBeforeDeadLetteringMessage)
                    .Append("\n\tMaxSizeInMB: ").Append(queue.MaxSizeInMB)
                    .Append("\n\tMessageCount: ").Append(queue.MessageCount)
                    .Append("\n\tScheduledMessageCount: ").Append(queue.ScheduledMessageCount)
                    .Append("\n\tStatus: ").Append(queue.Status)
                    .Append("\n\tTransferMessageCount: ").Append(queue.TransferMessageCount)
                    .Append("\n\tLockDurationInSeconds: ").Append(queue.LockDurationInSeconds)
                    .Append("\n\tTransferDeadLetterMessageCount: ").Append(queue.TransferDeadLetterMessageCount);

            Utilities.Log(builder.ToString());
        }
```

这是打印 Topic 的属性的方法：

```csharp
        public static void Print(ITopic topic)
        {
            StringBuilder builder = new StringBuilder()
                    .Append("Service bus topic: ").Append(topic.Id)
                    .Append("\n\tName: ").Append(topic.Name)
                    .Append("\n\tResourceGroupName: ").Append(topic.ResourceGroupName)
                    .Append("\n\tCreatedAt: ").Append(topic.CreatedAt)
                    .Append("\n\tUpdatedAt: ").Append(topic.UpdatedAt)
                    .Append("\n\tAccessedAt: ").Append(topic.AccessedAt)
                    .Append("\n\tActiveMessageCount: ").Append(topic.ActiveMessageCount)
                    .Append("\n\tCurrentSizeInBytes: ").Append(topic.CurrentSizeInBytes)
                    .Append("\n\tDeadLetterMessageCount: ").Append(topic.DeadLetterMessageCount)
                    .Append("\n\tDefaultMessageTtlDuration: ").Append(topic.DefaultMessageTtlDuration)
                    .Append("\n\tDuplicateMessageDetectionHistoryDuration: ").Append(topic.DuplicateMessageDetectionHistoryDuration)
                    .Append("\n\tIsBatchedOperationsEnabled: ").Append(topic.IsBatchedOperationsEnabled)
                    .Append("\n\tIsDuplicateDetectionEnabled: ").Append(topic.IsDuplicateDetectionEnabled)
                    .Append("\n\tIsExpressEnabled: ").Append(topic.IsExpressEnabled)
                    .Append("\n\tIsPartitioningEnabled: ").Append(topic.IsPartitioningEnabled)
                    .Append("\n\tDeleteOnIdleDurationInMinutes: ").Append(topic.DeleteOnI-
```

dleDurationInMinutes)
        .Append("\n\tMaxSizeInMB: ").Append(topic.MaxSizeInMB)
        .Append("\n\tScheduledMessageCount: ").Append(topic.ScheduledMessageCount)
        .Append("\n\tStatus: ").Append(topic.Status)
        .Append("\n\tTransferMessageCount: ").Append(topic.TransferMessageCount)
        .Append("\n\tSubscriptionCount: ").Append(topic.SubscriptionCount)
        .Append("\n\tTransferDeadLetterMessageCount: ").Append(topic.TransferDeadLetterMessageCount);

        Log(builder.ToString());
}
```

还有更多 Example 代码，可以下载下来，详见 https://github.com/Azure/azure-service-bus，如图 6-34 所示。

图 6-34 此样例代码可以 Clone 或者下载

用 Git 客户端程序下载这些代码。打开 azure-service-bus\samples\DotNet\GettingStarted\BasicSendReceiveQuickStart 文件夹，如图 6-35 所示。

打开此目录中的 .sln 文件，则代码大概是这样的：

```
//Program.cs 大概的样子：
using Microsoft.Azure.ServiceBus;
using System;
using System.Text;
using System.Threading;
using System.Threading.Tasks;

namespace BasicSendReceiveQuickStart
```
```

Service Bus

图 6 - 35 基本收发的样例程序

```
{
    class Program
    {
        static IQueueClient queueClient;

        static void Main(string[] args)
        {
            string ServiceBusConnectionString = "";
            string QueueName = "";

            for (int i = 0; i < args.Length; i++)
            {
                var p = new Program();
                if (args[i] == " - ConnectionString")
                {
                    Console.WriteLine($"ConnectionString:{args[i+1]}");
                    ServiceBusConnectionString = args[i+1]; // Alternatively enter your connection string here. 可以使用你的链接串
                }
                elseif(args[i] == " - QueueName")
                {
                    Console.WriteLine($"QueueName:{args[i+1]}");
                    QueueName = args[i+1]; // Alternatively enter your queue name here. 可以输入你的 queue 的名字
                }
            }

            if (ServiceBusConnectionString != ""&& QueueName != "")
                MainAsync(ServiceBusConnectionString, QueueName).GetAwaiter().Ge-
```

tResult();
            else
            {
                Console.WriteLine("Specify - Connectionstring and - QueueName to execute the example.");
                Console.ReadKey();
            }
        }

        staticasync Task MainAsync (string ServiceBusConnectionString, string QueueName)
        {
            constint numberOfMessages = 10;
            queueClient = new QueueClient(ServiceBusConnectionString, QueueName);

            Console.WriteLine("===============================");
            Console.WriteLine("Press any key to exit after receiving all the messages.");
            Console.WriteLine("===============================");

            // Register QueueClient's MessageHandler and receive messages in a loop 注
            // 册 Queue-Client，MessageHanlder 没有起作用的话
            RegisterOnMessageHandlerAndReceiveMessages();

            // Send Messages
            await SendMessagesAsync(numberOfMessages);

            Console.ReadKey();

            await queueClient.CloseAsync();
        }

        static void RegisterOnMessageHandlerAndReceiveMessages()
        {
            // Configure the MessageHandler Options in terms of exception handling,
            // number of concurrent messages to deliver etc. 配置 MessageHandler Options
            var messageHandlerOptions = new MessageHandlerOptions(ExceptionReceivedHandler)
            {
                // Maximum number of Concurrent calls to the callback 'ProcessMessages-
                // Async', set to 1 for simplicity. 最大同时 call 此处理消息的方法-
                // ProcessMessagesAsync, 设置为 1
                // Set it according to how many messages the application wants to process
                // in parallel. 显示地设置为 1
                MaxConcurrentCalls = 1,

```csharp
            // Indicates whether MessagePump should automatically complete the
            // messages after returning from User Callback.指出是否 MessagePump
            // 应该自动 complete 消息
            // False below indicates the Complete will be handled by the User Call-
            // back as in 'ProcessMessagesAsync' below.False 表示不自动 complete
            // 消息,将由用户代码自行处理
            AutoComplete = false
        };

        // Register the function that will process messages 注册此 function 将处理
        // 消息
        queueClient.RegisterMessageHandler(ProcessMessagesAsync, messageHandlerOptions);
    }

    static async Task ProcessMessagesAsync(Message message, CancellationToken token)
    {
        // Process the message 这是处理消息的方法,处理消息
        Console.WriteLine($"Received message: SequenceNumber:{message.SystemProperties.SequenceNumber} Body:{Encoding.UTF8.GetString(message.Body)}");

        // Complete the message so that it is not received again.
        // This can be done only if the queueClient is created in ReceiveMode.Peek-
        // Lock mode (which is default).将消息标记成 complete
        await queueClient.CompleteAsync(message.SystemProperties.LockToken);

        // Note:Use the cancellationToken passed as necessary to determine if the
        // queueClient has already been closed.
        // If queueClient has already been Closed, you may chose to not call
        // CompleteAsync() or AbandonAsync() etc. calls
        // to avoid unnecessary exceptions.如果需要,可以调用 AbandonAsync()方法
        // 来放弃 complete 消息,这样这个消息可以再次被处理
    }

    // Use this Handler to look at the exceptions received on the MessagePump
    static Task ExceptionReceivedHandler(ExceptionReceivedEventArgs exceptionReceivedEventArgs)
    {
        Console.WriteLine($"Message handler encountered an exception {exceptionReceivedEventArgs.Exception}.");
        var context = exceptionReceivedEventArgs.ExceptionReceivedContext;
        Console.WriteLine("Exception context for troubleshooting:");
        Console.WriteLine($"- Endpoint: {context.Endpoint}");
```

```
            Console.WriteLine($" - Entity Path: {context.EntityPath}");
            Console.WriteLine($" - Executing Action: {context.Action}");
            return Task.CompletedTask;
        }

        staticasync Task SendMessagesAsync(int numberOfMessagesToSend)
        {
            try
            {
                for (var i = 0; i < numberOfMessagesToSend; i ++ )
                {
                    // Create a new message to send to the queue
                    string messageBody = $"Message {i}";
                    var message = new Message(Encoding.UTF8.GetBytes(messageBody));

                    // Write the body of the message to the console
                    Console.WriteLine($"Sending message: {messageBody}");

                    // Send the message to the queue
                    await queueClient.SendAsync(message);
                }
            }
            catch (Exception exception)
            {
                Console.WriteLine($"{DateTime.Now} :: Exception: {exception.Message}");
            }
        }
    }
}
```

6.4 Service Bus Queue 和 Storage Queue 的区别

Service Bus Queue 和 Storage Queue 的区别如表 6-1～表 6-7 所列。

表 6-1 基础功能比较

Comparison Criteria	Storage Queue	Service Bus Queue
Ordering guarantee	No	Yes-First-In-First-Out (FIFO) (through the use of messaging sessions)
Delivery guarantee	At-Least-Once	At-Least-Once At-Most-Once

续表 6-1

Comparison Criteria	Storage Queue	Service Bus Queue
Transaction support	No	Yes (through the use of local transactions)
Receive behavior	Non-blocking (completes immediately if no new message is found)	Blocking with/without timeout (offers long polling, or the "Comet technique") Non-blocking (through the use of .NET managed API only)
Receive mode	Peek & Lease	Peek & Lock Receive & Delete
Exclusive access mode	Lease-based	Lock-based
Lease/Lock duration	30 seconds (default) 7 days (maximum)	60 seconds (default) 5 minutes (maximum)
Lease/Lock granularity	Message level (each message can have a different timeout value)	Queue level (each queue has a lock granularity applied to all of its messages, fixed for the lifetime of the queue)
Batched receive	Yes (explicitly specifying message count when retrieving messages, up to a maximum of 32 messages)	Yes (implicitly enabling a pre-fetch property or explicitly through the use of transactions)
Batched send	No	Yes (through the use of transactions or client-side batching)

表 6-2 高级功能比较

Comparison Criteria	Storage Queue	Service Bus Queue
Scheduled delivery	Yes	Yes
Automatic dead lettering	No	Yes
Message deferral	Yes (via in-place update of visibility timeout)	Yes (provided via a dedicated API function)
Poison message support	Yes	Yes
In-place update	Yes	No
Server-side transaction log	Yes	No

续表 6-2

Comparison Criteria	Storage Queue	Service Bus Queue
Storage metrics	Yes	No
Purge queue function	Yes	No
Message groups	No	Yes (through the use of messaging sessions)
Duplicate detection	No	Yes (configurable on the sender side)
WCF integration	No	Yes (offers out-of-the-box WCF bindings)
WF integration	Custom (requires building a custom WF activity)	Native (offers out-of-the-box WF activities)

表 6-3 容量和配额的比较

Comparison Criteria	Storage Queue	Service Bus Queue
Maximum message size	64 KB (48 KB when using Base64 encoding)	256 KB (including both header and body, maximum header size: 64 KB)
Maximum queue size	100 TB (limited to a single storage account capacity)	1, 2, 3, 4 or 5 GB (defined upon creation of a queue)
Maximum message TTL	7 days	Unlimited
Maximum number of queues	Unlimited	10,000 (per service namespace, can be increased)
Maximum number of concurrent clients	Unlimited	Unlimited (100 concurrent connection limit only applies to TCP protocol-based communication)

表 6-4 管理和操作的比较

Comparison Criteria	Storage Queue	Service Bus Queue
Management protocol	REST over HTTP/HTTPS	REST over HTTPS

续表 6-4

Comparison Criteria	Storage Queue	Service Bus Queue
Runtime protocol	REST over HTTP/HTTPS	REST over HTTPS TCP with TLS
.NET Managed API	Yes (.NET managed Storage Client API)	Yes (.NET managed brokered messaging API)
Java API	Yes	Yes
PHP API	Yes	Yes
Node.js API	Yes	No
Arbitrary metadata support	Yes	No
Queue naming rules	Up to 63 characters long [3,63] (letters in a queue name must be lowercase)	Up to 260 characters long (queue names are case-insensitive)
Get queue length function	Yes (approximate value)	Yes (exact, point-in-time value)
Peek function	Yes	No

表 6-5　性能的比较

Comparison Criteria	Storage Queue	Service Bus Queue
Maximum throughput	Up to 2,000 messages per second	Up to 2,000 messages per second (based on benchmark with 1 KB messages)
Average latency	10 ms (with TCP Nagle disabled)	100 ms
Throttling behavior	Reject with HTTP 503 code (throttled requests are not treated as billable)	Reject with exception/HTTP 503 (throttled requests are not treated as billable)

表 6-6　授权和认证的比较

Comparison Criteria	Storage Queue	Service Bus Queue
Authentication	Symmetric key	ACS claims
Role-based access control	No	Yes (through the use of ACS roles)
Identity provider federation	No	Yes

表6-7 编程的比较

Comparison Criteria	Storage Queue	Service Bus Queue
Message Class Name	CloudQueueMessage	BrokeredMessage
Send Method Name	CloudQueue.AddMessage()	QueueClient.Send()
Queue Size	ApproximateMessageCount	

How to Create Queue	//Retrieve storage account from con-
	// nection string.
	CloudStorageAccount storageAccount
	=CloudStorageAccount.Parse(
	　CloudConfigurationManager.Get-
	Setting("StorageConnectionString"));
	//Create the queue client.
	CloudQueueClient queueClient = stor-
	ageAccount.CreateCloudQueueClient();
	// Retrieve a reference to a queue.
	CloudQueue queue = queueClient.
	GetQueueReference("myqueue");
	// Create the queue if it doesn't already
	// exist.
	queue.CreateIfNotExists();
		QueueDescription qd = new QueueDe-
		scription("TestQueue");
		qd.MaxSizeInMegabytes=5120;
		qd.DefaultMessageTimeToLive = new
		TimeSpan(0, 1, 0);
		// Create a new Queue with custom
		//settings
		string connectionString=CloudConfig-
		urationManager.GetSetting("Micro-
		soft.ServiceBus.ConnectionString");
		var namespaceManager=
		NamespaceManager.CreateFromCon-
		nectionString(connectionString);
		if (! namespaceManager.QueueExists
		("TestQueue"))
		{
		　namespaceManager.CreateQueue(qd);
		}

How to Send message	// Create a message and add it to the
	// queue.
	CloudQueueMessage message = new
	CloudQueueMessage("Hello, World");
	queue.AddMessage(message);
		string connectionString=CloudConfig-
		urationManager.GetSetting("Micro-
		soft.ServiceBus.ConnectionString");
		QueueClient Client=QueueClient.Cre-
		ateFromConnectionString(connection-
		String,"TestQueue");
		Client.Send(new BrokeredMessage());

How to Receive Mes-
sage
	// Get the next message
	CloudQueueMessage retrievedMessage
	=queue.GetMessage();
	// Process the message in less than 30
	// seconds, and then delete the message
	// queue.DeleteMessage(retrievedMes-
	// sage);
		BrokeredMessage message = Client.
		Receive();
		if (message!=null)
		{
		　try
		　{
		　　// Remove message from queue
		　　message.Complete();
		　}
		　catch (Exception)
		　{
		　　// Indicate a problem, unlock
		　　// message in queue
		　　message.Abandon();
		　}
		}

Azure Storage Queue 的流程如图 6-36 所示。

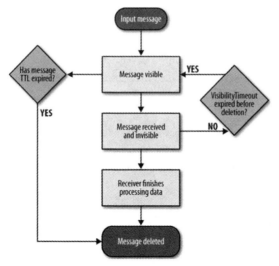

图 6-36　Azure Storage Queue 的流程

Azure Service Bus Queue 的流程如图 6-37 所示。

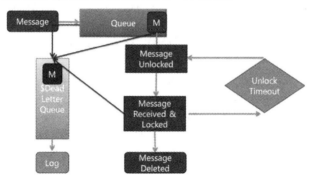

图 6-37　Azure Service Bus Queue 的流程

6.5　用 Visual Studio 开发 Service Bus Topic

Topic 和 Subscription 是一对多的关系，一个 Topic 之下可以有多个 Subscription，如图 6-38 所示。

针对每一个消息，每一个 Subscription 都会收到一份消息的拷贝。

这可以还用前面下载的样例代码。还是打开 azure-service-bus\samples\DotNet\GettingStarted\Microsoft.Az-

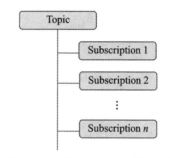

图 6-38　Topic 和 Subscription 的一对多关系

ure.ServiceBus 目录下的 sln 文件,可以看到 BasicSendReceiveUsingTopicSubscriptionClient 工程。

```csharp
// Program.cs 文件源码:
// Copyright (c) Microsoft. All rights reserved.
// Licensed under the MIT license. See LICENSE file in the project root for full license
// information.

namespace BasicSendReceiveUsingTopicSubscriptionClient
{
    using Microsoft.Azure.ServiceBus;
    using System;
    using System.Text;
    using System.Threading;
    using System.Threading.Tasks;

    classProgram
    {
// Connection String for the namespace can be obtained from the Azure portal
// under the 请用你的 Azure Portal 中给的 Service Bus 链接串替换这里的链接串
// 'Shared Access policies' section.访问 'Shared Access policies' 菜单,点击其中
// 的 rule 条目,来获得链接串
conststring ServiceBusConnectionString = "{ServiceBus connection string}";
conststring TopicName = "{Topic Name}";
conststring SubscriptionName = "{Subscription Name}";
static ITopicClient topicClient; //注意:这里用的是 ITopicClient,用来发消息
        static ISubscriptionClient subscriptionClient; //注意:这里用的是 ISubscriptionClient,用来作为订阅者读取消息

        static void Main(string[] args)
        {
            MainAsync().GetAwaiter().GetResult();
        }

        static async Task MainAsync()
        {
            const int numberOfMessages = 10;
            topicClient = new TopicClient(ServiceBusConnectionString, TopicName);
            subscriptionClient = new SubscriptionClient(ServiceBusConnectionString, TopicName, SubscriptionName);

            // Ensure default rule exists
```

```csharp
                await subscriptionClient.RemoveRuleAsync(RuleDescription.DefaultRuleName);
                await subscriptionClient.AddRuleAsync(new RuleDescription(RuleDescription.DefaultRuleName, new TrueFilter()));

                Console.WriteLine("================================");
                Console.WriteLine("Press any key to exit after receiving all the messages.");
                Console.WriteLine("================================");

                // Register Subscription's MessageHandler and receive messages in a loop 注
                // 册订阅的对应 MessageHandler 并在循环中接收消息
                RegisterOnMessageHandlerAndReceiveMessages();

                // Send Messages 发送消息
                await SendMessagesAsync(numberOfMessages);

                Console.ReadKey();

                await subscriptionClient.CloseAsync();
                await topicClient.CloseAsync();
        }

        static void RegisterOnMessageHandlerAndReceiveMessages()
        {
                // Configure the MessageHandler Options in terms of exception handling,
                // number of concurrent messages to deliver etc.
                var messageHandlerOptions = new MessageHandlerOptions(ExceptionReceivedHandler)
                {
                        // Maximum number of Concurrent calls to the callback 'ProcessMessagesAsync',
                        // Async', set to 1 for simplicity. 最大同时调用 ProcessMessagesAsync
                        // 方法,显示地设置成 1
                        // Set it according to how many messages the application wants to
                        // process in parallel. 根据应用的需求来设置并行处理消息的个数
                        MaxConcurrentCalls = 1,

                        // Indicates whether MessagePump should automatically complete the
                        // messages after returning from User Callback. 指示是否 MessagePump
                        // 应该自动 complete 消息,在用户调用返回之后
                        // False below indicates the Complete will be handled by the User Callback as in 'ProcessMessagesAsync' below. False 表示不自动 complete
                        // 消息,用户调用方法 ProcessMessagesAsync 会手动地将消息 complete
```

```
                AutoComplete = false
            };

            // Register the function that will process messages 注册处理消息的方法，调用的同时需
            // 要传入相应的 option
            subscriptionClient.RegisterMessageHandler(ProcessMessagesAsync, messageHandlerOptions);
        }

        // ProcessMessagesAsync 是具体处理消息的方法，传入的是消息 Message
        static async Task ProcessMessagesAsync(Message message, CancellationToken token)
        {
            // Process the message 处理此消息，此处可以加入具体的商业代码
            Console.WriteLine($"Received message: SequenceNumber:{message.SystemProperties.SequenceNumber} Body:{Encoding.UTF8.GetString(message.Body)}");

            // Complete the message so that it is not received again. 完成此消息，所以
            // 此消息不会被再 receive 一次
            // This can be done only if the subscriptionClient is created in Receive-
            // Mode.PeekLock mode (which is default). 注意，此 CompleteAsync 方法只能
            // 在 PeekLock 模式下才可用。如果是 Receive 和 Delete 的模式，则不必调用
            // CompeleteAsync 方法
            await subscriptionClient.CompleteAsync(message.SystemProperties.LockToken);


            // Note: Use the cancellationToken passed as necessary to determine if the
            // subscriptionClient has already been closed. 注：cancellationToken 参数
            // 用来检测 subscriptionClient 是否已经关闭
            // If subscriptionClient has already been Closed, you may chose to not call
            // CompleteAsync() or AbandonAsync()etc. calls 如果 subscriptionClient 已
            // 经是关闭的状态，你最好不调用 CompleteAsync()或者 AbandonAsync()，这样
            // 可避免不必要的 Exception
            // to avoid unnecessary exceptions.
        }

        // Use this Handler to look at the exceptions received on the MessagePump 用
        // handler 来处理 MessagePump 收到的消息
        static Task ExceptionReceivedHandler(ExceptionReceivedEventArgs exceptionReceivedEventArgs)
        {
```

```csharp
            Console.WriteLine($"Message handler encountered an exception {exceptionReceivedEventArgs.Exception}.");
            var context = exceptionReceivedEventArgs.ExceptionReceivedContext;
            Console.WriteLine("Exception context for troubleshooting:");
            Console.WriteLine($"- Endpoint: {context.Endpoint}");
            Console.WriteLine($"- Entity Path: {context.EntityPath}");
            Console.WriteLine($"- Executing Action: {context.Action}");
            return Task.CompletedTask;
        }

        static async Task SendMessagesAsync(int numberOfMessagesToSend)
        {
            try
            {
                for (var i = 0; i < numberOfMessagesToSend; i++)
                {
                    // Create a new message to send to the topic 创建消息
                    string messageBody = $"Message {i}";
                    var message = new Message(Encoding.UTF8.GetBytes(messageBody));

                    // Write the body of the message to the console 将消息体打印出来
                    Console.WriteLine($"Sending message: {messageBody}");

                    // Send the message to the topic 发送消息到 topic
                    await topicClient.SendAsync(message);
                }
            }
            catch (Exception exception)
            {
                Console.WriteLine($"{DateTime.Now} :: Exception: {exception.Message}");
            }
        }
    }
}
```

6.6 用 Visual Studio 开发 Service Bus Relay On

中继(Relay On)提供了一种可利用现有应用的方法。一般,现有应用是在本地的,而需要跟部署在云端的应用进行连接。中继就是这种情况的解决方案。在这种混合本地应用和云应用的场合下，可以通过 Point-to-site VPN，将本地一台机器与云端的一个站点（site）连接起来。本地网路中只有这一台机器与 Azure 云端的网络连接，所以可以按照 Point-to-site VPN 的方式在本地此机器上安装 VPN 的客户端软件。此 VPN 客户端软件可以支持的操作系统包括：

- Windows 7 (32 bit、64 bit)；
- Windows Server 2008 R2 (64 bit)；
- Windows 8 (32 bit、64 bit)；
- Windows 8.1 (32 bit、64 bit)；
- Windows Server 2012 (64 bit)；
- Windows Server 2012 R2 (64 bit)；
- Windows 10。

其网络结构大体如图 6-39 所示。

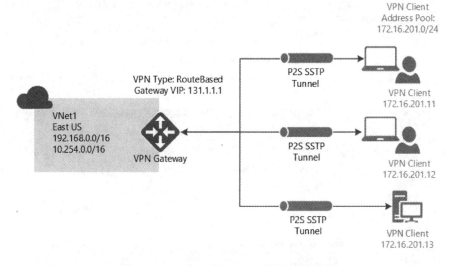

图 6-39　Point to site VPN 网络的结构图

详细情况可以参考：https://docs.microsoft.com/en-us/azure/vpn-gateway/vpn-gateway-howto-point-to-site-classic-azure-portal。

除了 Point-to-site VPN，另外一种就是可以用 Site-to-site VPN，即用本地网络和 Azure 云环境的网络建立一个 VPN，双方之间有 VPN Gateway，网络通信跨越两个子网络的，就得通过 VPN Gateway。其网络结构图如图 6-40 所示。

左边的是 Azure 云环境网络，右边的是本地网络。可以见到其用的协议与 Point-to-site VPN 是不同的。详细情况可参考：https://docs.microsoft.com/en-us/azure/vpn-gateway/vpn-gateway-howto-site-to-site-classic-portal。

在用上述方法搭建好 VPN 后，即可进行下一步。

在 Service Bus 中建立一个 Shared Access Policy，让客户端可以访问此

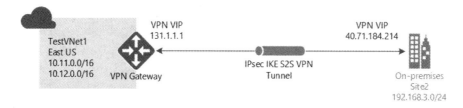

图 6-40　Site-to-site VPN 网络的结构图

Service Bus。

Service Bus Relay On 可以支持本地网络中的 WCF 服务。之前开发的 WCF 服务，是旧的资产，不能马上就淘汰，所以可以将其与云环境中的应用进行集成。

假设：本地已有一个 WCF 服务，部署在一台服务器上。下面要做的就是：

① 将此 WCF 服务注册到 Service Bus Relay On 上。

② 客户端从公网上通过 Service Bus 的 Shared Access Policy 来访问注册的 WCF 服务。

详见图 6-41。

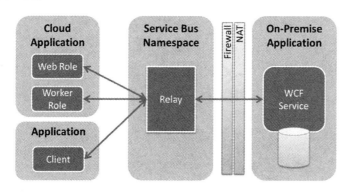

图 6-41　Service Bus Relay 服务网络图

这个图很好地说明了 Service Bus Relay 所起的作用，Relay 可以给客户端应用和云应用起到一个桥梁的作用，让这些应用可以访问到公司本地网中的 WCF 服务，从中获得数据。

第一步：创建一个 console 类型的程序。程序文件 Program.cs 的代码如下：

```
using System;
using System.Collections.Generic;
using System.Linq;
using System.Text;
using System.Threading.Tasks;
using System.ServiceModel;
using Microsoft.ServiceBus;
```

```csharp
using System.ServiceModel.Description;

namespace ServiceBusRelayServer
{
    class Program
    {
        static void Main(string[] args)
        {
            // 注:这里用的是默认的 Policy,可以另外创建一个 Shared Access Policy
            // 注:此处要求 policy 有
            var tokenProvider = TokenProvider.CreateSharedAccessSignatureTokenProvider("RootManageSharedAccessKey", " + LSd5njWikUZEi + RY/NpkkR/GL3PlpVRGCv23gwKrM0 = ");

            // 注:将 WCF Service 的类型 YourWCFServiceType 注册
            var serviceHost = new ServiceHost(typeof(YourWCFServiceType));

            // 注:将 WCF Service 的接口类型 IWCFService 注册成为 Service Endpoint
            // 注:URL sb://your_service_bus_namespace.servicebus.windows.net/process 中的 /process 中有全部的信息。这样注册后,以后只要访问/process URL,就会将客户的请求发送到公司本地网路中的 WCF 服务。
            var serviceEndpoint = serviceHost.AddServiceEndpoint(
                typeof(IWCFService),
                new NetTcpRelayBinding(),
                "sb://your_service_bus_namespace.servicebus.windows.net/process");

            serviceEndpoint.Behaviors.Add(new TransportClientEndpointBehavior(tokenProvider));

            serviceHost.Open();
            Console.WriteLine("Press ENTER to close");
            Console.ReadLine();
            serviceHost.Close();
        }
    }
}
```

第二步:写客户端代码,来调用此 Service Bus Relay On 服务。

```csharp
using Microsoft.ServiceBus;
using System;
using System.Collections.Generic;
using System.Linq;
```

```csharp
using System.ServiceModel;
using System.Text;
using System.Threading.Tasks;
using System.Diagnostics;


namespace ServiceBusRelayClient
{
    class Program
    {
        staticvoid Main(string[] args)
        {
            var binding = new NetTcpRelayBinding();

            var channelFactory = new ChannelFactory <IEchoChannel>
            (
                binding,
                new EndpointAddress("sb://your_service_bus_namespace.servicebus.windows.net/process")
            );

            channelFactory.Endpoint.EndpointBehaviors.Add( new TransportClientEndpointBehavior { TokenProvider = TokenProvider.CreateSharedAccessSignatureTokenProvider ("RootManageSharedAccessKey", " + LSd5njWikUZEi + RY/NpkkR/GL3PlpVRGCv23gwKrM0 = ") });

            using (var ch = channelFactory.CreateChannel())
            {
                string returnValue = ch.GetValue();
                Console.WriteLine("Get {0} From WCF Server", returnValue);
            }
            Console.ReadLine();
        }
    }
}
```

第 7 章

Service Fabric

7.1 基 础

Azure Service Fabric 是一款分布式系统平台,可方便用户轻松打包、部署和管理可缩放的可靠微服务和容器。开发人员和管理员不需解决复杂的基础结构问题,只需专注于实现苛刻的任务关键型工作负荷,即那些可缩放、可靠且易于管理的工作负荷即可。Service Fabric 代表用于生成和管理这些企业级的一级云规模应用程序的下一代中间件平台。Azure Service Fabric 可以作为服务的容器平台,是如同 Docker 一样的容器,它专为支持微服务而生。它可以支持跨平台的集群,同时支持多种编程语言、多种平台,语言如 Java、C♯、Php、Python、Node.js 等,平台如 Windows、Linux;支持诸如 Web 微服务、Web 应用,或者 tcp/ip 服务等。它还提供了全面的应用程序管理功能,用于设置、部署、监视、升级/修补和删除部署的应用程序。

7.1.1 开发环境准备

以下这些操作系统支持开发 Azure Service Fabric:
- Windows 7;
- Windows 8/Windows 8.1;
- Windows Server 2012 R2;
- Windows 10。

如果是 Visual Studio 2015,则需要安装:
- Servie Fabric Tools For Visual Studio 2015;
- Service Fabric SDK。

要求 Visual Studio 2015 update 2 或者之后的版本,才能安装以上两个安装包。需要到 MSDN 网站上搜寻安装包,下载了以后来安装。

本书默认都使用 Visual Studio 2017,在安装时需要选择合适的包,如图 7-1 所示。

在图 7-1 中,安装时选择 Azure development 就可以了,Service Fabric Tools 已经包含在其中。安装好之后,Visual Studio 2017 就会有支持 Service Fabric 的开

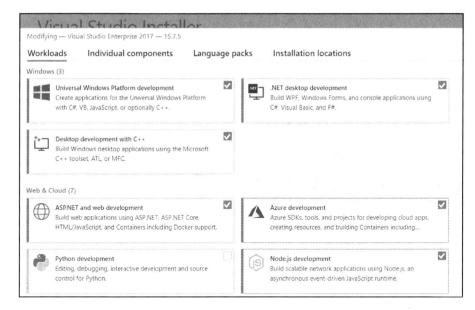

图 7-1 安装包

发了。可以创建 Service Fabric 类型的工程。

另外还需要安装 Azure Service Fabric SDK，详见 http://www.microsoft.com/web/handlers/webpi.ashx?command=getinstallerredirect&appid=MicrosoftAzure-ServiceFabric-CoreSDK。在此链接可以下载 Service Fabric SDK 的安装包。这些安装包可以帮我们建立一个本地的开发环境。

创建好 Service Fabric Local Cluster Manager 之后，就会有一个 System tray 的图标，如图 7-2 所示。

这个 Service Fabric Local Cluster Manager 在本地运行了一个模拟 Azure 环境的 Service Fabric 部署空间。可以将开发出来的应用程序部署到此 Service Fabric Local Cluster Manager 中。

图 7-2 Service Fabric Local Cluster Manager 在任务栏的图标

它有 1 个节点(node)的模式，还有 5 个节点的模式。我们可以在这两个模式间切换。选择了 5 个节点的模式后，会弹出一个请求改动计算机设置的对话框，单击 Yes 按钮，允许对计算机做出改动，如图 7-3 所示。

然后一个 Powershell 窗口弹出，它会运行脚本将 Service Fabric Local Cluster Manager 设置好。过一段时间会收到已设置好的信息。

设置好之后，就可以对本地的 Service Fabric Local Cluster 进行管理了，如图 7-4 所示。

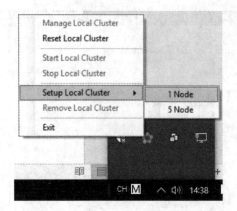

图7-3 选择 Setup Local Cluster，
1 Node 或者 5 Node

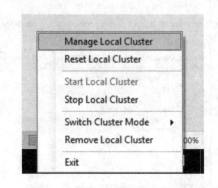

图7-4 停止 Local Cluster 或者
开始 Local Cluster

选择第一项，Manage Local Cluster，打开了一个浏览器窗口，如图7-5所示。

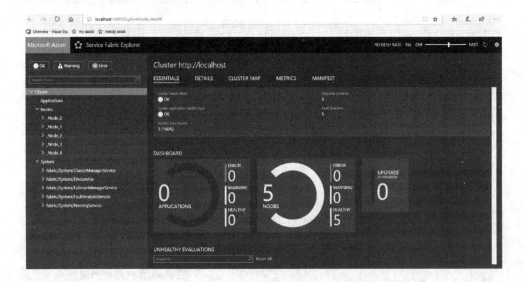

图7-5 Service Fabric Explorer 查看 Local Cluster 的状态

可以看到已经启动了5个节点。目前还没有应用部署到这个 Local Cluster Manager。

因为 Service Fabric 项目的部署和升级是基于 Power Shell 的，因此需要执行如下命令用于执行 Power Shell 脚本。

使用管理员权限打开 Power Shell，执行如下命令：

Set-ExecutionPolicy -ExecutionPolicy Unrestricted -Force -Scope CurrentUser

这样允许当前登录计算机的用户执行任何命令。

7.1.2 基本概念和俯瞰图

Service Fabric 服务:它在容器中运行的一系列程序来服务外部客户,有可能需要调用第三方微服务,或者需要访问一些存储,或者访问数据库,完成请求后再将结果返回给客户端。服务有无状态和有状态之分。无状态的服务可以由任何一个节点来响应用户的请求。无状态服务的节点不保存任何状态信息。同一个客户端的多个请求可以由多个节点来响应。

无状态的 Service Fabric 如图 7-6 所示。

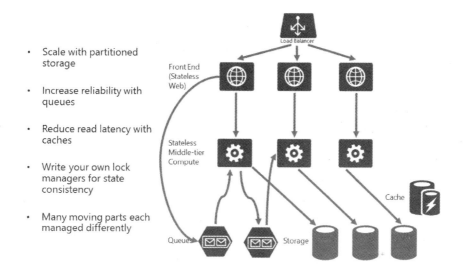

图 7-6 无状态的 Service Fabric

有状态的服务会有数据保存到可靠的存储当中。这些可靠的存储都是由 Service Fabric 来管理的。有状态的服务也可以由任何一个节点来服务客户端。Service Fabric 会智能地调整由哪些节点来响应用户的请求。每个响应用户请求的节点都会多一个步骤,需要将状态数据读出,在完成请求时再将改动写回可靠存储。

有状态的 Service Fabric 如图 7-7 所示。

Service Fabric 行动者:它是在服务之下运行的单线程的独立运行单元。行动者与服务类似,可以用来完成一些相对比较独立的工作,其区别主要是行动者没有对第三方的微服务的调用,或者进行一些耗时的存储访问,相对比较独立的计算单元,只要调用,传入相应的参数,行动者就执行程序计算。最后将执行结果返回给客户端。

这些技术合作的关系图如图 7-8 所示。

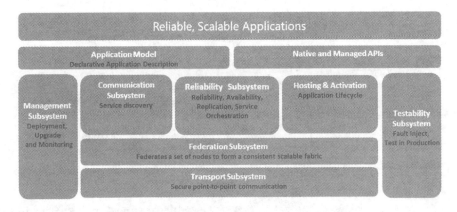

图 7-7 有状态的 Service Fabric

图 7-8 Service Fabric 各技术的关系图

7.2 Stateless Service 服务的创建

可以打开 Visual Studio,新建一个工程,File→New→Project。

在后面的窗口中选择,如图 7-9 所示。

Cloud→Service Fabric Application。

输入其工程名字,比如 Service Fabric Application1,然后输入一个保存目录。单击 OK 按钮,进入下一页,如图 7-10 所示。

从此窗口可以看出,.NET Core 2.0 和.NET Framework 是专为对应的技术准备的。分有状态和无状态的服务,以及有状态和无状态的 ASP.NET Core 程序等。Hosted Containersand Applications 是为别的编程语言以及开发平台准备的。

Service Fabric 7

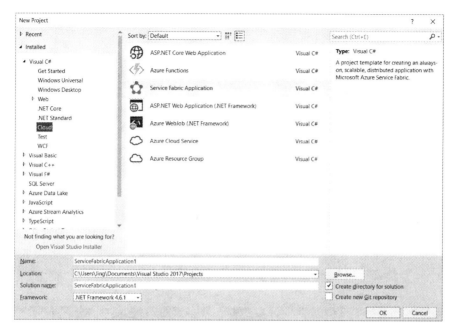

图 7-9 新建 Service Fabric Application

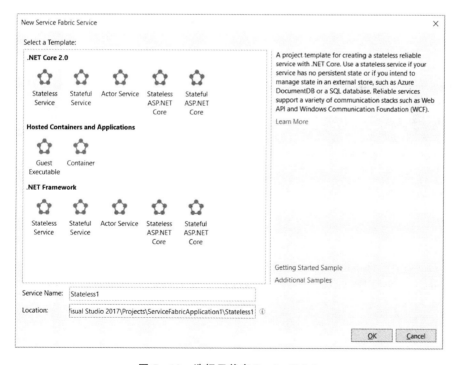

图 7-10 选择无状态 Service Fabric

在此界面输入 Service Name 和 Location(保存路径),单击 OK 按钮,等一会儿就创建好了,如图 7-11 所示。

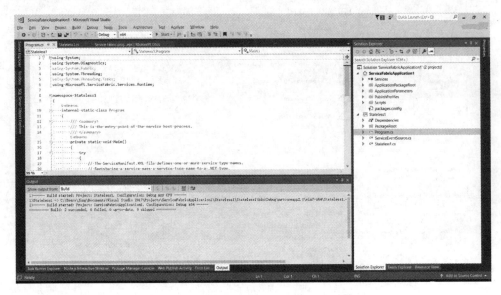

图 7-11　创建好无状态 Service Fabric 后的样子

此工程是一个 Stateless 的 Service,即无状态的 Service。

//Program.cs 文件
using System;
using System.Diagnostics;
using System.Fabric;
using System.Threading;
using System.Threading.Tasks;
using Microsoft.ServiceFabric.Services.Runtime;

namespace Stateless1
{
　　internal static class Program
　　{
　　　　/// <summary>
　　　　/// This is the entry point of the service host process.
　　　　/// </summary>
　　　　private static void Main()
　　　　{
　　　　　　try
　　　　　　{
　　　　　　　　// The ServiceManifest.XML file defines one or more service type names.

```
            // Registering a service maps a service type name to a .NET type.
            // When Service Fabric creates an instance of this service type,
            // an instance of the class is created in this host process.

            ServiceRuntime.RegisterServiceAsync("Stateless1Type",
                context => new
Stateless1(context)).GetAwaiter().GetResult();

ServiceEventSource.Current.ServiceTypeRegistered(Process.GetCurrentProcess().Id, typeof
(Stateless1).Name);

            // Prevents this host process from terminating so services keep running.
            Thread.Sleep(Timeout.Infinite);
        }
        catch (Exception e)
        {
ServiceEventSource.Current.ServiceHostInitializationFailed(e.ToString());
            throw;
        }
    }
  }
}
```

此代码用于注册此无状态的服务。注册完之后就无限休眠,让无状态的 Stateless1 类保持继续运行。服务 Stateless1 会被部署到所有的节点上。比如有 50 个节点,这 50 个节点都部署了 Stateless1 服务。Stateless1 是无状态的服务,不会保存用户的临时数据在服务中,所以客户端请求可以由任何一个节点来处理。这样就带来了便利性,可以由任何一个节点来响应客户端的请求。客户端的前后多个请求,可以由多个不同的节点来响应。

```
//Stateless1.cs 文件
using System;
using System.Collections.Generic;
using System.Fabric;
using System.Linq;
using System.Threading;
using System.Threading.Tasks;
using Microsoft.ServiceFabric.Services.Communication.Runtime;
using Microsoft.ServiceFabric.Services.Runtime;

namespace Stateless1
{
```

```csharp
/// <summary>
/// An instance of this class is created for each service instance by the Service Fabric
/// runtime.
/// </summary>
internal sealed class Stateless1 : StatelessService
{
    public Stateless1(StatelessServiceContext context)
        : base(context)
    { }

    /// <summary>
    /// Optional override to create listeners (e.g., TCP, HTTP) for this service replica to
    /// handle client or user requests.
    /// </summary>
    /// <returns> A collection of listeners. </returns>
    protected override IEnumerable <ServiceInstanceListener> CreateServiceInstanceListeners()
    {
        return new ServiceInstanceListener[0];
    }

    /// <summary>
    /// This is the main entry point for your service instance.
    /// </summary>
    /// <param name = "cancellationToken"> Canceled when Service Fabric needs to
    /// shut down this service instance. </param>
    protected override async Task RunAsync(CancellationToken cancellationToken)
    {
        // TODO: Replace the following sample code with your own logic
        //       or remove this RunAsync override if it's not needed in your service.

        long iterations = 0;

        while (true)
        {
            cancellationToken.ThrowIfCancellationRequested();

            ServiceEventSource.Current.ServiceMessage(this.Context, "Working-{0}", ++iterations);

            await Task.Delay(TimeSpan.FromSeconds(1), cancellationToken);
        }
```

                }
            }
        }

注：Stateless1 继承了 Stateless Service 类，将由 Service Fabric 的容器来创建 Stateless1 类的实例，并且容器负责管理 Stateless1 的生命周期。这个容器就是 Service Fabric Runtime。在创建 Stateless1 的实例时，Service Fabric Runtime 传入一个 Stateless Service Context，作为 Stateless1 的上下文。

重写 Create Service Instance Listeners 方法创建一个侦听器的集合。可以通过侦听器来侦听服务器端和客户端的通信。这个侦听器的集合不是必需的，所以可以不必重写 Create Service Instance Listeners 方法。

Protected Override Async Task RunAsync(Cancellation Token Cancellation Token)是 Stateless Service 的一个必须覆盖的方法，用来表示此服务的自定义行为。当部署到节点上运行的时候，都是这个方法响应客户端的请求。

```
//ServiceEventSource.cs 文件
using System;
using System.Collections.Generic;
using System.Diagnostics.Tracing;
using System.Fabric;
using System.Linq;
using System.Text;
using System.Threading.Tasks;
using Microsoft.ServiceFabric.Services.Runtime;

namespace Stateless1
{
    [EventSource(Name = "MyCompany-ServiceFabricApplication1-Stateless1")]
    internal sealed class ServiceEventSource : EventSource
    {
        public static readonly ServiceEventSource Current = new ServiceEventSource();

        // Instance constructor is private to enforce singleton semantics
        private ServiceEventSource() : base() { }

        #region Keywords
        // Event keywords can be used to categorize events.
        // Each keyword is a bit flag. A single event can be associated with multiple key-
        // words(via EventAttribute.Keywords property).
        // Keywords must be defined as a public class named 'Keywords' inside EventSource
        /// that uses them.
```

```csharp
public static classKeywords
{
    public const EventKeywords Requests = (EventKeywords)0x1L;
    public const EventKeywords ServiceInitialization = (EventKeywords)0x2L;
}
#endregion

#region Events
// Define an instance method for each event you want to record and apply an [Event]
// attribute to it.
// The method name is the name of the event.
// Pass any parameters you want to record with the event (only primitive integer
// types, DateTime, Guid & string are allowed).
// Each event method implementation should check whether the event source is
// enabled, and if it is, call WriteEvent() method to raise the event.
// The number and types of arguments passed to every event method must exactly match
// what is passed to WriteEvent().
// Put [NonEvent] attribute on all methods that do not define an event.
// For more information see https://msdn.microsoft.com/en-us/library/system.
// diagnostics.tracing.eventsource.aspx

[NonEvent]
public void Message(string message, paramsobject[] args)
{
    if (this.IsEnabled())
    {
        string finalMessage = string.Format(message, args);
        Message(finalMessage);
    }
}


private const int MessageEventId = 1;
[Event(MessageEventId, Level = EventLevel.Informational, Message = "{0}")]
public void Message(string message)
{
    if (this.IsEnabled())
    {
        WriteEvent(MessageEventId, message);
    }
}


[NonEvent]
```

```csharp
        public void ServiceMessage(StatelessServiceContext serviceContext, string mes-
sage, paramsobject[] args)
        {
            if (this.IsEnabled())
            {
                string finalMessage = string.Format(message, args);
                ServiceMessage(
                    serviceContext.ServiceName.ToString(),
                    serviceContext.ServiceTypeName,
                    serviceContext.InstanceId,
                    serviceContext.PartitionId,
                    serviceContext.CodePackageActivationContext.ApplicationName,
                    serviceContext.CodePackageActivationContext.ApplicationTypeName,
                    serviceContext.NodeContext.NodeName,
                    finalMessage);
            }
        }

        // For very high-frequency events it might be advantageous to raise events using
        // WriteEventCore API.
        // This results in more efficient parameter handling, but requires explicit alloca-
        // tion of EventData structure and unsafe code.
        // To enable this code path, define UNSAFE conditional compilation symbol and turn
        //on unsafe code support in project properties.
        privateconstint ServiceMessageEventId = 2;
        [Event(ServiceMessageEventId, Level = EventLevel.Informational, Message = "{7}")]
        private
#if UNSAFE
        unsafe
#endif
        void ServiceMessage(
            string serviceName,
            string serviceTypeName,
            long replicaOrInstanceId,
            Guid partitionId,
            string applicationName,
            string applicationTypeName,
            string nodeName,
            string message)
        {
#if !UNSAFE
            WriteEvent(ServiceMessageEventId, serviceName, serviceTypeName, replicaOrIn-
```

```csharp
stanceId, partitionId, applicationName, applicationTypeName, nodeName, message);
#else
            const int numArgs = 8;
            fixed (char * pServiceName = serviceName, pServiceTypeName = service-
TypeName, pApplicationName = applicationName, pApplicationTypeName = applicationTypeName,
pNodeName = nodeName, pMessage = message)
            {
                EventData * eventData = stackalloc EventData[numArgs];
                eventData[0] = new EventData { DataPointer = (IntPtr) pServiceName,
Size = SizeInBytes(serviceName) };
                eventData[1] = new EventData { DataPointer = (IntPtr) pService-
TypeName, Size = SizeInBytes(serviceTypeName) };
                eventData[2] = new EventData { DataPointer = (IntPtr)
(&replicaOrInstanceId), Size = sizeof(long) };
                eventData[3] = new EventData { DataPointer = (IntPtr) (&partitionId),
Size = sizeof(Guid) };
                eventData[4] = new EventData { DataPointer = (IntPtr) pApplication-
Name, Size = SizeInBytes(applicationName) };
                eventData[5] = new EventData { DataPointer = (IntPtr) pApplication-
TypeName, Size = SizeInBytes(applicationTypeName) };
                eventData[6] = new EventData { DataPointer = (IntPtr) pNodeName, Size =
SizeInBytes(nodeName) };
                eventData[7] = new EventData { DataPointer = (IntPtr) pMessage, Size =
SizeInBytes(message) };

                WriteEventCore(ServiceMessageEventId, numArgs, eventData);
            }
#endif
        }


        private const int ServiceTypeRegisteredEventId = 3;
        [Event(ServiceTypeRegisteredEventId, Level = EventLevel.Informational, Message =
"Service host process {0} registered service type {1}", Keywords = Keywords.ServiceInitial-
ization)]
        public void ServiceTypeRegistered(int hostProcessId, string serviceType)
        {
            WriteEvent(ServiceTypeRegisteredEventId, hostProcessId, serviceType);
        }

        private const int ServiceHostInitializationFailedEventId = 4;
        [Event(ServiceHostInitializationFailedEventId, Level = EventLevel.Error, Message =
"Service host initialization failed", Keywords = Keywords.ServiceInitialization)]
```

```csharp
        public void ServiceHostInitializationFailed(string exception)
        {
            WriteEvent(ServiceHostInitializationFailedEventId, exception);
        }


        // A pair of events sharing the same name prefix with a "Start"/"Stop" suffix
        // implicitly marks boundaries of an event tracing activity.
        // These activities can be automatically picked up by debugging and profiling tools,
        // which can compute their execution time, child activities,
        // and other statistics.
        private const int ServiceRequestStartEventId = 5;
        [Event(ServiceRequestStartEventId, Level = EventLevel.Informational, Message = "Service request '{0}' started", Keywords = Keywords.Requests)]
        public void ServiceRequestStart(string requestTypeName)
        {
            WriteEvent(ServiceRequestStartEventId, requestTypeName);
        }

        private const int ServiceRequestStopEventId = 6;
        [Event(ServiceRequestStopEventId, Level = EventLevel.Informational, Message = "Service request '{0}' finished", Keywords = Keywords.Requests)]
        public void ServiceRequestStop(string requestTypeName, string exception = "")
        {
            WriteEvent(ServiceRequestStopEventId, requestTypeName, exception);
        }
        #endregion

        #region Private methods
#if UNSAFE
        private int SizeInBytes(string s)
        {
            if (s == null)
            {
                return 0;
            }
            else
            {
                return (s.Length + 1) * sizeof(char);
            }
        }
#endif
        #endregion
```

```
    }
}
```

注：Service Event Source.cs 定义了一个单例模式的类 Service Event Source，用来帮助我们向 Trace 系统发送 Event 事件和 Log。如果部署的节点服务器的操作系统是 Windows，则可在 System Event 中看到这些应用程序产生的事件和消息。

//PackageRoot/ServiceManifest.cs 文件的源码

```xml
<?xmlversion="1.0"encoding="utf-8"?>
<ServiceManifest Name="Stateless1Pkg"
                 Version="1.0.0"
                 xmlns="http://schemas.microsoft.com/2011/01/fabric"
                 xmlns:xsd="http://www.w3.org/2001/XMLSchema"
                 xmlns:xsi="http://www.w3.org/2001/XMLSchema-instance">
    <ServiceTypes>
        <!-- This is the name of your ServiceType.
             This name must match the string used in the RegisterServiceAsync call in Program.cs. -->
        <StatelessServiceTypeServiceTypeName="Stateless1Type" />
    </ServiceTypes>

    <!-- Code package is your service executable. -->
    <CodePackage Name="Code"Version="1.0.0">
        <EntryPoint>
            <ExeHost>
                <Program> Stateless1.exe </Program>
            </ExeHost>
        </EntryPoint>
    </CodePackage>


    <!-- Config package is the contents of the Config directoy under PackageRoot that contains an
         independently-updateable and versioned set of custom configuration settings for your service. -->
    <ConfigPackage Name="Config"Version="1.0.0" />


    <Resources>
        <Endpoints>
            <!-- This endpoint is used by the communication listener to obtain the port on which to listen. Please note that if your service is partitioned, this port is shared with
```

replicas of different partitions that are placed in your code. -->
            <Endpoint Name = "ServiceEndpoint" />
        </Endpoints>
    </Resources>
</ServiceManifest>
```

注：Stateless1 Type 要和 Program.cs 中注册的 Service Type 名字一致。

Code Package 节的信息 Stateless1.exe 是编译后的产出文件。我们可以看到 Stateless1 工程的属性如图 7-12 所示。

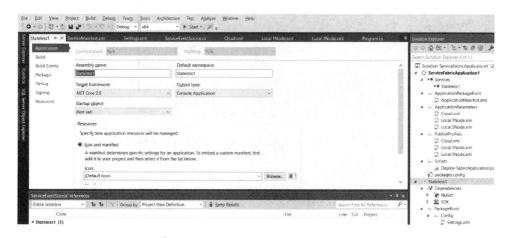

图 7-12  Stateless1 工程的属性

它是编译成 Stateless1.exe 文件的，并且是没有图形界面的 console 程序。

Endpoints 节是定义端点 Service Endpoint，用来加入侦听器 Listener 的时候使用。

```
//PackageRoot/Config/Settings.xml
<? xmlversion = "1.0"encoding = "utf - 8" ? >
<Settingsxmlns:xsd = "http://www.w3.org/2001/XMLSchema"xmlns:xsi = "http://www.w3.org/2001/XMLSchema - instance"xmlns = "http://schemas.microsoft.com/2011/01/fabric">
    <! -- Add your custom configuration sections and parameters here -->
    <! --
    <Section Name = "MyConfigSection">
    </Section>
    -->
</Settings>
```

注：这个文件用来配置自定义的配置节和自定义的参数，如注释的 MyConfigSection 节。自定义参数如 MyParameter。
```

部署相关的设置

这个 Stateless 服务将会被部署到一个容器中去,这个容器的节点数目是由容器来决定的。所以针对部署环境,应有一个对应的部署配置文件。

与部署有关的部分如图 7-13 所示。

图 7-13 部署 Service Fabric 的 Powershell 文件

其中,主要是在 Service Fabric Application1 工程中的这些文件。

先来看看部署脚本文件:

// Deploy - FabricApplication.ps1 文件

```
<#
.SYNOPSIS
Deploys a Service Fabric application type to a cluster.


.DESCRIPTION
This script deploys a Service Fabric application type to a cluster.  It is invoked by
Visual Studio when deploying a Service Fabric Application project.


.NOTES
WARNING: This script file is invoked by Visual Studio.  Its parameters must not be al-
tered but its logic can be customized as necessary.


.PARAMETER PublishProfileFile
Path to the file containing the publish profile.


.PARAMETER ApplicationPackagePath
Path to the folder of the packaged Service Fabric application.


.PARAMETER DeployOnly
Indicates that the Service Fabric application should not be created or upgraded after
registering the application type.
```

.PARAMETER ApplicationParameter
Hashtable of the Service Fabric application parameters to be used for the application.

.PARAMETER UnregisterUnusedApplicationVersionsAfterUpgrade
Indicates whether to unregister any unused application versions that exist after an upgrade is finished.

.PARAMETER OverrideUpgradeBehavior
Indicates the behavior used to override the upgrade settings specified by the publish profile.
'None' indicates that the upgrade settings will not be overridden.
'ForceUpgrade' indicates that an upgrade will occur with default settings, regardless of what is specified in the publish profile.
'VetoUpgrade' indicates that an upgrade will not occur, regardless of what is specified in the publish profile.

.PARAMETER UseExistingClusterConnection
Indicates that the script should make use of an existing cluster connection that has already been established in the PowerShell session. The cluster connection parameters configured in the publish profile are ignored.

.PARAMETER OverwriteBehavior
Overwrite Behavior if an application exists in the cluster with the same name. Available Options are Never, Always, SameAppTypeAndVersion. This setting is not applicable when upgrading an application.
'Never' will not remove the existing application. This is the default behavior.
'Always' will remove the existing application even if its Application type and Version is different from the application being created.
'SameAppTypeAndVersion' will remove the existing application only if its Application type and Version is same as the application being created.

.PARAMETER SkipPackageValidation
Switch signaling whether the package should be validated or not before deployment.

.PARAMETER SecurityToken
A security token for authentication to cluster management endpoints. Used for silent authentication to clusters that are protected by Azure Active Directory.

.PARAMETER CopyPackageTimeoutSec
Timeout in seconds for copying application package to image store.

.EXAMPLE

    . Scripts\Deploy-FabricApplication.ps1 -ApplicationPackagePath 'pkg\Debug'

Deploy the application using the default package location for a Debug build.

    .EXAMPLE
    . Scripts\Deploy-FabricApplication.ps1 -ApplicationPackagePath 'pkg\Debug' -DoNot-CreateApplication

Deploy the application but do not create the application instance.

    .EXAMPLE
    . Scripts\Deploy-FabricApplication.ps1 -ApplicationPackagePath 'pkg\Debug' -ApplicationParameter @{CustomParameter1 = 'MyValue'; CustomParameter2 = 'MyValue'}

Deploy the application by providing values for parameters that are defined in the application manifest.
    #>

Param
(
    [String]
    $PublishProfileFile,

    [String]
    $ApplicationPackagePath,

    [Switch]
    $DeployOnly,

    [Hashtable]
    $ApplicationParameter,

    [Boolean]
    $UnregisterUnusedApplicationVersionsAfterUpgrade,

    [String]
    [ValidateSet('None', 'ForceUpgrade', 'VetoUpgrade')]
    $OverrideUpgradeBehavior = 'None',

    [Switch]
    $UseExistingClusterConnection,

```powershell
        [String]
        [ValidateSet('Never','Always','SameAppTypeAndVersion')]
        $OverwriteBehavior = 'Never',

        [Switch]
        $SkipPackageValidation,

        [String]
        $SecurityToken,

        [int]
        $CopyPackageTimeoutSec
)

function Read-XmlElementAsHashtable
{
    Param(
        [System.Xml.XmlElement]
        $Element
    )

    $hashtable = @{}
    if( $Element.Attributes)
    {
        $Element.Attributes |
            ForEach-Object {
                $boolVal = $null
                if([bool]::TryParse($_.Value, [ref]$boolVal)) {
                    $hashtable[$_.Name] = $boolVal
                }
                else {
                    $hashtable[$_.Name] = $_.Value
                }
            }
    }

    return $hashtable
}

functionRead-PublishProfile
{
    Param(
```

```powershell
        [ValidateScript({Test-Path $_ -PathType Leaf})]
        [String]
        $PublishProfileFile
    )

    $publishProfileXml = [Xml](Get-Content $PublishProfileFile)
    $publishProfile = @{}

    $publishProfile.ClusterConnectionParameters = Read-XmlElementAsHashtable $publishProfileXml.PublishProfile.Item("ClusterConnectionParameters")
    $publishProfile.UpgradeDeployment = Read-XmlElementAsHashtable $publishProfileXml.PublishProfile.Item("UpgradeDeployment")
    $publishProfile.CopyPackageParameters = Read-XmlElementAsHashtable $publishProfileXml.PublishProfile.Item("CopyPackageParameters")

    if($publishProfileXml.PublishProfile.Item("UpgradeDeployment"))
    {
        $publishProfile.UpgradeDeployment.Parameters = Read-XmlElementAsHashtable $publishProfileXml.PublishProfile.Item("UpgradeDeployment").Item("Parameters")
        if($publishProfile.UpgradeDeployment["Mode"])
        {
            $publishProfile.UpgradeDeployment.Parameters[$publishProfile.UpgradeDeployment["Mode"]] = $true
        }
    }

    $publishProfileFolder = (Split-Path $PublishProfileFile)
    $publishProfile.ApplicationParameterFile = [System.IO.Path]::Combine($PublishProfileFolder, $publishProfileXml.PublishProfile.ApplicationParameterFile.Path)

    return $publishProfile
}

$LocalFolder = (Split-Path $MyInvocation.MyCommand.Path)

if(!$PublishProfileFile)
{
    $PublishProfileFile = "$LocalFolder\..\PublishProfiles\Local.xml"
}

if(!$ApplicationPackagePath)
{
```

```powershell
        $ApplicationPackagePath = "$LocalFolder\..\pkg\Release"
    }

    $ApplicationPackagePath = Resolve-Path $ApplicationPackagePath

    $publishProfile = Read-PublishProfile $PublishProfileFile

    if(-not $UseExistingClusterConnection)
    {
        $ClusterConnectionParameters = $publishProfile.ClusterConnectionParameters
        if($SecurityToken)
        {
            $ClusterConnectionParameters["SecurityToken"] = $SecurityToken
        }

        try
        {
            [void](Connect-ServiceFabricCluster @ClusterConnectionParameters)
        }
        catch [System.Fabric.FabricObjectClosedException]
        {
            Write-Warning "Service Fabric cluster may not be connected."
            throw
        }
    }

    $RegKey = "HKLM:\SOFTWARE\Microsoft\Service Fabric SDK"
    $ModuleFolderPath = (Get-ItemProperty -Path $RegKey -Name FabricSDKPSModulePath).FabricSDKPSModulePath
Import-Module "$ModuleFolderPath\ServiceFabricSDK.psm1"

$IsUpgrade = ($publishProfile.UpgradeDeployment -and $publishProfile.UpgradeDeployment.Enabled -and $OverrideUpgradeBehavior -ne 'VetoUpgrade') -or $OverrideUpgradeBehavior -eq 'ForceUpgrade'

$PublishParameters = @{
    'ApplicationPackagePath' = $ApplicationPackagePath
    'ApplicationParameterFilePath' = $publishProfile.ApplicationParameterFile
    'ApplicationParameter' = $ApplicationParameter
    'ErrorAction' = 'Stop'
}
```

```powershell
if( $publishProfile.CopyPackageParameters.CopyPackageTimeoutSec)
{
    $PublishParameters['CopyPackageTimeoutSec'] = $publishProfile.CopyPackageParameters.CopyPackageTimeoutSec
}

if( $publishProfile.CopyPackageParameters.CompressPackage)
{
    $PublishParameters['CompressPackage'] = $publishProfile.CopyPackageParameters.CompressPackage
}

# CopyPackageTimeoutSec parameter overrides the value from the publish profile
if( $CopyPackageTimeoutSec)
{
    $PublishParameters['CopyPackageTimeoutSec'] = $CopyPackageTimeoutSec
}

if( $IsUpgrade)
{
    $Action = "RegisterAndUpgrade"
    if( $DeployOnly)
    {
    $Action = "Register"
    }

    $UpgradeParameters = $publishProfile.UpgradeDeployment.Parameters

    if( $OverrideUpgradeBehavior -eq 'ForceUpgrade')
    {
    # Warning: Do not alter these upgrade parameters. It will create an inconsistency with Visual Studio's behavior.
        $UpgradeParameters = @{ UnmonitoredAuto = $true; Force = $true }
    }

    $PublishParameters['Action'] = $Action
    $PublishParameters['UpgradeParameters'] = $UpgradeParameters
    $PublishParameters['UnregisterUnusedVersions'] = $UnregisterUnusedApplicationVersionsAfterUpgrade

    Publish-UpgradedServiceFabricApplication @PublishParameters
}
```

```
else
{
    $Action = "RegisterAndCreate"
    if( $DeployOnly)
    {
        $Action = "Register"
    }

    $PublishParameters['Action'] = $Action
    $PublishParameters['OverwriteBehavior'] = $OverwriteBehavior
    $PublishParameters['SkipPackageValidation'] = $SkipPackageValidation

    Publish-NewServiceFabricApplication @PublishParameters
}
```

注：这个脚本提供了部署到一个 Azure Service Fabric Cluster 的所有步骤的脚本。

通常部署需要先建立一个 Azure Service Fabric Cluster，为这个 Cluster 定义好节点数、每个节点的服务器类型、操作系统。这些前提都准备好之后，我们还需要获得这个 Cluster 的连接信息，比如它的服务器证书的信息、其端点的 URL、建立连接的用户名等信息，改好对应的 Publish 目录中的文件。一般可以用 Visual Studio 来运行此 Powershell 脚本。Visual Studio 可以帮助我们运行正确的命令，省一些时间。另外一种方法就是脱离 Visual Studio 来运行此部署 Powershell 脚本，针对编译路径，以及输入 Publish 的文件，可以参考 Visual Studio 运行时打印出来的命令行改成正确的命令行，然后在 Powershell 窗口中运行。

下面来看 PublishProfiles/Cloud.xml 的内容：

```
<?xml version="1.0" encoding="utf-8"?>
<PublishProfile xmlns="http://schemas.microsoft.com/2015/05/fabrictools">
<!-- ClusterConnectionParameters allows you to specify the PowerShell parameters to use when connecting to the Service Fabric cluster.
    Valid parameters are any that are accepted by the Connect-ServiceFabricCluster cmdlet.

    For a remote cluster, you would need to specify the appropriate parameters for that specific cluster.
      For example: <ClusterConnectionParameters ConnectionEndpoint="mycluster.westus.cloudapp.azure.com:19000" />

    Example showing parameters for a cluster that uses certificate security:
```

```xml
    <ClusterConnectionParameters ConnectionEndpoint = "mycluster.westus.cloud-app.azure.com:19000"
            X509Credential = "true"
ServerCertThumbprint = "0123456789012345678901234567890123456789"
            FindType = "FindByThumbprint"
FindValue = "9876543210987654321098765432109876543210"
            StoreLocation = "CurrentUser"
            StoreName = "My" />

Example showing parameters for a cluster that uses Azure Active Directory (AAD) security:
    <ClusterConnectionParameters
ConnectionEndpoint = "mycluster.westus.cloudapp.azure.com:19000"
            AzureActiveDirectory = "true"
ServerCertThumbprint = "0123456789012345678901234567890123456789" />
    -->
    <ClusterConnectionParametersConnectionEndpoint = "" />
    <ApplicationParameterFilePath = "..\ApplicationParameters\Cloud.xml" />
    <CopyPackageParametersCompressPackage = "true" />
</PublishProfile>
```

注：这个文件需要填入正确的 Cluster Connection Parameters 参数，有如下几个：

- ConnectionEndpoint＝"mycluster.westus.cloudapp.azure.com:19000"；
- X509Credential＝"true"；
- ServerCertThumbprint＝"0123456789012345678901234567890123456789"；
- FindType＝"FindByThumbprint"；
- FindValue＝"9876543210987654321098765432109876543210"；
- StoreLocation＝"CurrentUser"；
- StoreName＝"My"。

第一个参数 ConnectionEndpoint 是创建 Cluster 时的 URL 和端口号。这可以在 Azure Portal 中获得。

后面几个参数都是关于这个端点的服务器证书的。必须要获得其服务器证书，保存到本地，用证书管理程序保存到当前用户的一个证书存储器中（Certificate Store）。同时要指定其证书指纹。下面来讲详细操作。

先在 Azure Portal 上建立一个 Service Fabric Cluster，可以单击 Create Resource 菜单，然后搜索 Service Fabric，出现 Service Fabric Cluster，单击它，出现如图 7-14 所示界面。

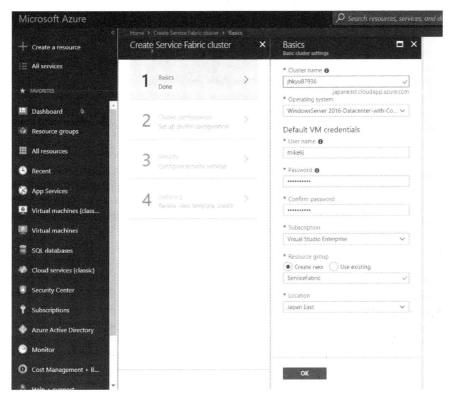

图 7 - 14　在 Azure Portal 中建立一个 Service Fabric Cluster

注：填入 Cluster 的名字，只有名字与别人的 Cluster 名字不重复才有可能通过。然后就选每个节点的操作系统，点击下拉列表选择一个合适的，如图 7 - 15 所示。

主要是 Windows Server 的不同版本和 Linux 的两个版本。通常. net framework 的 Service Fabric 必须选择 Windows Server 的操作系统，如果是. net core 的 ServiceFabric，可以选择 Windows Server 或者 Linux；其他编程语言的，如 PHP、Java、node. js 的，可以选择 Linux 的操作系统。下面是选择 Node types，如图 7 - 16 所示。

选择节点类型个数。根据实际需要来选，最多可以有三个节点类型。需要为每个节点类型设置类型名、VM 类型（包括 CPU、RAM、DISK 之类的内容），如图 7 - 17 所示。

选择一个 VM 类型之后，这个费用是要乘以个数的，比如个数是 5，选择的 VM 类型是 75.89 美元/月，那么就是 5 乘以 75.89 美元/月。这样的费用也需要精打细算，根据需要来动态调整。比如一开始个数弄少一点，就弄单节点的 Cluster，那就勾上 Single node cluster 的选项。

如果后面压力逐渐增大，可以将 Single node cluster 的勾去掉，然后填写节点个

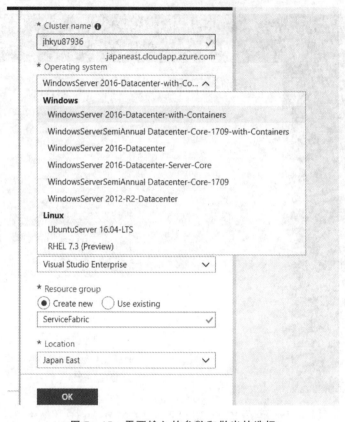

图 7-15 需要输入的参数和做出的选择

图 7-16 选择 Node types

Service Fabric 7

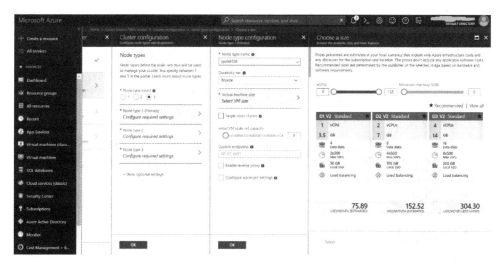

图 7-17 为每一个 Node types 选择具体的 VM 类型

数,为 2、3、4 的个数,逐渐增加个数。

选项 Enablereverse proxy 允许开发者打开反向代理功能,这样 Service Fabric 就可以有一个反向代理的手段来进行负载均衡的操作,让内部多个节点均衡地获得客户端的请求,并对客户端进行响应。

如图 7-18 所示,勾上 Configure advanced settings 选项后会显示高级设置的选项。

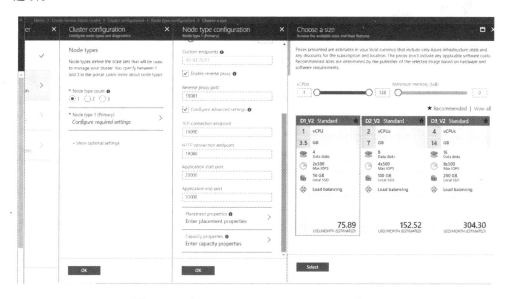

图 7-18 打开 Configure Advanced Settings 选项

它可以用来自定义各端点的端口号,其当前显示的是默认的端口号。

Placement properties 可以定义应用程序需要的一些设置项,用于应用程序的正常运行。

Capacity properties 可以配置将一些能力指标作为检测的对象。在其中选择能力指标。

单击 Select 按钮选择 VM 机型,然后单击 OK 按钮,则这个 Node type configuration 就配置好了。单击 OK 按钮进入下一界面,如图 7-19 所示。

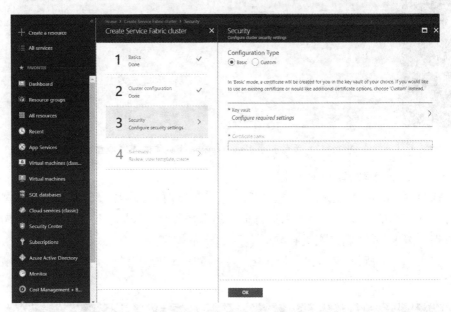

图 7-19 为 Service Fabric 设置 Security 选项

接下来,需要创建一个 Key vault,用来保存安全证书,如图 7-20 所示。

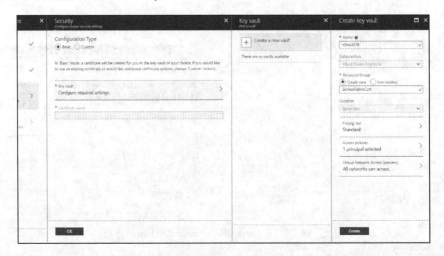

图 7-20 创建一个 Key vault

可点击 Pricing Tier，有两种选择：Standard、Premium，这两个价格区别不大。但是 Premium 有 HSM backed key，建议选 Premium。

再检查 Access Policies 访问策略，其中默认已将 Azure 账号加入，还可以将其他用户的账号加入，则这些用户也可以访问此 Service Fabric。这些用户是在 Azure Active Directory 中管理那些账号。要么是从已有的 Windows Active Directory 中导入的，要么是手工添加的。如图 7-21 所示，是从已有的 Windows Active Directory 中导入的。

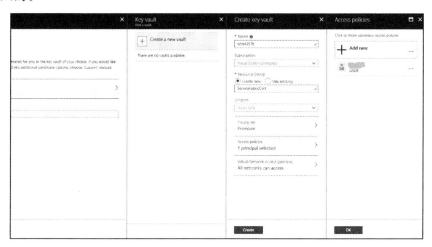

图 7-21　加入一些用户，允许访问此 Key vault

可访问的网络，默认是所有网络都可以访问。如果需要，可以改变默认的设置，只让特定的网络可以访问。

确认以后就可单击 Create 按钮，等待一段时间，则 Key vault 就建立好了。然后需要编辑 Access Policies，如图 7-22 中警告所示。

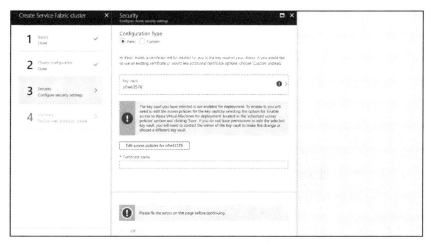

图 7-22　编辑 Access Policies

按照它的指示来设置,如图 7-23 所示。

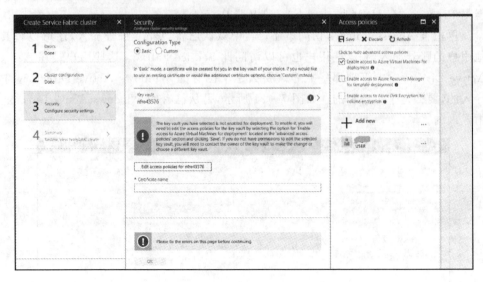

图 7-23 按照警告指示编辑 Access Policy,并勾上 Enable access to Azure Virtual Machine for deployment

单击 Save 按钮后,警告信息还在,需要再选择一遍此 Key vault,警告信息就会消失。

然后输入 Certificate name,如图 7-24 所示。

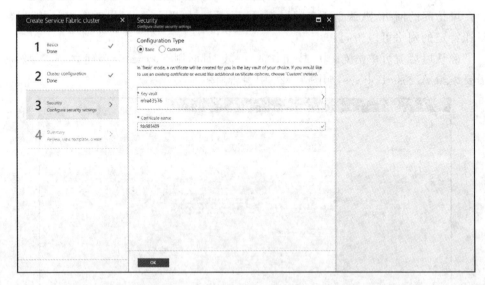

图 7-24 输入安全证书的名字

单击 OK 按钮,出现了 Summary 确认页面,如图 7-25 所示。

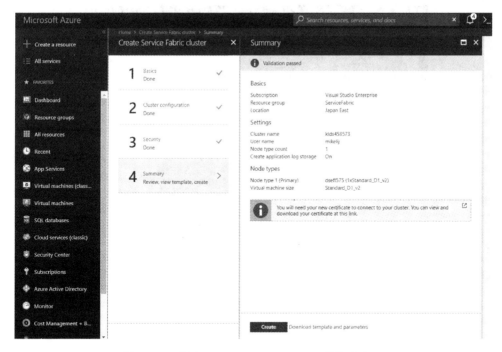

图 7-25　创建 Service Fabric Cluster 之前的确认页

可以点击如图 7-26 中所示的链接,去下载证书。

单击 Download as a certificate 按钮,将证书保存到本地。后面部署要用到。

都确认后,可以单击 Create 按钮创建这个 Service Fabric Cluster。等待一定时间,则此 Cluster 就建立好了,如图 7-27 所示。

然后可以从此 Cluster 中获得前面配置文件所需的配置信息,包括 Service Fabric Endpoint,以及服务器证书的信息。把服务器证书的指纹信息(将服务器证书放入一个本地 Store)、查找证书的 Store 信息等都填入上述配置文件。将 ClusterConnectionParameters 中的如下参数都填上相应的值:

- ConnectionEndpoint="mycluster.westus.cloudapp.azure.com:19000";
- X509Credential="true";
- ServerCertThumbprint="0123456789012345678901234567890123456789";
- FindType="FindByThumbprint";
- FindValue="9876543210987654321098765432109876543210";
- StoreLocation="CurrentUser";
- StoreName="My"。

然后 PublishFiles/Local.1Node.xml。

<?xmlversion="1.0"encoding="utf-8"?>
<PublishProfilexmlns="http://schemas.microsoft.com/2015/05/fabrictools"

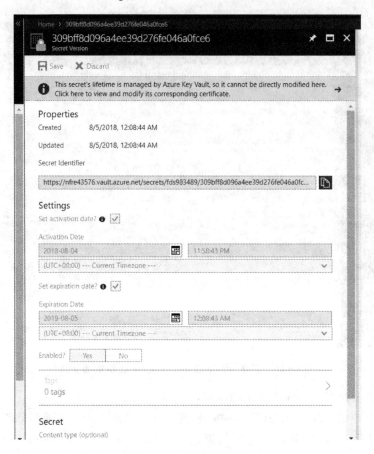

图 7 - 26　下载安全证书

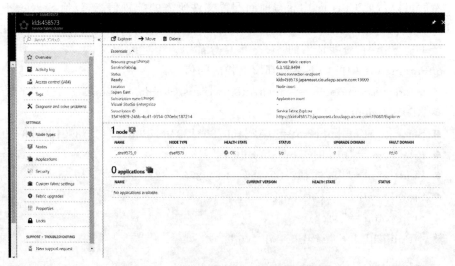

图 7 - 27　创建好了 Service Fabric Cluster

```xml
<!-- ClusterConnectionParameters allows you to specify the PowerShell parameters to use when connecting to the Service Fabric cluster.
     Valid parameters are any that are accepted by the Connect-ServiceFabricCluster cmdlet.

     For a local cluster, you would typically not use any parameters.
     For example: <ClusterConnectionParameters />
-->
<ClusterConnectionParameters />
<ApplicationParameterFilePath = "..\ApplicationParameters\Local.1Node.xml" />
</PublishProfile>
```

这用来配置应用程序的参数,可以给所有节点都配置一致的参数。
ApplicationParameters\Local.1Node.xml 文件如下:

```xml
<?xmlversion = "1.0"encoding = "utf-8"?>
<Applicationxmlns:xsd = "http://www.w3.org/2001/XMLSchema" xmlns:xsi = "http://www.w3.org/2001/XMLSchema-instance" Name = "fabric:/ServiceFabricApplication1" xmlns = "http://schemas.microsoft.com/2011/01/fabric">
    <Parameters>
        <ParameterName = "Stateless1_InstanceCount"Value = "1" />
    </Parameters>
</Application>
```

下面是配置具体参数的文件。
ApplicationPackageRoot\ ApplicationManifest.xml 文件如下:

```xml
<?xmlversion = "1.0"encoding = "utf-8"?>
<ApplicationManifestxmlns:xsd = "http://www.w3.org/2001/XMLSchema" xmlns:xsi = "http://www.w3.org/2001/XMLSchema-instance"
ApplicationTypeName = "ServiceFabric-Application1Type"
ApplicationTypeVersion = "1.0.0"
xmlns = "http://schemas.microsoft.com/2011/01/fabric">
    <Parameters>
        <ParameterName = "Stateless1_InstanceCount"DefaultValue = "-1" />
    </Parameters>
    <!-- Import the ServiceManifest from the ServicePackage. The ServiceManifestName and ServiceManifestVersion
         should match the Name and Version attributes of the ServiceManifest element defined in the
         ServiceManifest.xml file. -->
    <ServiceManifestImport>
        <ServiceManifestRefServiceManifestName = "Stateless1Pkg"ServiceManifestVersion = "1.0.0" />
```

```
            <ConfigOverrides />
        </ServiceManifestImport>
        <DefaultServices>
            <!-- The section below creates instances of service types, when an instance of this
                 application type is created. You can also create one or more instances of service type using the
                 ServiceFabric PowerShell module.

                 The attribute ServiceTypeName below must match the name defined in the imported ServiceManifest.xml file. -->
            <Service Name = "Stateless1"ServicePackageActivationMode = "ExclusiveProcess">
                <StatelessServiceServiceTypeName = "Stateless1Type" InstanceCount = "[Stateless1_InstanceCount]">
                    <SingletonPartition />
                </StatelessService>
            </Service>
        </DefaultServices>
    </ApplicationManifest>
```

这是此 Service Fabric 服务的部署描述文件。ServiceManifestName 和 ServiceManifestVersion 需要与 ServiceManufest.xml 文件中定义的值一致。

架构场景：

一个无状态的服务类似如图 7-28 所示。

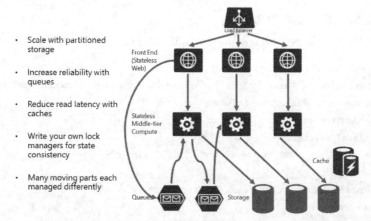

图 7-28　无状态 Service Fabric

这些无状态的 Web 服务,将一些数据缓存到 Cache 中,还有无状态的中间件服务,将数据写到消息队列和 Cache 服务中,这样可以方便进行 Scale 伸缩,任何一个无状态的服务器都可以服务用户请求。如果一个无状态服务器损坏,其他服务器也可处理客户端的请求。

7.3 Stateful 服务的创建

选择菜单 File→New Project，选择 Cloud→Service Fabric Application，输入工程名字，比如 ServiceFabricA1，然后选择 Stateful Service，如图 7-29 所示。

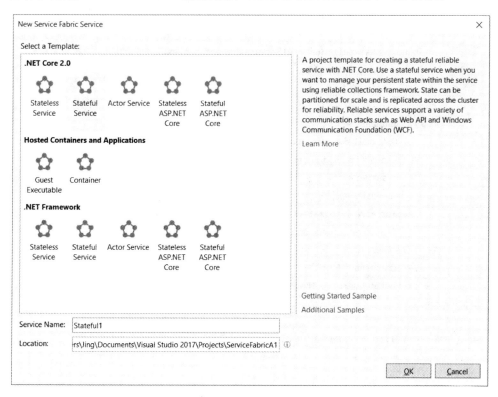

图 7-29 创建 Stateful Service Fabric

创建后，大部分文件都类似，只有 Stateful1.cs 文件与无状态的代码不同。

```
//Stateful1.cs 文件
using System;
using System.Collections.Generic;
using System.Fabric;
using System.Linq;
using System.Threading;
using System.Threading.Tasks;
using Microsoft.ServiceFabric.Data.Collections;
using Microsoft.ServiceFabric.Services.Communication.Runtime;
using Microsoft.ServiceFabric.Services.Runtime;
```

```csharp
namespace Stateful1
{
    /// <summary>
    /// An instance of this class is created for each service replica by the Service
    /// Fabric runtime.
    /// </summary>
    internal sealed class Stateful1 : StatefulService
    {
        public Stateful1(StatefulServiceContext context)
            : base(context)
        { }

        /// <summary>
        /// Optional override to create listeners (e.g., HTTP, Service Remoting, WCF,
        /// etc.) for this service replica to handle client or user requests.
        /// </summary>
        /// <remarks>
        /// For more information on service communication, see https://aka.ms/service-
        /// fabric servicecommunication
        /// </remarks>
        /// <returns> A collection of listeners. </returns>
        protected override IEnumerable <ServiceReplicaListener> CreateServiceReplicaListeners()
        {
            return new ServiceReplicaListener[0];
        }

        /// <summary>
        /// This is the main entry point for your service replica.
        /// This method executes when this replica of your service becomes primary and
        /// has write status.
        /// </summary>
        /// <param name = "cancellationToken"> Canceled when Service Fabric needs to
        /// shut down this service replica. </param>
        protected override async Task RunAsync(CancellationToken cancellationToken)
        {
            // TODO: Replace the following sample code with your own logic
            //       or remove this RunAsync override if it's not needed in your service.

            var myDictionary = await this.StateManager.GetOrAddAsync <IReliableDictionary <string, long>> ("myDictionary");
```

```csharp
                while (true)
                {
                        cancellationToken.ThrowIfCancellationRequested();

                        using (var tx = this.StateManager.CreateTransaction())
                        {
                                var result = await myDictionary.TryGetValueAsync(tx, "Counter");

                                ServiceEventSource.Current.ServiceMessage(this.Context, "Current Counter Value: {0}",
                                        result.HasValue ? result.Value.ToString() : "Value does not exist.");

                                await myDictionary.AddOrUpdateAsync(tx, "Counter", 0, (key, value) => ++value);

                                // If an exception is thrown before calling CommitAsync, the
                                // transaction aborts, all
                                // changes are discarded, and nothing is saved to the secondary
                                // replicas.
                                await tx.CommitAsync();
                        }

                        await Task.Delay(TimeSpan.FromSeconds(1), cancellationToken);
                }
            }
        }
}
```

在 RunSync 方法中使用 StateManager 属性。这是 StatefulService 提供的一个服务，提供了 ReliableService 来存储应用程序需要保存的数据。它有一系列的 Reliable(可靠的)存储结构，比如 IReliableDictionary 结构。容器会负责管理这些应用程序交付的数据的存储和传输，在多个节点同步。

```csharp
var myDictionary = await this.StateManager.GetOrAddAsync<IReliableDictionary<string, long>>("myDictionary");
```

得到一个 IReliableDictionary 的 dictionary，后面往此 dictionary 中加入元素。如这句：

```csharp
await myDictionary.AddOrUpdateAsync(tx, "Counter", 0, (key, value) => ++value);
```

同时 StateManager 支持事务性的操作。

要开始改变的时候,可以调用:

this.StateManager.CreateTransaction()

结束的时候,可以调用:

await tx.CommitAsync();

另外可以看一个样例,见 https://azure.microsoft.com/en-us/resources/samples/service-fabric-dotnet-core-getting-started/。下载下来以后,解压,到路径 service-fabric-dotnet-core-getting-started-master\Services\CounterService\src\CounterServiceApplication\CounterService,打开工程。

```csharp
//CounterService.cs 文件
using System;
using System.Collections.Generic;
using System.Fabric;
using System.Linq;
using System.Threading;
using System.Threading.Tasks;
using Microsoft.ServiceFabric.Data.Collections;
using Microsoft.ServiceFabric.Services.Communication.Runtime;
using Microsoft.ServiceFabric.Services.Runtime;
using Microsoft.ServiceFabric.Services.Remoting.Runtime;
using Microsoft.ServiceFabric.Data;
using CounterService.Interfaces;
namespace CounterService
{
    /// <summary>
    /// An instance of this class is created for each service replica by the Service
    /// Fabric runtime.
    /// </summary>
    internal sealed class CounterService : StatefulService, ICounterService
    {
        private string myName;
        private long myId;

        public CounterService(StatefulServiceContext context)
            : base(context)
        {
            this.myName = context.CodePackageActivationContext.ApplicationName;
            this.myId = context.ReplicaOrInstanceId;
        }

        /// <summary>
```

Service Fabric

```csharp
/// Optional override to create listeners (e.g., HTTP, Service Remoting, WCF,
/// etc.) for this service replica to handle client or user requests.
/// </summary>
/// <remarks>
/// For more information on service communication, see https://aka.ms/service-
/// fabric servicecommunication
/// </remarks>
/// <returns> A collection of listeners. </returns>
protected override IEnumerable <ServiceReplicaListener> CreateServiceReplicaListeners()
{
    return this.CreateServiceRemotingReplicaListeners();
}

/// <summary>
/// This is the main entry point for your service replica.
/// This method executes when this replica of your service becomes primary and
/// has write status.
/// </summary>
/// <param name = "cancellationToken"> Canceled when Service Fabric needs to
/// shut down this service replica. </param>
protected override async Task RunAsync(CancellationToken cancellationToken)
{
    // TODO: Replace the following sample code with your own logic
    //       or remove this RunAsync override if it's not needed in your service.

    var myDictionary = await this.StateManager.GetOrAddAsync < IReliableDictionary <string, long>> ("myDictionary");

    while (true)
    {
        cancellationToken.ThrowIfCancellationRequested();

        using (var tx = this.StateManager.CreateTransaction())
        {
            var result = await myDictionary.TryGetValueAsync(tx, "Counter");

            var counterValue = await myDictionary.GetOrAddAsync(tx, "Counter", 0);

            Console.WriteLine(" Incrementing counter... {0} {1} : {2}", this.myName, this.myId, counterValue);
            ServiceEventSource.Current.Message(" Incrementing counter from replica : " + this.myId + " : current value : " + counterValue);
```

```
            await myDictionary.AddOrUpdateAsync(tx, "Counter", 0, (key, val-
ue) => ++value);

            // If an exception is thrown before calling CommitAsync, the
            transaction aborts, all
            // changes are
            // discarded, and nothing is saved to the secondary replicas.
            await tx.CommitAsync();
        }

            await Task.Delay(TimeSpan.FromSeconds(1), cancellationToken);
        }
    }
    public async Task <long> GetValueAsync()
    {
        ConditionalValue <long> result;
        var myDictionary = await this.StateManager.GetOrAddAsync <IReliableDic-
tionary <string, long>> ("myDictionary");
        using (var tx = this.StateManager.CreateTransaction())
        {
            result = await myDictionary.TryGetValueAsync(tx, "Counter");
            await tx.CommitAsync();
        }
        if (result.HasValue)
        {
            return result.Value;
        }
        else
        {
            return -1;
        }
    }
}
```

可以看到类似的代码对 State Manager 的利用,以及开始事务、结束事务、IReliableDirectionry 的使用。

7.4 调试 Service Fabric 应用

要调试 Service Fabric 应用,需要配置好本地调试环境。
① 确保 Windows Service 的正常运行。

② 为了保证各节点间互相引用使用正常,需要使用机器名来作为本地开发 Cluster 的名字。在反向代理模式中查找服务也是需要使用机器名。

a. 打开一个 Powershell 窗口,用 Administrator 权限。

b. 进入 SDK Cluster 的目录,C:\Program Files\Microsoft SDKs\Service Fabric\ClusterSetup。

c. 输入命令,C:\Program Files\Microsoft SDKs\Service Fabric\ClusterSetup\DevClusterSetup.ps1 -UseMachineName。

注:此命令默认创建 5 个节点的 Cluster,如果要创建一个节点的 Cluster,则可以加上参数 CreateOneNodeCluster。

用本地 Service Fabric Cluster 调试容器的局限性列表。

用本地容器当中无法支持使用 Localhost 来做 ClusterFODNorIP 选项的 DNS 查询。解决方法就是用机器名来配置本地的 Service Fabric Cluster。

在虚拟机里运行 Windows 10 不会得到 DNS 回复到容器。解决方法是:在虚拟机的网卡 NIC 的 IPv4 协议中不允许 UDP Checksum(检查和)Offload。注:这会降低此虚拟机网络的性能。参考页面如下:

https://github.com/Azure/service-fabric-issues/issues/1061。

如果因这个应用是用 DockerCompose 部署的,则在同一个应用程序当中使用 DNS Service Name 查找服务,在 Windows 10 上不能正常工作。解决办法是:使用 servicename.applicationname 属性来查找 Service Endpoint。参考页面如下:

https://github.com/Azure/service-fabric-issues/issues/1062。

如果为 ClusterFODNorIP 选项使用 IP 地址,则改变主机的主 IP 将使 DNS 功能失效。解决方法是:使用新的主 IP 来重建这些本地 Service Fabric Cluster。或者使用机器名来重建本地 Service Fabric Cluster。这是按设计实现的功能。

如果 Cluster 中 FQDN 不能在网络上查找到,则 DNS 查找功能将失败。解决方法是:用主机的主 IP 来重建本地 Service Fabric Cluster。这是按设计实现的功能。

当调试一个容器时,Docker Logs 将只会在 Visual Studio 的 Output 窗口,不会通过 Service Fabric 的 API,也不会在 Service Fabric Explorer 中出现。

调试一个在 Docker 容器里运行的.NET 应用使用的步骤,作为 Administrator 来运行 Visual Studio。

打开已有应用程序的工程文件,或者是创建一个新的.NET 应用。

右键单击工程,在弹出的菜单里选择 Add→Container Orchestrator Support→Service Fabric。然后按 F5 键来调试启动此应用程序。

Visual Studio 支持 Console and ASP.NET Project Types for .NET and .NET Core,然后就可以在应用程序代码中打端点,等端点命中时,程序就会暂停下来,可以做价差变量的值、单步、跳过等操作。调试操作在 Visual Studio 中讲得比较详细,这里就不详细叙述了。

第 8 章

Media Service

8.1 基础

8.1.1 什么是 Media Service

Media Service 是一个可扩展的云平台服务,用于进行多媒体流播放、live 广播,允许开发者创建可伸缩的媒体管理和交付应用。Media Service 基于 RESTFul API,允许开发者安全地上传、存储、编码,并且能够安全地上传、存储、编码和打包视频或音频内容,以供点播和直播流传输到各种客户端(例如,电视、PC 和移动设备)。

可以使用完全媒体服务来构建端到端工作流,还可以选择为工作流程的某些部分使用第三方组件。例如,使用第三方编码器进行编码,然后使用媒体服务上传、保护、打包、递送。

8.1.2 用 Media Service 能做什么

Azure 媒体服务能够构建可扩展、成本有效、端到端的媒体分发解决方案,可以上传、编码、打包和流媒体到 Windows、iOS、Android、Adobe Flash 及其他设备和平台。

媒体服务对传统媒体工作流构建方法所带来的好处如下:
- 允许开发人员轻松创建、管理和维护自定义媒体工作流的 API。
- 它是一种标准化的工作流程,当参与创建和管理的内容有多个时,可以提高协调性和生产率。
- 通过使用全局数据中心来自动转换可扩展性,而不必计划容量尖峰或担心空闲数据中心。
- 成本效率通过编码媒体一次,然后使用动态包装以多种格式传递它。

8.1.3 Media Service 概览

1. 媒体服务提供您需要的一切,来轻松地建立和操作三个标准媒体解决方案
   - 视频点播(VOD)服务。媒体服务提供了将 VOD 服务操作到多个设备和平

- 在线视频平台（OVP）。媒体服务将与您的 OVP 和内容管理系统（CMS）集成，以帮助您的平台通过使用 Azure 数据中心的全球足迹来获得规模，而不必计划容量尖峰或担心空闲数据中心。
- 端到端解决方案。媒体服务可以用来轻松地在 Azure 中完全构建安全的端到端媒体工作流，从内容摄取到编码、打包和保护。这有助于降低与集成多个产品和提供者相关联的成本。

不管开发场景如何，分享视频内容的第一步都是选择用户的媒体体验。

2. 选择 Azure 媒体服务视频体验

分享视频内容的第一步是决定你想要什么样的用户体验。这可以通过提出一些问题来回答：

- 你的用户将如何观看视频内容？
- 你的用户会连接到互联网吗？
- 你的用户希望视频内容在高清频道中吗？
- 你的用户会在计算机或手持设备上观看视频内容吗？

提供这些问题的答案将有助于给用户最好的体验。

必须做出的另一个决定是选择支持的查看设备类型。表8-1 概述了媒体服务支持的查看设备，以及这些设备的推荐查看方法。

表8-1 媒体服务支持的查看设备

设 备	描 述	推荐查看方法
Web 浏览器	Web 浏览器可以在桌面 PC、智能手机和平板电脑上运行。当在桌面 PC 上运行时，可以利用大屏幕尺寸和大存储容量，允许浏览高清视频	离线查看、渐进式下载和流式传输
智能手机	智能手机屏幕小，存储容量小	流媒体传输
平板电脑	平板电脑比智能手机具有更大的屏幕，但通常具有更小的存储容量	流媒体传输。具有更大存储容量的平板电脑可以利用离线观看和逐步下载的优势
Xbox	Xbox 控制台具有大屏幕和大存储容量的优点	离线查看、渐进式下载和流式传输
机顶盒与电视连线	这些设备通常具有大屏幕，但存储容量最小	流媒体传输

3. Azure 媒体服务视频处理流程的组织

图8-1 显示了使用媒体服务处理媒体时的标准工作流的高级概述。

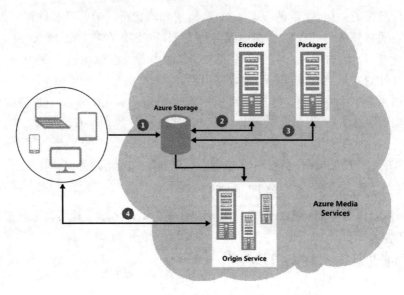

图 8-1 Azure Media Services 的工作流

4. 标准媒体服务工作流的高级概述

工作流程中涉及的步骤如下：
- 媒体被上传到媒体服务并存储在 Azure Blob 中。
- 使用 Azure 媒体编码器对上传的媒体进行编码，将编码的媒体存储在 Azure 存储中。
- 编码媒体由 Azure 媒体打包器打包，结果存储在 Azure 存储中。
- 客户端应用程序回放位于 URL 上的媒体，而 Media Service 处理从存储到客户端应用的出站流。

5. 向 Azure 媒体服务上传视频

必须将内容上传到 Azure 媒体服务中，以便能够编码、管理和消费它。媒体服务使用 Azure 存储来存储媒体以进行处理和查看。可以使用媒体服务 REST API 或可用的客户端 SDK 之一编程地上传内容。这些 API 允许一次上传一个文件或执行批量上传操作。媒体服务还允许执行内容的安全上传和存储。存储加密将在把内容上传到 Azure 存储之前将其本地加密，在那里它将以加密的形式存储。

媒体服务的基本内容是一种资产。资产包含一个或多个文件，如视频、音频、闭幕字幕文件和有关文件的元数据。每个资产包含一个或多个资产文件，每个资产文件包含关于特定媒体文件的元数据。一旦通过上传文件创建了一个资产，它就可以在媒体服务工作流中使用，例如编码和流媒体。

每个资产都映射到 Azure 存储账户中的 Blob 容器，资产中的文件作为容器中的 Blob 存储。Blob 容器要分组。一组 Blob 就像文件夹对一组文件进行分组一样。它们作为访问控制的边界点被用在媒体服务中。Azure 存储账户可以包含无限数量的 Blob 容器,容器可以存储无限数量的 Blob。

媒体服务账户与一个或多个 Azure 存储账户相关联。每个账户可以包含无限数量的 Blob 容器,并且只受基础账户的限制。媒体服务提供 SDK 工具来管理多个存储账户,并在上传期间执行资产分布的负载平衡。

6. Azure 媒体服务中支持的文件类型

可以将各种视频、音频和图像文件类型上传到媒体服务账户,而对使用媒体服务 SDK 上传的文件的类型或格式没有限制。然而,Azure 管理门户限制上传到 Azure 媒体编码器支持的格式。这些导入格式包括 MPEG-1、MPEG-2、MPEG-4、Windows Media 视频编码、MP3、WAVE 和 Windows 媒体音频编码音频,以及 BMP、JPEG 和 PNG 编码图像。Azure 媒体编码器可以导出数据作为 Windows Media Video、Windows Media Audio、MP4 和流畅的流式文件格式。

7. 使用微软 Azure 媒体服务处理媒体

在媒体服务中,媒体处理涉及获取媒体处理器实例、编码、打包和保护媒体文件。

8. 获取媒体处理器实例

媒体服务提供了许多能够处理视频的媒体处理器。媒体处理器处理特定的处理任务,例如编码、格式转换、加密或解密媒体内容。编码视频是最常见的媒体服务处理任务,它由 Azure 媒体编码器执行。

9. 编码视频

编码是将视频转换成可被用户消费的格式的过程。用户可以使用各种设备观看视频,包括台式计算机、智能手机、平板电脑、Xbox 控制台、机顶盒或互联网连接的电视。这些设备都具有影响所需编码的特性。例如,智能手机具有小屏幕和小存储空间,而台式计算机具有较大的屏幕和更大的存储空间。此外,智能手机的带宽可能比台式计算机更有限。因此,当选择如何编码一个视频时,必须记住用户将消耗视频的各种设备。

使用编码器预置字符串配置媒体编码器,每个预置字符串指定编码器所需的一组设置。编码器预置分为两组:通用预置和设备特定预置。由通用预设编码的视频可以被任何支持所需文件格式的设备使用。由特定设备预置的视频被设计成由特定设备如智能手机使用。对于所有预置列表,请参阅链接:https://docs.microsoft.com/en-us/azure/media-services/previous/media-services-mes-presets-overview。

10. 智能手机的编码

当为智能电话编码视频时,应该选择与目标设备的分辨率、支持的编解码器和支持的文件格式相匹配的编码预置。例如,对于支持高达 1 080 P 的 H.264 视频的 Windows 电话,应该使用"H264 平滑流式 Windows Phone 7 系列"预置。

iPhone 5 以 HLS 格式支持 H.264 视频到 1 080 P。然而,媒体服务不支持直接将视频编码到 HLS,但可以对 MP4 进行编码,然后使用静态打包将视频转换成 HLS。或者,可以以平滑流或 MP4 格式编码,并使用动态包装来将视频实时转换为 HLS。因此,如果想为 iPhone 5 编码视频到 1 080 P,将使用"H264 自适应比特率 MP4 集 1 080 P 只用于 iOS 蜂窝"预置。同样,对于支持 426X360 H.264 视频的 Android 手机,可以使用"H264 自适应比特率 MP4 集 SD4X3 只用于 iOS 蜂窝"预置。动态包装将用于实时转换视频到 HLS。

11. Xbox 编码

当为 Xbox 编码视频时,可以在 VC-1 和 H.264 之间选择分辨率高达 1 080 P 的平滑流。例如,使用 H.264 将视频编码到 720 P,将使用"H264 平滑流 720P Xbox Live ADK"预置。

12. 其他设备和平台的编码

通用预置可用于对各种设备进行编码,包括台式计算机、平板电脑和机顶盒。要选择合适的编码预设,必须确定用户将如何查看你的内容,以及他们的查看设备支持什么分辨率、比特率、编解码器和文件格式。

表 8-2 列出了每种类型的设备和由媒体服务支持的客户端技术。

表 8-2 每种类型的设备和由媒体服务支持的客户端技术

设备	技术
Windows 8	流畅的流媒体、渐进式下载、MPEG-DASH
Windows RT	流畅的流式和渐进式下载
Windows Phone	流畅的流式和渐进式下载
Web 浏览器	通过微软提供的附加 SDK 和播放器框架来支持流畅的流式传输。 通过 HTML5 视频元素在浏览器中支持渐进式下载。 Internet Explorer 11 和 Chrome 都通过使用媒体源扩展(MSE)来支持 MPEG-DASH
Xbox	流畅的流式和渐进式下载
麦金塔	苹果的 HLS 和渐进式下载
iOS	流畅的流、苹果的 HLS 和逐步下载
Android	流畅的流媒体、渐进式下载、苹果的 HLS
机顶盒,连接电视	流畅的流媒体、渐进式下载、苹果的 HLS

13. Azure 媒体服务包装视频

一旦一个视频被编码,它将被放置在一个输出资产中,之后它可以被放置到各种文件容器中。这个过程被称为包装。例如,可以使用 Azure 媒体打包器将 MP4 文件转换成流畅的流内容,以将编码的内容放入不同的文件容器中。

媒体服务允许用户决定他们是否会用媒体处理器预先包装视频(称为静态包装),或按需包装视频(称为动态包装)。

Azure Media Packager 是一种能够进行静态包装的媒体处理器。静态包装包括按用户所需的格式创建内容的副本。例如,如果用户需要两种格式,MP4 文件都可以转换成流畅的流媒体内容。这将导致存在的内容有两个副本,每个副本有不同的格式。

动态打包不是由媒体处理器执行的,而是由源服务器执行的。当客户端应用程序请求特定的视频格式时,OrthServer 将源媒体打包,允许只对视频进行一次编码,并将其实时转换为客户端应用程序所请求的格式。使用动态包装,则视频通常被存储为自适应比特率 MP4 的文件集。当客户端应用程序请求视频时,它指定所需的格式。然后,源服务器将 MP4 自适应比特率文件转换为客户端实时请求的格式。这确保了只有一个拷贝的视频必须被存储,因此降低了存储成本。

动态包装是发布视频的首选方法。

14. 利用 Azure 媒体服务保护视频

为了在发布媒体时保护媒体,媒体服务支持基于 Play Ready 及示例的普通加密和 AES 128 位 CBC 信封加密。Play Ready 是由微软开发的数字版权管理系统。DRM 允许控制谁可以访问您的内容。当用户试图观看 Play Ready 保护内容时,客户端应用程序请求来自媒体服务的内容。媒体服务将客户端重定向到授权服务器,该授权服务器认证和授权用户对内容的访问。客户端应用程序可以安全地下载解密密钥,该密钥允许内容被解密和查看。

AES 信封加密提供内容加密,但不允许复杂的数字版权管理或安全密钥传递(仅由 SSL 提供)。内容所有者应该信任他们的客户,如果他们选择 AES 128 位信封加密,那么对于不信任的恶意终端用户来说,获取和重新分配密钥要容易得多。

15. 从 Azure 媒体服务传送视频

媒体服务提供了传输媒体资产的不同机制,这些媒体资产已经上载到媒体服务。它可以用来传递仅仅存储在媒体服务中的内容,或者它还可以包括以不同方式处理或编码的内容。

用户通常可以使用四种方法来访问视频:
- 离线观看。
- 渐进式下载。
- 流式传输。
- 自适应比特率流。

离线观看涉及用户将整个视频下载到他们的计算机或设备上。因为视频可能相当大,下载可能需要一些时间才能完成,并且设备必须有足够的存储空间来保存整个视频。然而,这种方法的好处是,一旦下载了视频,就不需要网络连接来查看视频。

渐进式下载允许连接到互联网的用户在下载整个视频之前开始观看视频。然而,它确实要求观看设备具有足够的存储空间来保存整个视频。

流式传输还需要互联网连接,但与渐进式下载不同,因为它只下载少量视频,一旦被显示就丢弃它。这种方法的好处是它在观看设备上几乎不需要存储。

自适应比特率流允许客户端应用程序确定网络条件,并将视频内容的数据速率适配到可用的网络带宽。当网络通信降级时,客户端可以自动选择内容的较低比特率版本,因此允许用户继续观看视频,尽管质量较差。当网络条件改善时,客户端可以自动切换到更高的比特率,提高了视频质量。这种方法的好处是,视频播放器可以自动反映播放过程中带宽的变化,从而提供更好的用户体验。

媒体服务支持三种自适应比特率流技术:

① Smooth Streaming 流畅流。这是微软公司开发的一种自适应比特率流技术。

② HTTP 实时流(HLS)。这是苹果公司开发的一种自适应比特率流技术。

③ MPEG-DASH。这是一个由运动图像专家组(MPEG)创建的自适应比特率流协议,并且是一个国际标准。

16. 从 Azure 媒体服务中处理出站视频

媒体服务源服务处理对内容的请求。它从 Azure 存储中检索文件,并将它们直接提供给内容分发网络(CDN)或客户端应用程序。源服务器具有允许它们每秒响应几百个请求,并提供动态加密和动态打包服务的特性。

17. 在 Azure 媒体服务中提供视频访问

访问媒体服务中的内容需要一个定位器,它提供一个入口点来访问包含在一个资产中的文件。Access 策略用于定义客户端访问给定资产的权限和持续时间。多个定位器可以共享访问策略,使得不同的定位器可以提供不同的开始和停止时间,同时使用访问策略提供的相同的权限和持续时间设置。

有两种类型的定位器:

● 共享访问签名定位器。

● 按需原点定位器。

共享访问签名定位器授予 Azure 存储中媒体资产的底层 Blb 容器的访问权限。通过指定共享访问签名,可以给具有 URL 访问权限的用户授予特定资源的指定时间段。还可以指定能够通过共享访问签名定位器访问的资源执行哪些操作。

应使用按需原点定位器来授予对流式内容的访问权。点播源定位器由媒体服务原点服务公开,它从 Azure 存储中提取内容并将其传送给客户端。按需原点定位器混淆了基础资产的 Blb 容器和存储账户 URL。相反,它们总是指向媒体服务源服务,因此允许先进的方案,如 IP 限制、缓存控制和 CDN 认证。

18. 从 Azure 媒体服务中消费视频

媒体服务为创建在不同设备和平台上运行的媒体播放器应用提供支持,包括PC、Macintosh、Windows Phone、iOS 设备和 Android 设备。微软还提供了许多不同的 SDK 和播放器框架,允许创建从媒体服务中消耗流媒体的应用程序。

8.2 视频点播 Video-On-Demand

1. 启动流端点

使用 Azure 媒体服务时最常见的场景之一是通过自适应比特率流传输视频。媒体服务能够提供动态包装。通过动态打包,可以在媒体服务支持的实时流格式中提供自适应比特率的 MP4 编码内容。例子包括苹果 HTTP 直播流(HLS)、微软平滑流和动态自适应流 HTTP(DASH,也称为 MPEG-DASH)。通过使用媒体服务自适应比特率流,可以在不存储每个流格式的预打包版本的情况下传递视频。

注:当创建媒体服务账户时,在停止状态下将默认流端点(streaming endpoint)添加到账户中。若要启动流内容,并利用动态打包和动态加密的优点,则要使从中流内容的流端点必须处于运行状态。

启动流端点的步骤如下:
① 登录 Azure 门户网站。
② 选择设置→流端点。
③ 选择默认流端点,出现默认流端点详细信息窗口。
④ 选择 Start 图标。
⑤ 选择 Save 按钮。

2. 上传文件

通过使用媒体服务流视频,上传源视频,将它们编码成多个比特率,然后发布结果。第一部分将在本节中介绍。

① 在 Azure 门户网站中,选择 Azure 媒体服务账户 Media Services Account。
② 选择设置→资产 Assets。然后,选择上传 Upload 按钮,如图 8-2 所示。
这时,上传一个视频资产的窗口出现了。
注:媒体服务不限制上传视频文件的大小。
③ 在你的计算机上选择想上传的视频,然后单击 OK 按钮。
上传开始了。可以看到文件名下的进度。
上传完成后,新资产将显示在"资产"窗格中。

3. 编码资产

要利用动态打包,必须将源文件编码成多比特率流的 MP4 文件集。编码步骤在本节中演示。

图 8-2　Upload 按钮，用于上传多媒体资源

4. 在门户中编码资产

在 Azure 门户中使用媒体编码器标准对内容进行编码：

① 在 Azure 门户网站中，选择 Azure 媒体服务账户。

② 选择设置→资产，选择要编码的资产。

③ 选择 Encode 按钮，如图 8-3 所示。

④ 在对资产窗格进行编码时，选择媒体编码器标准处理器和预置。有关预置的信息，请参见自动生成比特率梯形和媒体编码器标准的任务预置。选择最适合输入视频的预置是很重要的。例如，如果知道输入的视频有 1 920×1 080 像素的分辨率，则可以选择 H264 多比特率 1 080 P 预置。如果有一个低分辨率(640×360)的视频，则不应该使用 H264 多比特率 1 080 P 的预置。

为了帮助管理资源，可以编辑输出资产的名称和作业的名称。

⑤ 选择 Create 按钮。

5. 监控作业进度的编码

要监控编码作业的进度，请在页面顶部选择设置，然后选择作业，如图 8-4 所示。

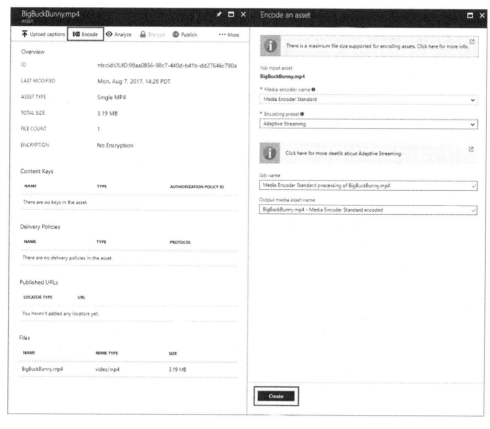

图 8-3 Encode 按钮帮助将资源文件进行编码

图 8-4 查看作业

6. 发布内容

为了给用户提供一个可以用来流或下载内容的 URL,首先必须通过创建一个定位器来发布你的资产。定位器提供对资产中的文件的访问。Azure 媒体服务支持两种定位器:

① 流(On Demand Origin)定位器。流定位器用于自适应流式传输。自适应流的例子包括 HLS、平滑流和 MPEG-DASH。要创建流式定位器,资产必须包含.ISM 文件。

② 渐进式(共享访问签名 Shared Access Signature)定位器。渐进式定位器用于通过渐进式下载来传送视频。

要构建一个 HLS 流式 URL,应在 URL 上附加(format=M3U8AAPL):

{streaming endpoint name-media services account name}.streaming.mediaservices.windows.net/{locator ID}/{file name}.ism/Manifest(format=m3u8-aapl)

若要生成流式 URL 以播放流畅的流式资产,请使用以下 URL 格式:

{streaming endpoint name-media services account name}.streaming.mediaservices.windows.net/{locator ID}/{file name}.ism/Manifest

要构建 MPEG-DASH 流式 URL,请将 URL(Frase=MPD 时间 CSF)附加到 URL:

{streaming endpoint name-media services account name}.streaming.mediaservices.windows.net/{locator ID}/{file name}.ism/Manifest(format=mpd-time-csf)

共享访问签名 URL 具有以下格式:

{blob container name}/{asset name}/{file name}/{shared access signature}

若要更新定位器上的过期日期,可以使用 REST API 或.NET API。

注:当更新共享访问签名定位器的过期日期时,URL 会发生更改。

7. 使用门户发布资产

① 在 Azure 门户网站中,选择 Azure 媒体服务账户。

② 选择设置→资产 Assets。选择要发布的资产。

③ 选择 Publish 按钮,如图 8-5 所示。

④ 选择 Locator type。

⑤ 选择 Add 按钮。

URL 被添加到发布的 URL 列表中。

8. 从门户播放内容

可以在 Azure Portal 的内容播放器上测试你的视频。

选择视频,然后选择播放按钮,如图 8-6 所示。

Media Service 8

图 8-5 Publish 资源

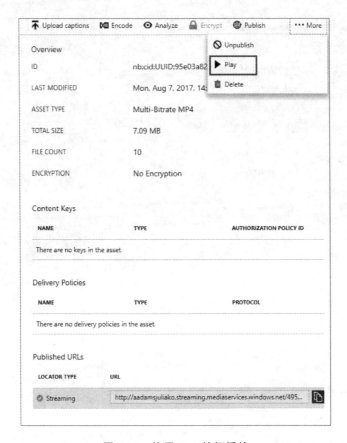

图 8-6　使用 Play 按钮播放

9. Azure 点播媒体编码器的综述与比较

编码概述

Azure 媒体服务为云中媒体的编码提供了多种选择。

当开始使用媒体服务时,了解编解码器和文件格式之间的差异是很重要的。编解码器是实现压缩/解压缩算法的软件,而文件格式是保存压缩视频的容器。

媒体服务提供动态的包装,使流媒体服务支持的格式提供自适应码率 MP4 或平滑的流媒体编码内容(MPEG DASH,HLS,平滑流),而不必重新包装成这些流媒体格式。

注:当你的系统账户创建一个默认的流端点处于停止状态并添加到你的账户时,开始流你的内容,并利用动态包装和动态加密,流的终点及你想流的内容必须在运行状态。

媒体服务支持需求的编码器,本章介绍如下几种:

- Media Encoder Standard。
- Media Encoder Premium Workflow。

本章简要介绍了按需媒体编码器,并提供链接的文章,提供更详细的信息。该主题还提供了编码器的比较。

注:默认情况下,每个媒体服务账户一次可以有一个活动编码任务。可以保留编码单元,允许同时运行多个编码任务,每个购买的编码保留单元都有一个编码任务。有关信息,请参阅缩放编码单元。

10. 使用.NET 编码媒体编码器标准的资产

编码作业是媒体服务中最常见的处理操作之一。创建编码作业以将媒体文件从一种编码转换为另一种编码。当编码时,可以使用媒体服务内置媒体编码器,还可以使用由媒体服务伙伴提供的编码器;第三方编码器可通过 Azure 市场获得。

本书展示了如何使用.NET 以媒体编码器标准(MES)来编码资产,使用这里描述的编码器预置之一配置媒体编码器标准。

建议始终将源文件编码为自适应比特率的 MP4 集,然后使用动态打包 Dynamic Packaging 将该集合转换为所需格式。

如果输出资产是存储加密的,则必须配置资产传递策略。有关更多信息,请参见配置资产传递策略 Configuring asset delivery policy。

注:MES 生成一个输出文件,该文件包含输入文件名的前 32 个字符。该名称基于预设文件中指定的内容。例如,"文件名":"{BaseNe}}{{index }{Ext}"。{BaseNe}被输入文件名的前 32 个字符替换。

11. MES(Media Encoder Standard) 格式

输入容器/文件格式如表 8-3 所列。

表 8-3 输入容器/文件格式

文件格式(文件扩展名)	是否支持
FLV(用 H.264 和 AAC 编解码器)(.FLV)	是
MXF(.MXF)	是
GXF(.GXF)	是
MPEG2-PS,MPEG2-TS,3GP(.TS,PS,3GP,.3GPP,.MPG)	是
Windows Media Video(WMV)/ASF(.WMV,ASF)	是
AVI(未压缩 8 位/10 位)(.AVI)	是
MP4(.MP4,.M4A,.M4V)/ISMV(.ISMA,.ISMV)	是
微软数字视频录制(DVR MS)(.DVR MS)	是
MatROSKA/WebM(.MKV)	是
Wave/wav(.wav)	是
QuiTimeTo(.MOV)	是

注:以上是更常见的文件扩展名。媒体编码器标准确实支持许多其他标准(例如.M2TS、.MPEG2VIEW、.QT)。如果您尝试对文件进行编码,就会得到不支持格式的错误信息。

12. 输入容器中的音频格式

媒体编码标准支持输入集装箱携带以下的音频格式:

MXF、GXF 和 QuickTime 文件,其中有音频轨道交错的立体或 5.1 样本。MXF、GXF 和 QuickTime 文件的音频进行单独的 PCM 轨道,但信道映射(立体声或5.1)可以从文件元数据导出。

显式/用户提供的信道映射将在不久的将来提供支持。

输入视频编解码器如表 8−4 所列。

表 8−4 输入视频编解码器

输入视频编解码器	是否支持
AVC 8 位/10 位,最多 4∶2∶2,包括 AVCITRA	8 位 4∶2∶0 和 4∶2∶2
Avid DNxHD(在 MXF 中)	是
DVCPRO/DVCPROHD(在 MXF 中)	是
数字视频(DV)(在 AVI 文件中)	是
JPEG 2000	是
MPEG-2(高达 422 的配置文件和高级别;包括 XDCAM、XDCAM HD、XDCAM IMX、CABLELABS ® 和 D10 等变型)	高达 422 的配置文件
MPEG-1	是
VC-1/WMV9	是
Canopus HQ/HQX	否
MPEG-4 第 2 部分	是
Theora	是
YV420 未压缩,或夹层	是
苹果 ProRes 422	是
苹果 ProRes 422 LT	是
苹果 ProRes 422HQ	是
苹果 ProRes Proxy	是
苹果 ProRes 4444	是
苹果 ProRes 4444XQ	是
HEVC/H.265	主要配置文件

输入音频编解码器如表 8−5 所列。

表 8-5 输入音频编解码器

输入音频编解码器	是否支持
AAC（AAC-LC、AAC-HE 和 AAC-HEv2；up to 5.1）	是
MPEG Layer 2	是的
MP3（MPEG-1 Audio Layer 3）	是
Windows Media Audio	是
WAV/PCM	是
FLAC	是
Opus	是
Vorbis	是
AMR（adaptive multi-rate）	否
Dolby® E	否
Dolby® Digital（AC3）	否
Dolby® Digital Plus（E-AC3）	否

13. 输出格式和编解码器

表 8-6 列出了文件格式、视频编码器、音频编码器。

表 8-6 编解码和文件格式、支持出口

文件格式	视频编码器	音频编码器
MP4（including multi-bitrate MP4 containers）	H.264（High、Main 和 Baseline Profiles）	AAC-LC、HE-AAC v1、HE-AAC v2
MPEG2-TS	H.264（High、Main 和 Baseline Profiles）	AAC-LC、HE-AAC v1、HE-AAC v2

14. MES 的预置

媒体编码器标准（MES）配置了一个编码器预置。参考下面的链接查看编码器预置列表：https://docs.microsoft.com/en-us/azure/media-services/previous/media-services-dotnet-encode-with-media-encoder-standard。

输入和输出的元数据

当对输入的资产使用 MES 进行编码时，在编码任务完成时得到输出资产。输出资产包含视频、音频、缩略图、宣言等，基于使用的编码器预置。

输出资产的元数据 XML 文件包含关于输入资产的元数据信息。元数据 XML 文件的名字符合如下格式：<资产 ID>_metadata.xml（例如：41114ad3-eb5e-4c57-8d92-5354e2b7d4a4_metadata.xml），其中<资产>是输入资产的资产 ID。此 XML 文件的 schema 在下面的链接中描述，请查阅：https://docs.microsoft.com/en-us/azure/media-services/previous/media-services-input-metadata-schema。

输出资产中也有一个 XML 文件含有输出资产的元数据信息。输出元数据 XML 文件的名字是如下格式：<源文件名>_manifest.xml（例如：BigBuckBunny_manifest.xml）。输出元数据 XML 文件的 schema 在下面的链接中描述，请查阅：https://docs.microsoft.com/en-us/azure/media-services/previous/media-services-output-metadata-schema。

如果想仔细查看这两个元数据文件，可以创建一个 SAS 定位器并下载此文件到本地计算机。在 Media Services .NET SDK 扩展工具中有一个样例演示如何创建一个 SAS 定位器，并下载一个文件到本地计算机。

Media Service On-Demand 样例代码的下载地址如下：

https://azure.microsoft.com/documentation/samples/media-services-dotnet-on-demand-encoding-with-media-encoder-standard/。

下载了之后，解压如图 8-7 所示。

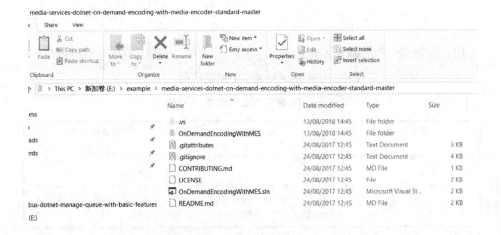

图 8-7 样例程序解压后的目录结构

```
//Program.cs

using Microsoft.WindowsAzure.MediaServices.Client;
using System;
using System.Collections.Generic;
using System.Configuration;
using System.IO;
using System.Linq;
using System.Threading;
```

```csharp
namespace OnDemandEncodingWithMES
{
    class Program
    {
        // Read values from the App.config file.
        static string _AADTenantDomain =
            ConfigurationManager.AppSettings["AMSAADTenantDomain"];
        static string _RESTAPIEndpoint =
            ConfigurationManager.AppSettings["AMSRESTAPIEndpoint"];

        private static readonly string _mediaFiles =
                Path.GetFullPath(@"../..\Media");

        private static readonly string _presetFiles =
            Path.GetFullPath(@"../..\Presets");

        // Field for service context.
        private static CloudMediaContext _context = null;

        static void Main(string[] args)
        {
            try
            {
                AzureAdTokenCredentials tokenCredentials = new AzureAdTokenCredentials(_AADTenantDomain, AzureEnvironments.AzureCloudEnvironment);
                AzureAdTokenProvider tokenProvider = new AzureAdTokenProvider(tokenCredentials);

                _context = new CloudMediaContext(new Uri(_RESTAPIEndpoint), tokenProvider);
                // If you want to secure your high quality input media files with
                // strong encryption at rest on disk,
                // use AssetCreationOptions.StorageEncrypted instead of AssetCre-
                // ationOptions.None.

                Console.WriteLine("Upload a file.\n");
                IAsset inputAsset =
                    UploadFile(Path.Combine(_mediaFiles, @"BigBuckBunny.mp4"), AssetCreationOptions.None);

                Console.WriteLine("Generate thumbnails and get URLs.\n");
```

```csharp
            IAsset thumbnailAsset = GenerateThumbnail(inputAsset, AssetCreationOptions.None);
            PublishAssetGetURLs(thumbnailAsset, false, ".bmp");

            Console.WriteLine("Encode to audio and get an on demand URL.\n");

            IAsset audioOnly = EncodeToAudioOnly( inputAsset, AssetCreationOptions.None);
            PublishAssetGetURLs(audioOnly);

            Console.WriteLine("Encode to adaptive bitraite MP4s and get on demand URLs.\n");

            // If you want to secure your high quality encoded media files with
            // strong encryption at rest on disk,
            // use AssetCreationOptions.StorageEncrypted instead of AssetCre-
            // ationOptions.None.
            //
            // If your asset is AssetCreationOptions.StorageEncrypted,
            // make sure to call ConfigureClearAssetDeliveryPolicy defined below.

            IAsset encodedAsset =
                EncodeToAdaptiveBitrateMP4s( inputAsset, AssetCreationOptions.None);

            // If your want to delivery a storage encrypted asset,
            // you must configure the asset's delivery policy.
            // Before your asset can be streamed,
            // the streaming server removes the storage encryption and
            //streams your content using the specified delivery policy.

            // ConfigureClearAssetDeliveryPolicy(encodedAsset);

            PublishAssetGetURLs(encodedAsset);
        }
        catch (Exception exception)
        {
            // Parse the XML error message in the Media Services response and
            // create a new exception with its content.
            exception = MediaServicesExceptionParser.Parse(exception);
```

```csharp
                Console.Error.WriteLine(exception.Message);
            }
            finally
            {
                Console.ReadLine();
            }
        }

        static public IAsset UploadFile(string fileName, AssetCreationOptions options)
        {
            IAsset inputAsset = _context.Assets.CreateFromFile(
                fileName,
                options,
                (af, p) =>
                {
                    Console.WriteLine("Uploading '{0}' - Progress: {1:0.##}%", af.Name, p.Progress);
                });

            Console.WriteLine("Asset {0} created.", inputAsset.Id);

            return inputAsset;
        }

        static public IAsset EncodeToAdaptiveBitrateMP4s(IAsset asset, AssetCreationOptions options)
        {

            // Prepare a job with a single task to transcode the specified asset
            // into a multi-bitrate asset.

            IJob job = _context.Jobs.CreateWithSingleTask(
                "Media Encoder Standard",
                "Adaptive Streaming",
                asset,
                "Adaptive Bitrate MP4",
                options);

            Console.WriteLine("Submitting transcoding job...");
```

```csharp
            // Submit the job and wait until it is completed.
            job.Submit();

            job = job.StartExecutionProgressTask(
                j =>
                {
                    Console.WriteLine("Job state: {0}", j.State);
                    Console.WriteLine("Job progress: {0:0.##}%", j.GetOverallProgress());
                },
                CancellationToken.None).Result;

            Console.WriteLine("Transcoding job finished.");

            IAsset outputAsset = job.OutputMediaAssets[0];

            return outputAsset;
        }

        static public IAsset GenerateThumbnail(IAsset asset, AssetCreationOptions options)
        {
            // Load the XML (or JSON) from the local file.
            string configuration = File.ReadAllText(Path.Combine(_presetFiles, @"ThumbnailPreset_JSON.json"));

            IJob job = _context.Jobs.CreateWithSingleTask(
                "Media Encoder Standard",
                configuration,
                asset,
                "Thumbnail",
                options);

            Console.WriteLine("Submitting transcoding job...");

            // Submit the job and wait until it is completed.
            job.Submit();

            job = job.StartExecutionProgressTask(
                j =>
                {
```

```csharp
                    Console.WriteLine("Job state: {0}", j.State);
                    Console.WriteLine("Job progress: {0:0.##}%", j.GetOverallProgress());
                },
                CancellationToken.None).Result;

            Console.WriteLine("Transcoding job finished.");

            IAsset outputAsset = job.OutputMediaAssets[0];

            return outputAsset;
        }

        static public IAsset EncodeToAudioOnly(IAsset asset, AssetCreationOptions options)
        {
            // Load the XML (or JSON) from the local file.
            string configuration = File.ReadAllText(Path.Combine(_presetFiles, @"AudioOnlyPreset_JSON.json"));

            IJob job = _context.Jobs.CreateWithSingleTask(
                "Media Encoder Standard",
                configuration,
                asset,
                "Audio only",
                options);

            Console.WriteLine("Submitting transcoding job...");

            // Submit the job and wait until it is completed.
            job.Submit();

            job = job.StartExecutionProgressTask(
                j =>
                {
                    Console.WriteLine("Job state: {0}", j.State);
                    Console.WriteLine("Job progress: {0:0.##}%", j.GetOverallProgress());
                },
                CancellationToken.None).Result;

            Console.WriteLine("Transcoding job finished.");
```

```csharp
            IAsset outputAsset = job.OutputMediaAssets[0];

            return outputAsset;
        }

        static public void PublishAssetGetURLs(IAsset asset, bool onDemaindURL = true,
string fileExt = "")
        {
            // Publish the output asset by creating an Origin locator for adaptive
            // streaming,
            // and a SAS locator for progressive download.

            if (onDemaindURL)
            {
                _context.Locators.Create(
                    LocatorType.OnDemandOrigin,
                    asset,
                    AccessPermissions.Read,
                    TimeSpan.FromDays(30));

                // Get the Smooth Streaming, HLS and MPEG-DASH URLs for adaptive
                // streaming,
                // and the Progressive Download URL.
                Uri smoothStreamingUri = asset.GetSmoothStreamingUri();
                Uri hlsUri = asset.GetHlsUri();
                Uri mpegDashUri = asset.GetMpegDashUri();

                // Display  the streaming URLs.
                Console.WriteLine("Use the following URLs for adaptive streaming: ");
                Console.WriteLine(smoothStreamingUri);
                Console.WriteLine(hlsUri);
                Console.WriteLine(mpegDashUri);
                Console.WriteLine();
            }
            else
            {
                _context.Locators.Create(
                    LocatorType.Sas,
                    asset,
                    AccessPermissions.Read,
                    TimeSpan.FromDays(30));
```

```
                    IEnumerable <IAssetFile> assetFiles = asset
                        .AssetFiles
                        .ToList()
                        .Where(af => af.Name.EndsWith(fileExt,
StringComparison.OrdinalIgnoreCase));

                    // Get the URLs for progressive download for each specified file that
                    // was generated as a result of encoding.
                    List <Uri> sasUris = assetFiles.Select(af => af.GetSasUri()).
ToList();

                    // Display the URLs for progressive download.
                    Console.WriteLine("Use the following URLs for progressive down-
load.");
                    sasUris.ForEach(uri => Console.WriteLine(uri + "\n"));
                    Console.WriteLine();
                }
            }
        }
    }
```

下面的程序演示了如何使用 SDK 来上传媒体文件:

```
    UploadFile(Path.Combine(_mediaFiles, @"BigBuckBunny.mp4"), AssetCreationOptions.
None);
```

同时也有通过 Azure AD 登录 Media Service 的操作, 如:

```
            AzureAdTokenCredentials tokenCredentials = new
AzureAdTokenCredentials(_AADTenantDomain,
AzureEnvironments.AzureCloudEnvironment);
            AzureAdTokenProvider tokenProvider = new
AzureAdTokenProvider(tokenCredentials);
```

也有创建一个自适应编码率的 MP4 格式媒体文件的, 如:

```
IAsset encodedAsset =
    EncodeToAdaptiveBitrateMP4s(inputAsset, AssetCreationOptions.None);
```

还有将媒体文件产生概览视频的程序段, 如:

```
IAsset thumbnailAsset = GenerateThumbnail(inputAsset, AssetCreationOptions.None);
```

还有将视频转换成只有音频的程序, 如:

```
IAsset audioOnly = EncodeToAudioOnly(inputAsset, AssetCreationOptions.None);
```

最后还有将视频资源在 Azure 中进行发布的功能,如:

```
PublishAssetGetURLs(audioOnly);
```

8.3 视频直播 Live Streaming

1. 先决条件

完成教程需要以下内容:

① Azure 的账号。需要一个 Azure 免费试用账号,或者 MSDN 账号,或者付费账号。

② 媒体服务账户。若要创建媒体服务账户,则需要在 Portal 中创建媒体服务账户。

③ 网络摄像头。例如,市面上常用的网络摄像头,或者一个笔记本电脑内置的摄像头等。网络摄像头需要系统能够进行驱动,能在系统中正常使用。

2. 公共直播流场景

下面的步骤描述了创建公用流媒体应用程序所涉及的任务,这些应用程序使用配置为传递信号的通道。本教程介绍如何创建和管理通过通道和实况事件。

注:要确保从中流内容的流端点处于运行状态。

① 把摄像机/摄像头连接到计算机上。启动并配置一个现场直播编码器,输出多比特率 RTMP 或碎片 MP4 流。有关更多信息,请参阅 Azure 媒体服务 RTMP 支持和实时编码器。此步骤也可以在创建频道之后执行。

② 创建并启动一个通过通道。

③ 检索频道摄取 URL。

④ 实况编码器使用摄取 URL 将流发送到信道。

⑤ 检索频道预览 URL。

⑥ 使用此 URL 验证通道是否正确接收实况流。

⑦ 创建现场事件/程序。

⑧ 当使用 Azure 门户时,创建一个实况事件,也会创建一个资产。

⑨ 准备好启动和归档时启动事件/程序。

⑩ (可选)可以用活编码器发送信号以启动广告。广告插入到输出流中。

⑪ 每当想停止流和归档事件时,停止事件/程序。

⑫ 删除事件/程序(并可选地删除资产)。

3. 查看通知和错误

如果想查看 Azure 门户生成的通知和错误,请单击通知图标,如图 8-8 所示。

图8-8 通道建立好的通知

4. 创建并开始启动通过通道和事件

通道与事件/程序相关联,这些事件/程序能够控制实况流中的段的发布和存储。通道管理事件。

通过设置归档窗口长度,可以指定要为程序保留记录内容的小时数。这个值可以从最小 5 分钟设置到最大 25 小时。存档窗口长度也决定了客户端可以从当前的实时位置及时返回的最大时间量。事件可以在指定的时间内运行,但是落后于窗口长度的内容会被连续丢弃。此属性的值还决定客户端显示的增长时间。

每个事件都与一个资产相关联。要发布事件,必须为关联的资产创建 OnDebug 定位符。有了这个定位器,就可以构建一个流式 URL,并可以提供给客户机。

一个通道支持多达三个同时运行的事件,这样就可以创建同一个输入流的多个存档。允许根据需要发布和归档一个事件的不同部分。例如,您的业务需求是存档一个程序 6 个小时,但只需广播 10 分钟。要实现这一点,需要创建两个并发运行的程序:一个程序被设置为归档 6 小时的事件,但程序没有公布;另一个程序设置为存档 10 分钟,这个程序被公布。

不应重用现有的实况事件。相反,应为每个事件创建和启动一个新事件。

当准备开始流和归档时,启动事件。当想停止流和归档事件时,停止该程序。

若要删除归档内容,请停止并删除事件,然后删除关联的资产。如果事件被使用,则不能删除该资产,必须首先删除事件。

即使在停止和删除事件之后,只要不删除该资产,用户就可以根据需要将归档内容作为视频点播。

如果确实想要保留存档的内容,但不允许它用于流媒体,则删除流定位器。

① 在 Azure 门户网站中,选择 Azure 媒体服务账户。

② 在 Settings 窗口中,单击 Lives streaming(如图 8-9 所示),出现直播流窗口。

③ 单击快速创建,以 RTMP 摄取协议创建一个通过通道。创建一个新的频道窗口出现了。

④ 给新频道一个名字,然后单击创建。这样,通过 RTMP 摄取协议创建了一个通过通道。

5. 创建事件

选择要添加事件的通道。

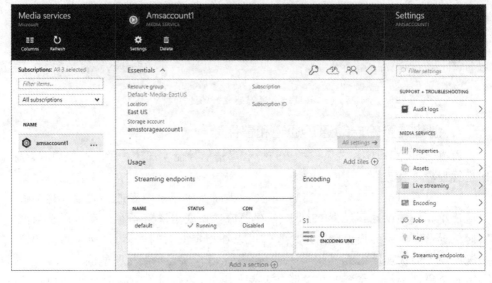

图 8-9　查看直播流

选择实时事件 Live event 按钮,如图 8-10 所示。

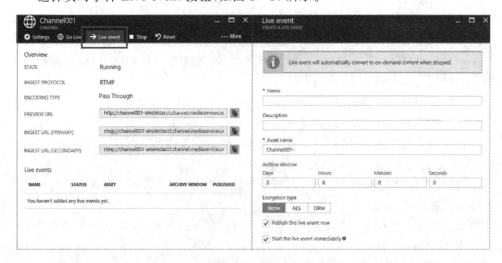

图 8-10　为通道添加实时事件 Live event

6. 获取摄取 URL

一旦创建了频道,就可以获得提供给实时编码器的摄取 URL。编码器使用这些 URL 输入直播流/实时流,如图 8-11 所示。

7. 观看事件

要观看事件,点击 Azure 门户中的手表或复制流式 URL 并使用所选择的播放器,如图 8-12 所示。

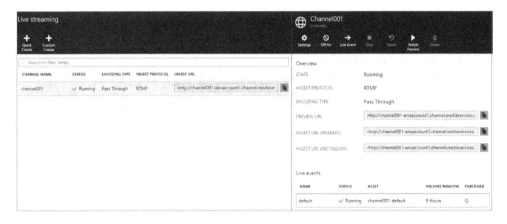

图 8-11 获得摄取 URL

图 8-12 在 Azure Portal 观看事件

当停止时,实况事件自动转换为所需内容。

8. 清 理

有关通过通道的更多细节,请参见创建多比特率流的内置编码器的实时流式传输。

只有当信道上的所有事件/程序都已停止时,才能停止信道。一旦信道停止,它就不承担任何费用。当需要重新启动时,它将具有相同的摄取 URL,因此不需要重新配置编码器。

只有当信道上的所有实况事件都被删除时,才能删除信道。

9. 查看归档内容

即使在停止和删除事件之后,只要不删除该资产,用户就可以根据需要将归档内容作为视频点播。如果事件被使用,则不能删除该资产,必须首先删除事件。

若要管理资产,请选择 Settings 并单击 Assets,如图 8-13 所示。

图 8-13 查看从通道归档的资源

10. 通过创建多比特率流的内置编码器实现流媒体直播

在 Azure 媒体服务中,频道表示用于处理流媒体内容的流水线。信道以两种方式之一接收实况输入流:

- 现场实况编码器发送多比特率 RTMP 或平滑流(片段化 MP4)流到不能使用媒体服务执行实时编码的信道。摄取的流通过通道不进行任何进一步的处理。这种方法称为传递。
- 实况编码器还可以将单个比特率流发送到未启用实况编码的信道,但不建议这样做。媒体服务将流传递给请求它的客户。

注:使用传递方法是进行流式传输最经济的方式。

现场实况编码器向信道发送单个比特率流,该信道能够以以下格式之一执行媒体编码的实时编码:RTMP 或平滑流(分段 MP4)。然后,信道对多比特率(自适应)视频流执行传入的单比特率流的实况编码。媒体服务将流传递给请求它的客户。

从媒体服务 2.10 发布开始,当创建一个频道时,可以指定希望频道如何接收输入流,还可以指定是否希望频道执行流的实况编码。有两种选择:

① 传递 Pass Through:如果计划使用具有多比特率流(通过流)的现场实况编码器作为输出,则指定此值。在这种情况下,传入流在没有任何编码的情况下传递到输出。这是 2.10 版本之前的一个频道的行为。本书详细介绍了使用这种通道的方法。

② 实况编码 Live Encoding:如果计划使用媒体服务将单比特率实况流编码为多比特流,则选择此值。在运行状态下留下一个实况编码信道会导致计费。我们建议立即停止运行频道后,完成直播流事件,以避免额外的小时收费。媒体服务将流传递给请求它的客户。

图 8-14 表示的是一个实时流式工作流,它使用一个 Office 实况编码器,有多比特率 RTMP 或分段 MP4(平滑流)流作为输出,如图 8-14 所示。

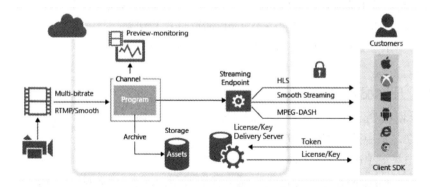

图 8-14 一个实时流式工作流

11. 公共直播流场景

下面的步骤描述创建公共流媒体应用程序所涉及的任务。

① 把摄像机连接到计算机上,启动并配置具有多比特率 RTMP 或分段 MP4 流(平滑流)的现场实况编码器作为输出。有关更多信息,请参阅 Azure 媒体服务 RTMP 支持和实时编码器。

还可以在创建频道之后执行此步骤。

② 创建并启动一个通道。

③ 检索频道摄取 URL。实况编码器使用摄取 URL 将流发送到信道。

④ 检索频道预览 URL。使用此 URL 验证通道是否正确接收实况流。

⑤ 创建一个程序。

当使用 Azure 门户时,创建一个程序,也会创建一个资产。

当使用.NETSDK 或 REST 时,需要创建一个资产并指定在创建程序时使用此资产。

⑥ 发布与程序关联的资产。

注:当创建了 Azure 媒体服务账户时,在停止状态下将默认流端点添加到账户中。要从中流内容的流端点必须处于运行状态。

⑦ 当准备好启动和归档时,启动程序。

⑧ (可选)可以激活编码器发送信号以启动广告。将广告插入到输出流中。

⑨ 当想停止流和归档事件时,停止该程序。

⑩ 删除程序(并可选地删除资产)。

12. 使用 Azure 门户配置内容保护策略

通过 Azure 媒体服务,可以确保存储、处理和传递信息的安全。可以使用媒体服务并通过使用 128 位加密密钥,以高级加密标准(AES)动态地加密您的内容。还可以使用 PlayReady 和/或 Widevine 数字版权管理(DRM)和苹果 FairPlay 使用普通加密(CCEC)。

媒体服务向授权客户提供 DRM 许可证和 AES 清晰密钥的服务。可以使用 Azure 门户为所有类型的加密创建一个密钥/许可证授权策略。

本书演示了如何使用门户配置内容保护策略，还展示了如何将动态加密应用到资产中。

13. 开始配置内容保护

若要使用 Portal 通过使用媒体服务账户配置全局内容保护，请采取以下步骤：
① 在门户中选择媒体服务账户。
② 选择 Settings→Context protection，如图 8 – 15 所示。

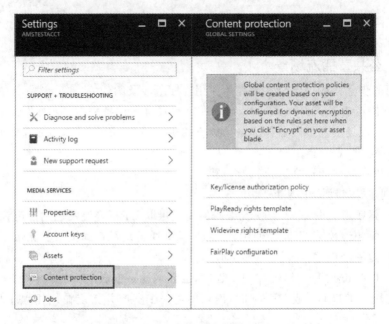

图 8 – 15　内容保护

14. 密钥/许可授权策略

媒体服务支持对密钥或许可证请求进行认证的多种方式。必须配置内容密钥授权策略。客户必须在密钥/许可证交付之前满足该策略。内容密钥授权策略可以具有一个或多个授权限制，即开放或令牌限制。

可以使用门户为所有类型的加密创建一个密钥/许可证授权策略。

(1) 开放授权

开放限制意味着系统将密钥传递给任何发出密钥请求的人。这种限制对于测试目的可能是有用的。

(2) 令牌授权

令牌限制策略必须伴随由安全令牌服务(STS)发出的令牌。媒体服务支持简单 Web 令牌(SWT)和 JSON Web 令牌(JWT)格式中的令牌。媒体服务不提供 STS。

可以创建自定义 STS 或使用 Azure 访问控制服务发出令牌。必须配置 STS 以创建用指定密钥签名的令牌,并发出令牌限制配置中指定的声明。如果令牌有效,令牌中的权利要求与密钥(或许可证)配置的请求相匹配,则媒体服务密钥传递服务将请求的密钥(或许可证)返回给客户端。

配置令牌限制策略时,必须指定主验证密钥、发行者和受众参数。主验证密钥包含令牌签署的密钥。发行者是发出令牌的安全令牌服务。受众(有时称为作用域)描述令牌或令牌授权访问的资源的意图。媒体服务密钥传递服务验证令牌中的这些值与模板中的值匹配,如图 8-16 所示。

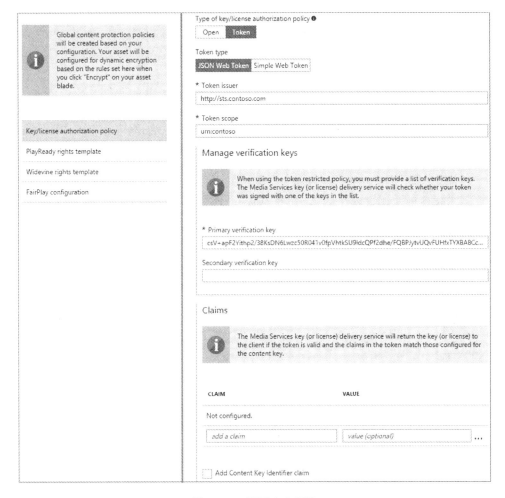

图 8-16 配置内容保护

15. PlayReady 授权模板

(1) 非持久性 Non persistent

如果将许可证配置为非持久性,则只在玩家使用许可证时才将其保存在内存中,如图 8-17 所示。

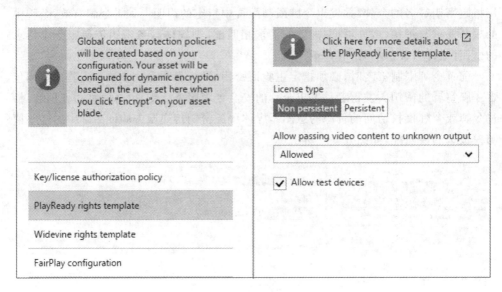

图 8-17　非持久化的授权

(2) 持久的 Persistent

如果将许可证配置为持久的,则将其保存在客户端的持久存储中,如图 8-18 所示。

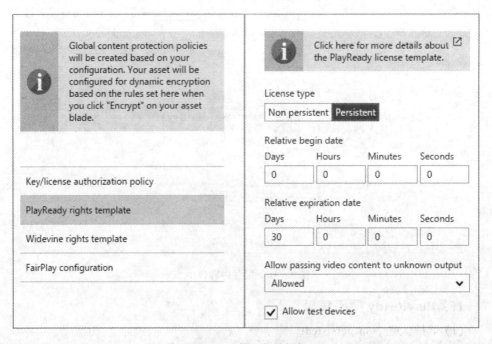

图 8-18　持久化的授权

16. Widevine 授权模板

Widevine 授权模板设置了在 Widevine 许可证上启用的功能。

(1) 基本的 Basic

当选择 Basic 时,模板将使用所有默认值创建。

(2) 先进的 Advanced

可以选择 Tracktype、Renew、Play 等选项。

17. FairPlay 配置

要启用 FalPlay 加密,请选择 FairPlay 配置。然后选择应用程序证书并输入应用程序密钥,如图 8-19 所示。

图 8-19　FairPlay 的配置

18. 将动态加密应用于您的资产

为了利用动态加密,应将源文件编码为一组自适应比特率 MP4 文件。

(1) 选择要加密的资产

要查看所有资产,请选择设置 Settings→资产 Assets,如图 8-20 所示。

(2) 用 AES 或 DRM 加密

当为资产选择加密时,会看到两种选择:AES 或 DRM。

① AES:在所有流协议上启用 AES 清除密钥加密:平滑流、HLS 和 MPEG-DASH,如图 8-21 所示。

② DRM:在选择 DRM 之后,会看到不同的内容保护策略(必须由这一点配置)和一组流协议:

- PlayReady and Widevine with MPEG-DASH,用 PlayReady 和 Widevine DRMS 动态地将 MPEG-DASH 流加密。
- PlayReady and Widevine with MPEG-DASH+FairPlay with HLS,用 PlayReady 和 Widevine DRMS 来进行 HLS 动态加密 MPEG-DASH 流。此选项还用 FairPlay 加密 HLS 流。

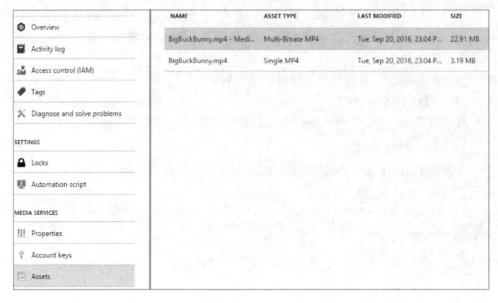

图 8-20 查看资源

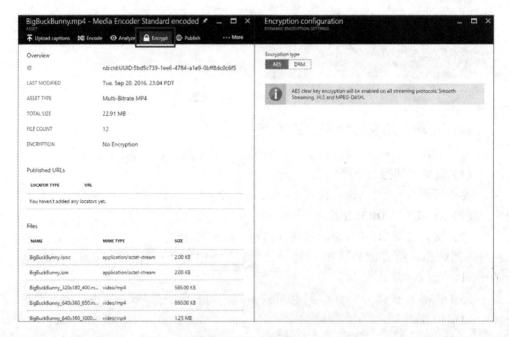

图 8-21 加密视频

- PlayReady only with Smooth Streaming，HLS，and MPEG-DASH，用 PlayReady 的 DRM 动态地加密。
- Widevine only with MPEG-DASH，用 Widevine DRM 动态地加密 MPEG-

DASH。

● FairPlay only with HLS，用 FairPlay 动态地加密 HLS 流。

为了启用 FairPlay 加密，在内容保护全局设置刀片上，选择 FairPlay 配置。然后选择应用程序证书，并输入应用程序密钥，如图 8-22 所示。

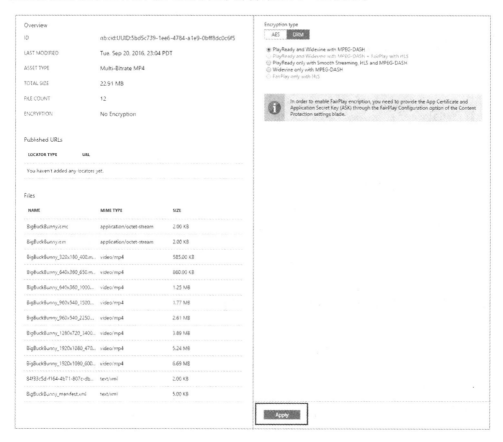

图 8-22　启用 DRM 版权保护

(3) 加密类型

在进行加密选择之后，选择 Apply 按钮。

第 9 章

HDInsight

9.1 简 介

Azure HDInsight 是一个完全管理、全频谱、开源分析服务的企业。HDInsight 是一种云服务,它使处理大量数据变得容易、快速和节约成本。HDInsight 还支持广泛的场景,如提取、转换和加载(ETL),数据仓库,机器学习和 IoT。

本章展示了如何创建集群、处理和分析大数据,并使用最流行的开源框架(如 Hadoop、Spark、HIVE、LLAP、Kafka、Storm 和微软机器学习服务器)开发解决方案。

9.2 用 HDInsight 创建 Hadoop

在本章中,将学习如何使用 Azure 门户在 HDInsight 中创建 Hadoop 集群,以及在 HDInsight 中运行 HIVE 作业。大多数 Hadoop 作业都是批处理作业。创建集群,运行一些作业,然后删除集群。在本章中,将执行所有三个任务。

目前 HDInsight 有七种不同的集群类型。每个集群类型支持不同的组件集。所有集群类型都支持 Hive。

每个 HDInsight 版本都是 HORTONWORKS 数据平台(HDP)版本的云分布。Hadoop 组件具有不同的 HDInsight 版本。

Azure HDISTION 支持多个 Hadoop 集群版本,可以在任何时候部署。每个版本选择创建 HDP 分布的特定版本和包含在该分布中的一组组件。截至 2017 年 4 月 4 日,Azure HDISTION 使用的默认集群版本为 3.6,基于 HDP 2.6。

与 HDISTIN 集群版本相关联的组件版本列在表 9-1 中。

注:HDInsight 服务的默认版本可能会在不另行通知的情况下更改。如果有版本依赖关系,那么当用 Azure Powershell 和 Azure CLI 创建.NET SDK 的集群时,应指定 HDInsight 版本。

1. 检查当前 Hadoop 组件版本信息

与 HDInsight 群集版本相关联的 Hadoop 生态系统组件版本可以随着 HDInsight 的更新而改变。若要检查 Hadoop 组件并验证正在使用哪些版本用于群集,请使用 AMBARI REST API。GetComponentInformation 命令检索有关服务组件的信息。有关详细信息,请参阅 AMBARI 文档,详见 https://github.com/apache/ambari/blob/trunk/ambari-server/docs/api/v1/index.md。

表 9-1　HDInsight 版本和对应部件版本

Component	HDInsight 3.6 (default)	HDInsight 3.5	HDInsight 3.4	HDInsight 3.3	HDInsight 3.2	HDInsight 3.1	HDInsight 3.0
Hortonworks Data Platform	2.6	2.5	2.4	2.3	2.2	2.1.7	2.0
Apache Hadoop and YARN	2.7.3	2.7.3	2.7.1	2.7.1	2.6.0	2.4.0	2.2.0
Apache Tez	0.7.0	0.7.0	0.7.0	0.7.0	0.5.2	0.4.0	—
Apache Pig	0.16.0	0.16.0	0.15.0	0.15.0	0.14.0	0.12.1	0.12.0
Apache Hive and HCatalog	1.2.1	1.2.1	1.2.1	1.2.1	0.14.0	0.13.1	0.12.0
Apache Hive2	2.1.0	—	—	—	—	—	—
Apache Tez Hive2	0.8.4	—	—	—	—	—	—
Apache Ranger	0.7.0	0.6.0	—	—	—	—	—
Apache HBase	1.1.2	1.1.2	1.1.2	1.1.1	0.98.4	0.98.0	—
Apache Sqoop	1.4.6	1.4.6	1.4.6	1.4.6	1.4.5	1.4.4	1.4.4
Apache Oozie	4.2.0	4.2.0	4.2.0	4.2.0	4.1.0	4.0.0	4.0.0
Apache Zookeeper	3.4.6	3.4.6	3.4.6	3.4.6	3.4.6	3.4.5	3.4.5
Apache Storm	1.1.0	1.0.1	0.10.0	0.10.0	0.9.3	0.9.1	—
Apache Mahout	0.9.0+	0.9.0+	0.9.0+	0.9.0+	0.9.0	0.9.0	—
Apache Phoenix	4.7.0	4.7.0	4.4.0	4.4.0	4.2.0	4.0.0.2.1.7.0-2162	—
Apache Spark	2.3.0, 2.2.0, 2.1.0	1.6.2, 2.0	1.6.0	1.5.2	1.3.1 (Windows only)	—	—
Apache Livy	0.4	0.3	0.3	0.2	—	—	—
Apache Kafka	1.0, 0.10.1	0.10.0	0.9.0	—	—	—	—
Apache Ambari	2.6.0	2.4.0	2.2.1	2.1.0	—	—	—
Apache Zeppelin	0.7.0	—	—	—	—	—	—
Mono	4.2.1	4.2.1	3.2.8	—	—	—	—
Apache Slider	0.92.0	—	—	—	—	—	—

对于 Windows 集群，检查组件版本的另一种方法是通过使用远程桌面登录到集群，并检查 C:\Appps\DIST\Directory 的内容。

Linux 是唯一在 HDInsight 版本 3.4 或更高版本上使用的操作系统。Windows 操作系统将从 HDInsight 3.4 以上版本退役。

2. 支持的 HDInsight 版本

表 9-2 列出了 HDInsight 的版本。对应于每个 HDInsight 版本的 HDP 版本连同产品发布日期一起列出。支持的到期和退休日期也提供。

表 9-2 HDInsight 版本以及其他部署方法

HDInsight version	HDP version	VM OS	Release date	Support expiration date	Retirement date	High availability	Availability on the Azure portal
HDInsight 3.6	HDP 2.6	Ubuntu 16.0.4 LTS	April 4, 2017			Yes	Yes

3. 可用版本

表 9-2 列出了 Azure 门户中可用的 HDInsight 版本以及其他部署方法，如 Powershell 和 .NETSDK。

注：在支持一个版本已经过期后，它可能无法通过微软 Azure 门户网站获得。但是，在 Windows Powershell New-AzureRmHDInsightCluster 命令和 .NETSDK 中的版本参数继续可用，直到版本退休日期为止，群集版本仍然可用。

4. 退休版本

表 9-3 列出了 Azure 门户中不可用的 HDInsight 版本。

表 9-3 Azure 门户中不可用的 HDInsight 版本

HDInsight version	HDP version	VM OS	Release date	Support expiration date	Retirement date	High availability	Availability on the Azure portal
HDInsight 3.5	HDP 2.5	Ubuntu 16.0.4 LTS	September 30, 2016	September 5, 2017	June 28, 2018	Yes	No
HDInsight 3.4	HDP 2.4	Ubuntu 14.0.4 LTS	March 29, 2016	December 29, 2016	January 9, 2018	Yes	No
HDInsight 3.3	HDP 2.3	Windows Server 2012 R2	December 2, 2015	June 27, 2016	July 31, 2018	Yes	No
HDInsight 3.3	HDP 2.3	Ubuntu 14.0.4 LTS	December 2, 2015	June 27, 2016	July 31, 2017	Yes	No

续表 9-3

HDInsight version	HDP version	VM OS	Release date	Support expiration date	Retirement date	High availability	Availability on the Azure portal
HDInsight 3.2	HDP 2.2	Ubuntu 12.04 LTS or Windows Server 2012 R2	February 18, 2015	March 1, 2016	April 1, 2017	Yes	No
HDInsight 3.1	HDP 2.1	Windows Server 2012 R2	June 24, 2014	May 18, 2015	June 30, 2016	Yes	No
HDInsight 3.0	HDP 2.0	Windows Server 2012 R2	February 11, 2014	September 17, 2014	June 30, 2015	Yes	No
HDInsight 2.1	HDP 1.3	Windows Server 2012 R2	October 28, 2013	May 12, 2014	May 31, 2015	Yes	No
HDInsight 1.6	HDP 1.1		October 28, 2013	April 26, 2014	May 31, 2015	No	No

5. 集群的默认节点配置和虚拟机大小

表 9-4 列出了 HDInsight 集群的默认虚拟机(VM)大小。

如果在群集中需要超过 32 个工作节点,则必须选择具有至少 8 个内核和 14 GB RAM 的节点大小。

所有支持区域(除了 Brazil South 和 Japan West)有效。

表 9-4 集群虚拟机类型

Cluster type	Hadoop	HBase	Interactive Query	Storm	Spark	ML Server
Head: default VM size	D3 v2	D3 v2	D13, D14	A3	D12 v2	D12 v2
Head: recommended VM sizes	D3 v2, D4 v2, D12 v2	D3 v2, D4 v2, D12 v2	D13, D14	A3, A4, A5	D12 v2, D13 v2, D14 v2	D12 v2, D13 v2, D14 v2
Worker: default VM size	D3 v2	D3 v2	D13, D14	D3 v2	Windows: D12 v2; Linux: D4 v2	Windows: D12 v2; Linux: D4 v2

续表 9-4

Cluster type	Hadoop	HBase	Interactive Query	Storm	Spark	ML Server
Worker: recommended VM sizes	D3 v2, D4 v2, D12 v2	D3 v2, D4 v2, D12 v2	D13, D14	D3 v2, D4 v2, D12 v2	Windows: D12 v2, D13 v2, D14 v2; Linux: D4 v2, D12 v2, D13 v2, D14 v2	Windows: D12 v2, D13 v2, D14 v2; Linux: D4 v2, D12 v2, D13 v2, D14 v2
ZooKeeper: default VM size		A3		A2		
ZooKeeper: recommended VM sizes		A3, A4, A5		A2, A3, A4		
Edge: default VM size						Windows: D12 v2; Linux: D4 v2
Edge: recommended VM size						Windows: D12 v2, D13 v2, D14 v2; Linux: D4 v2, D12 v2, D13 v2, D14 v2

表 9-5 列出了巴西南和日本西特殊的虚拟机设置。

表 9-5 巴西南和日本西特殊的虚拟机设置(与其他的地区不一样)

Cluster type	Hadoop	HBase	Interactive Query	Storm	Spark	ML Server
Head: default VM size	D3	D3	D13, D14	A3	D12	D12
Head: recommended VM sizes	D3, D4, D12	D3, D4, D12	D13, D14	A3, A4, A5	D12, D13, D14	D12, D13, D14
Worker: default VM size	D3	D3	D13, D14	D3	Windows: D12; Linux: D4	Windows: D12; Linux: D4
Worker: recommended VM sizes	D3, D4, D12	D3, D4, D12	D13, D14	D3, D4, D12	Windows: D12, D13, D14; Linux: D4, D12, D13, D14	Windows: D12, D13, D14; Linux: D4, D12, D13, D14

续表 9-5

Cluster type	Hadoop	HBase	Interactive Query	Storm	Spark	ML Server
ZooKeeper：default VM size		A2		A2		
ZooKeeper：recommended VM sizes		A2，A3，A4		A2，A3，A4		
Edge：default VM sizes						Windows：D12；Linux：D4
Edge：recommended VM sizes						Windows：D12，D13，D14；Linux：D4，D12，D13，D14

注：

Head 是 Storm 集群类型 Cluster type 之一的 Nimbus。

Worker 是 Storm 集群类型 Cluster type 之一的 Supervisor。

Worker 是 HBase 集群类型 Cluster type 之一的 Region。

6. 创建 Hadoop 集群

在本节中，将使用 Azure 门户在 HDInsight 中创建 Hadoop 集群。

登录 Azure 门户网站。在 Azure 门户中，选择 Create a resource→Data＋Analytics→HDInsight，如图 9-1 所示。

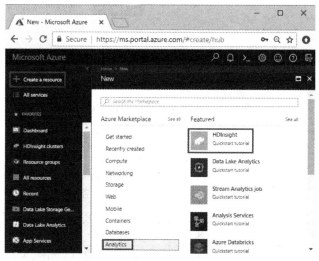

图 9-1 创建 HDInsight

在 HDInsight→Quick Create→Basics 之下，提供图 9-2 中建议的值。

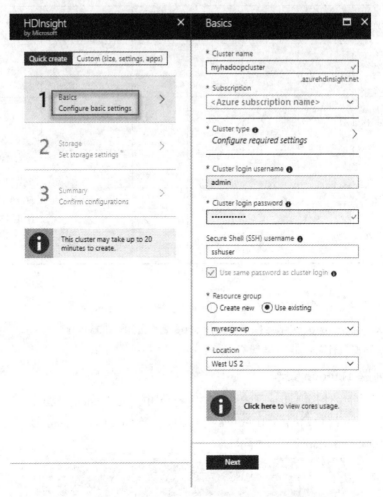

图 9-2 输入 HDInsight 的信息以创建它

输入或选择下列值，如表 9-6 所列。

表 9-6 创建 HDInsight 需要的信息

属　性	描　述
Cluster name	输入 Hadoop 集群的名称。因为 HDInsight 中的所有集群共享相同的 DNS 命名空间，所以这个名称应该是唯一的。该名称可以包括多达 59 个字符，包括字母、数字和连字符。名称的第一个和最后一个字符不能是连字符
Subscription	你用的订阅
Cluster Type	现在跳过这个。在该过程的下一步骤中提供此输入

续表 9-6

属 性	描 述
Cluster login username and password	默认登录名是 Admin。密码的长度必须至少为 10 个字符,并且必须包含至少一个数字、一个大写字母和一个小写字母、一个非字母数字字符(除了字符"")。确保不提供诸如"Pass@word1"这样的常用密码
SSH username	默认用户名是 SSHUSER。可以为 SSH 用户名提供另一个名称
Use same password as cluster login	选择此复选框,以便对 SSH 用户使用与您为集群登录用户提供的密码相同的密码
Resource group	创建资源组或选择现有资源组。资源组是 Azure 组件的容器。在这种情况下,资源组包含 HDInsight 群集和依赖 Azure 存储账户
Location	地点,可以选择一个距离较近的数据中心

选择群集类型,然后提供输入,如图 9-3 所示。

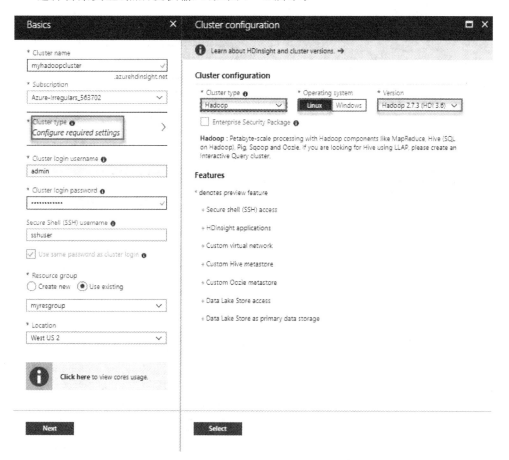

图 9-3 选择群集类型

选择以下值,如表9-7所列。

表9-7 输入的属性及其描述(1)

属 性	描 述
Cluster type	Select Hadoop
Operating system	Select Linux 或者 Windows,一般地建议用 Linux
Version	Select Hadoop 2.7.3 (HDI 3.6)

注:当然也可以选择其他版本。请根据实际需要来选择。

单击 Select 按钮,然后单击 Next 按钮。

在 Storage 选项卡中,提供如图9-4所示的输入。

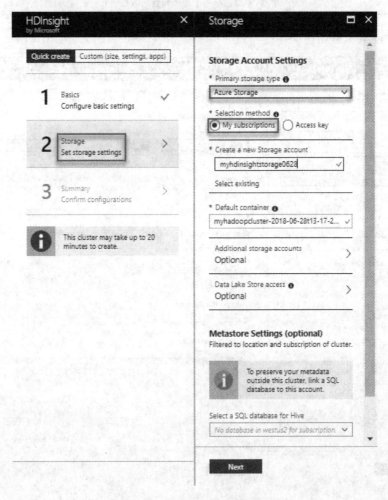

图9-4 HDInsight 设置 Storage

表 9-8 为 HDInsight 设置 Storage 的属性及描述。

表 9-8　HDInsight 设置 Storage 的属性及描述

属性	描述
Primary storage type	在本章中，选择 Azure 存储，使用 Azure 存储 BLB 作为默认存储账户。也可以使用 Azure 数据湖存储作为默认存储
Selection method	对于本章，请选择我的订阅从 Azure 订阅中使用存储账户。若要使用其他订阅的存储账户，请选择 Access 键，然后为该账户提供访问密钥
Create a new storage account	为存储账户提供名称

接受所有其他默认值，然后单击 Next 按钮。

在 Summary 选项卡中，验证在早期步骤中选择的值，如图 9-5 所示。

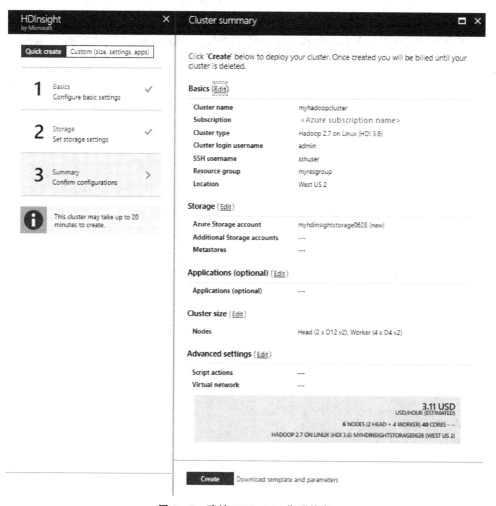

图 9-5　确认 HDInsight 集群信息

选择创建 Create，将看到一个新的瓦片，标题为"入口"仪表板上的 HDInsight 部署。创建一个集群需要大约 20 分钟，如图 9-6 所示。

图 9-6　HDInsight 正在创建中

一旦创建了集群，就可以看到 Azure 门户中的集群概览页面，如图 9-7 所示。

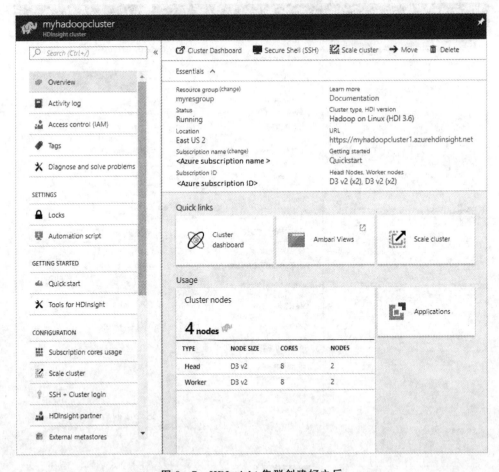

图 9-7　HDInsight 集群创建好之后

每个集群都有 Azure 存储账户或 Azure Data Lake 账户依赖项。它被称为默认存储账户。HDInsight 集群及其默认存储账户必须位于同一个 Azure 区域中。删除集群不会删除存储账户。

7. 运行 Hive 查询

Apache Hive 是 HDInsight 中最流行的组件。在 HDInsight 中有很多方法来运行 HIVE 工作。在本教程中，您使用门户中的 AMBARI HiVE 视图。对于提交 HIVE 作业的其他方法，请参见 HDInsight 使用 HIVE。

要打开 Ambari，从上一次截图选择集群仪表板。也可以浏览到 https://<ClusterName>.azurehdinsight.net，其中<ClusterName>是您在上一节中创建的集群，如图 9-8 所示。

图 9-8　点击 Cluster Dashboard

输入在创建集群时指定的 Hadoop 用户名和密码。默认用户名是管理员。打开 Hive 视图，如图 9-9 所示。

图 9-9　Hive 视图

在 QUERY 选项卡中，将下列 HIVEQL 语句粘贴到工作表中，如图 9-10 所示。选择执行 Execute。结果选项卡出现在查询选项卡下方，并显示有关作业的信息。

查询完成后，QUERY 选项卡显示操作的结果。你会看到一张名为 hivesam-

图 9-10 在 HIVE 中运行语句

pletable 的表格。这个样本 HIVE 表带有所有 HDInsight 集群,如图 9-11 所示。

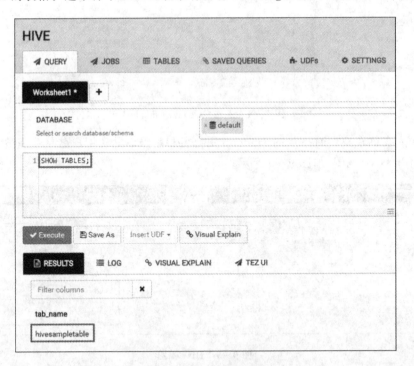

图 9-11 运行结果显示出来

运行以下查询:
SELECT * FROM hivesampletable;
还可以保存查询的结果。选择右侧的菜单按钮,并指定是希望将结果作为 CSV 文件下载,还是将其存储到与集群关联的存储账户,如图 9-12 所示。

图 9-12 hivesampletable 的数据

8. 清理资源

完成教程后,可能需要删除集群。使用 HDInsight,则数据存储在 Azure 存储中,因此当不使用时,可以安全地删除集群。另外,还需要为 HDInsight 集群付费,即使它不在使用中。由于集群的费用是存储费用的许多倍,因此当集群不使用时删除集群,在经济上是有意义的。

注:如果还需要继续使用,就不要马上删除 HDInsight 集群。

9. 删除集群和/或默认存储账户

回到浏览器选项卡,那里有 Azure 门户,您应该在集群概述页上。如果只想删除集群,但保留默认存储账户,请选择删除,如图 9-13 所示。

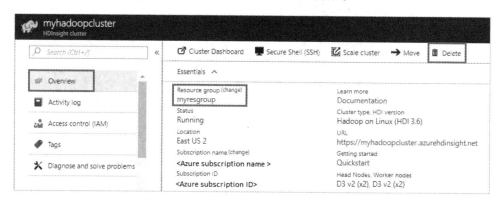

图 9-13 删除 HDInsight 集群

如果要删除集群以及默认存储账户,请选择资源组名称(在前面的屏幕快照中突出显示)以打开资源组页面。

选择"删除资源组"删除包含集群和默认存储账户的资源组。注意删除资源组删除存储账户。如果要保留存储账户,请选择只删除集群。

10. 使用 Apache Hive 在 Azure HDInsight 中提取、转换和加载数据（ETL）

在本教程中，获取原始 CSV 数据文件，将其导入 HDInsight 集群存储，然后使用 Azure HDInsight 上的 Apache Hive 转换数据。一旦数据被转换，就使用 Apache SQOOP 将数据加载到 Azure SQL 数据库中。在本教程中，使用公共可用的飞行数据。

其中包含以下任务：
- 下载示例飞行数据；
- 将数据上传到 HDInsight 集群；
- 用 Hive 变换数据；
- 在 Azure SQL 数据库中创建表；
- 使用 SQOOP 向 Azure SQL 数据库导出数据。

图 9-14 显示了一个典型的 ETL 应用程序流程。

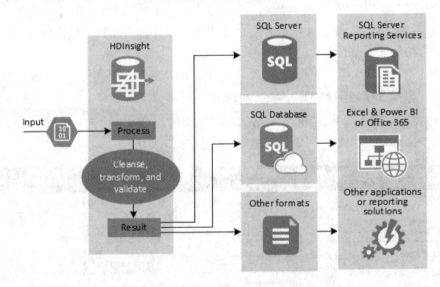

图 9-14 典型的 ETL 应用程序流程

(1) 先决条件

① 有一个基于 Linux 的 Hadoop 集群 HDInsight。请参阅在 HDInsight 开始使用 Hadoop 来创建一个新的基于 Linux 的 HDInsight 集群的步骤。

② 有 Azure SQL 数据库。使用 Azure SQL 数据库作为目的地数据存储区。如果没有 SQL 数据库，请参阅在 Azure 门户中创建 Azure SQL 数据库。

③ 有 Azure CLI 2。如果还没有安装 Azure CLI，请参阅安装 Azure CLI 以获得更多的步骤。

一个 SSH 客户端可以使用 SSH 连接到 HDInsight(Hadoop)。

(2) 下载飞行数据

访问链接：http://www.transtats.bts.gov/DL_SelectFields.asp? Table_ID=236&DB_Short_Name=On-Time，来下载飞行数据。

在页面上，选择以下值，如表 9-9 所列。

表 9-9 下载飞行数据时输入的值

名　称	值
Filter Year	2013
Filter Period	January
Fields	Year, FlightDate, UniqueCarrier, Carrier, FlightNum, OriginAirportID, Origin, OriginCityName, OriginState, DestAirportID, Dest, DestCityName, DestState, DepDelayMinutes, ArrDelay, ArrDelayMinutes, CarrierDelay, WeatherDelay, NASDelay, SecurityDelay, LateAircraftDelay.

清除所有其他字段。

选择下载，将得到一个 .zip 文件，其中包含所选择的数据字段。

11. 将数据上传到 HDInsight 集群

有许多方法将数据上传到与 HDInsight 群集相关的存储。在本节中，使用 SCP 来上传数据。要了解其他上传数据的方式，请参见将数据上传到 HDInsight 的方式。

打开命令提示符，并使用以下命令将 .zip 文件上载到 HDInsight 集群头节点：

scp <FILENAME>.zip <SSH-USERNAME>@<CLUSTERNAME>-ssh.azurehdinsight.net:<FILENAME.zip>

用 .zip 文件的名称替换<FILENAME>。用<SSH-USERNAME>替换 wHDInsight 集群的用户名。用 HDInsight 集群的名称替换<CLUSTERNAME>名字。

上传完成后，使用 SSH 连接到集群。在命令提示符上，输入以下命令：

ssh sshuser@clustername-ssh.azurehdinsight.net

使用以下命令解压缩 .zip 文件：

unzip FILENAME.zip

这个命令提取一个大约 60 MB 的 .CSV 文件。

使用以下命令在 HDISTION 存储上创建目录，然后将 .CSV 文件复制到目录：

hdfs dfs -mkdir -p /tutorials/flightdelays/data

hdfs dfs -put <FILENAME>.csv /tutorials/flightdelays/data/

12. 使用 Hive 查询转换数据

在 HDInsight 集群上运行蜂巢作业有很多方法。在本节中，使用直线运行 Hive 作业 job。

作为 HIVE 作业的一部分，将来自.CSV 文件的数据导入到一个名为 Delayes 的 HIVE 表中。

从 HDInsight 集群已有的 SSH 提示符中，使用以下命令创建和编辑名为 flight-delays.hql 的新文件：nano flightdelays.hql。

使用以下文本作为该文件的内容：

```
DROP TABLE delays_raw;
-- Creates an external table over the csv file
CREATE EXTERNAL TABLE delays_raw (
    YEAR string,
    FL_DATE string,
    UNIQUE_CARRIER string,
    CARRIER string,
    FL_NUM string,
    ORIGIN_AIRPORT_ID string,
    ORIGIN string,
    ORIGIN_CITY_NAME string,
    ORIGIN_CITY_NAME_TEMP string,
    ORIGIN_STATE_ABR string,
    DEST_AIRPORT_ID string,
    DEST string,
    DEST_CITY_NAME string,
    DEST_CITY_NAME_TEMP string,
    DEST_STATE_ABR string,
    DEP_DELAY_NEW float,
    ARR_DELAY_NEW float,
    CARRIER_DELAY float,
    WEATHER_DELAY float,
    NAS_DELAY float,
    SECURITY_DELAY float,
    LATE_AIRCRAFT_DELAY float)
-- The following lines describe the format and location of the file
ROW FORMAT DELIMITED FIELDS TERMINATED BY ','
LINES TERMINATED BY '\n'
STORED AS TEXTFILE
LOCATION '/tutorials/flightdelays/data';

-- Drop the delays table if it exists
DROP TABLE delays;
-- Create the delays table and populate it with data
-- pulled in from the CSV file (via the external table defined previously)
CREATE TABLE delays AS
```

```
SELECT YEAR AS year,
    FL_DATE AS flight_date,
    substring(UNIQUE_CARRIER, 2, length(UNIQUE_CARRIER) - 1) AS unique_carrier,
    substring(CARRIER, 2, length(CARRIER) - 1) AS carrier,
    substring(FL_NUM, 2, length(FL_NUM) - 1) AS flight_num,
    ORIGIN_AIRPORT_ID AS origin_airport_id,
    substring(ORIGIN, 2, length(ORIGIN) - 1) AS origin_airport_code,
    substring(ORIGIN_CITY_NAME, 2) AS origin_city_name,
    substring(ORIGIN_STATE_ABR, 2, length(ORIGIN_STATE_ABR) - 1)  AS origin_state_abr,
    DEST_AIRPORT_ID AS dest_airport_id,
    substring(DEST, 2, length(DEST) - 1) AS dest_airport_code,
    substring(DEST_CITY_NAME,2) AS dest_city_name,
    substring(DEST_STATE_ABR, 2, length(DEST_STATE_ABR) - 1) AS dest_state_abr,
    DEP_DELAY_NEW AS dep_delay_new,
    ARR_DELAY_NEW AS arr_delay_new,
    CARRIER_DELAY AS carrier_delay,
    WEATHER_DELAY AS weather_delay,
    NAS_DELAY AS nas_delay,
    SECURITY_DELAY AS security_delay,
    LATE_AIRCRAFT_DELAY AS late_aircraft_delay
FROM delays_raw;
```

若要保存文件,请按 Esc 键,然后输入 :x。

若要启动 Hive 并运行 FeleDelay.HQL 文件,请使用以下命令:

```
beeline -u 'jdbc:hive2://localhost:10001/;transportMode=http' -f flightdelays.hql
```

当接收到 jdbc:hive2://localhost:10001/> 提示时,使用以下查询从导入的航班延迟数据中检索数据:

```
INSERT OVERWRITE DIRECTORY '/tutorials/flightdelays/output'
ROW FORMAT DELIMITED FIELDS TERMINATED BY '\t'
SELECT regexp_replace(origin_city_name, '"', ''),
    avg(weather_delay)
FROM delays
WHERE weather_delay IS NOT NULL
GROUP BY origin_city_name;
```

此查询检索经历天气延迟的城市列表以及平均延迟时间,并将其保存到/tutorials/flightdelays/output。稍后,Sqoop 从该位置读取数据并将其导出到 Azure SQL 数据库。

退出 Beeline,进入!quit,在提示处退出。

13. 创建 SQL 数据库表

这里假设您已经创建了 Azure SQL 数据库。如果还没有 SQL 数据库,请使用 Azure 门户中创建 Azure SQL 数据库中的信息来创建一个 SQL 数据库。

如果已经有 SQL 数据库,则必须获取服务器名称。要找到服务器,请在 Azure 门户中指定名称,选择 SQL database.,然后筛选所使用的数据库的名称。服务器名称列在服务器名称列中,如图 9-15 所示。

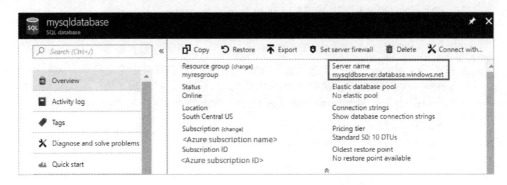

图 9-15 创建的 Azure SQL database,获得服务器名

注:连接 SQL 数据库并创建表的方法有很多种。下面的步骤使用 HDInsight 集群中的 FreeTDS。

要安装 FreeTDS,请使用 SSH 连接到集群的以下命令:

sudo apt-get --assume-yes install freetds-dev freetds-bin

安装完成后,使用以下命令连接到 SQL 数据库服务器。用 SQLServer 服务器名替换 serverName,用 SQL 数据库的登录名替换 Admin 登录和 Admin 口令。用数据库名称替换数据库名称。

TDSVER=8.0 tsql -H <serverName>.database.windows.net -U <adminLogin> -p 1433 -D <databaseName>

当提示时,输入 SQL 数据库管理员登录的密码。

接收与以下文本相似的输出:

```
locale is "en_US.UTF-8"
locale charset is "UTF-8"
using default charset "UTF-8"
Default database being set to sqooptest
1>
```

在 1>提示符下,输入以下几行:

CREATE TABLE [dbo].[delays](

```
[origin_city_name] [nvarchar](50) NOT NULL,
[weather_delay] float,
CONSTRAINT [PK_delays] PRIMARY KEY CLUSTERED
([origin_city_name] ASC))
GO
```

当进入 GO 语句时,对前面的语句进行评估。此查询创建一个名为延迟的表,使用聚集索引。

使用以下查询验证已创建的表:

```
SELECT * FROM information_schema.tables
GO
```

输出与下面的文本相似:

TABLE_CATALOG	TABLE_SCHEMA	TABLE_NAME	TABLE_TYPE
databaseName	dbo	delays	BASE TABLE

在 1> 提示中退出退出 TSQL 实用程序。

14. 使用 Sqoop 将数据导出到 SQL 数据库

在前面的部分中,复制了/TraceReals/Flash Time/Outlook 中的转换数据。这里,使用 Sqoop 将数据从/tutorials/flightdelays/output 导出到在 Azure SQL 数据库中创建的表。

使用以下命令验证 Sqoop 可以查看 SQL 数据库:

```
sqoop list-databases --connect jdbc:sqlserver://<serverName>.database.windows.net:1433 --username <adminLogin> --password <adminPassword>
```

此命令返回数据库列表,包括在其中创建延迟表的数据库。

使用以下命令将数据从 hivesampletable 导出到延迟表:

```
sqoop export --connect 'jdbc:sqlserver://<serverName>.database.windows.net:1433;database=<databaseName>' --username <adminLogin> --password <adminPassword> --table 'delays' --export-dir '/tutorials/flightdelays/output' --fields-terminated-by '\t' -m 1
```

Sqoop 连接到包含延迟表的数据库,并将/tutorials/flightdelays/output 目录中的数据导出到延迟表。

在 Sqoop 命令完成后,使用 TSQL 实用程序连接到数据库:

```
TDSVER=8.0 tsql -H <serverName>.database.windows.net -U <adminLogin> -P <adminPassword> -p 1433 -D <databaseName>
```

使用以下语句验证数据是否导出到 delays 表:

```
SELECT * FROM delays
```

GO

应该看到表中的数据列表。该表包括城市名称和该城市的平均航班延误时间。
键入 exit 以退出 TSQL 实用程序。

9.3 Apache Hive 简介

Apache Hive 蜂箱是 Hadoop 的数据仓库系统。Hive 实现数据的汇总、查询和数据分析。Hive 查询是用类似于 SQL 的查询语言 HiveQL 编写的。

Hive 允许在很大程度上在非结构化数据上构造结构。定义结构后,可以使用 HiveQL 查询数据,而不需要 Java 或 MapReduce。

HDInsight 提供了几种集群类型,它们可以针对特定的工作负载进行调整。下列集群类型最常用于蜂箱查询。

① 交互式查询:Hadoop 集群,提供低延迟分析处理(LLAP)功能,以改善交互式查询的响应时间。有关详细信息,请参阅 HDInsight 文档中的交互式查询的"开始"。

② Hadoop:用于批量处理工作负载的 Hadoop 集群。欲了解更多信息,请参见 Hadoop 在 HDInsight 文档中的"开始"。

③ Spark:Apache Spark 具有内置 Hive 的功能。有关更多信息,请参见在 HDInsight 文档上使用 Sark 的"开始"。

④ HBase:HiveQL 可以用来查询存储在 HBase 的数据。有关更多信息,请参见 HBase 在 HDInsight 文档上的"开始"。

1. 如何使用 Hive

使用表 9-10 查找与 HDInsight 一起使用 Hive 的不同方法。

表 9-10 与 HDInsight 一起使用 Hive 的不同方法

使用此方法	交互式查询	批处理	集群操作系统	客户端操作系统
HDInsight tools for Visual Studio Code	√	√	Linux	Linux、Unix、Mac OS X 或 Windows
HDInsight tools for Visual Studio	√	√	Linux 或 Windows	Windows
Hive View	√	√	Linux	Any (browser based)
Beeline client	√	√	Linux	Linux、Unix、Mac OS X 或 Windows
REST API	√	√	Linux 或 Windows	Linux、Unix、Mac OS X 或 Windows
Windows PowerShell	√	√	Linux 或 Windows	Windows

2. Hive 与数据结构

Hive 懂得如何使用结构化和半结构化数据,例如,字段中由特定字符分隔的文本文件。下面的 HiveQL 语句在空间分隔数据上创建表:

```
CREATE EXTERNAL TABLE log4jLogs (
    t1 string,
    t2 string,
    t3 string,
    t4 string,
    t5 string,
    t6 string,
    t7 string)
ROW FORMAT DELIMITED FIELDS TERMINATED BY ' '
STORED AS TEXTFILE LOCATION '/example/data/';
```

对于复杂或不规则结构的数据,Hive 还支持自定义序列化器/反序列化器。

3. Hive 内部表与外部表

可以用 Hive 创建两种类型的表。

① 内部:数据存储在蜂箱数据仓库中。数据仓库位于集群的默认存储区/HIV/仓库/默认位置。

当下列条件之一适用时使用内部表:
- 数据是暂时的。
- 希望用 Hive 管理表和数据的生命周期。

② 外部:数据存储在数据仓库之外。数据可以存储在集群可访问的任何存储区中。

当下列条件之一适用时,使用外部表:
- 数据也被用于 Hive 之外。例如,数据文件由另一个进程(不锁定文件)进行更新。
- 数据需要保留在基础位置,即使在删除表之后。
- 需要自定义位置,例如非默认存储账户。
- 用除了 Hive 之外的程序管理数据格式、位置等。

4. 自定义函数

蜂巢还可以通过用户定义的函数(UDF)来扩展。UDF 允许实现在 HiveQL 中不容易建模的功能或逻辑。

因为 Java 语言用得比较多,就用 Java 语言举例来说明如何用 Java 语言来定义一个滴定仪函数。

(1) 在 HDInsight 中使用 Java UDF

HDInsight 集群:

Linux 是唯一的操作系统,用于 HDInsight 版本 3.4 或更高版本。

本文档中的大多数步骤都基于 Windows 和基于 Linux 的集群。但是,将编译后的 UDF 上传到集群并运行它的步骤是基于 Linux 的集群所特有的,链接提供给可用于基于 Windows 的集群的信息。

(2) Java JDK 8 或更高版本(或等效的,如 OpenJDK)

● Apache Maven;

● 文本编辑器或 Java IDE。

如果在 Windows 客户端上创建 Python 文件,则必须使用 LF 作为行结束的编辑器。如果不确定编辑器是否使用 LF 或 CRLF,请参阅排除 CR 字符的步骤的故障排除部分。

(3) 创建一个示例 Java UDF

从命令行中,使用以下内容创建一个新的 Maven 项目:

mvn archetype:generate -DgroupId=com.microsoft.examples -DartifactId=ExampleUDF -DarchetypeArtifactId=maven-archetype-quickstart -DinteractiveMode=false

注:如果使用 Powershell,则必须在参数周围加引号。例如,

mvn archetype:generate "-DgroupId=com.microsoft.examples" "-DartifactId=ExampleUDF" "-DarchetypeArtifactId=maven-archetype-quickstart" "-DinteractiveMode=false"

此命令创建名为 exampleudf 的目录,该目录包含 Maven 项目。

创建项目后,删除作为项目的一部分创建的 exampleudf/src/test 目录。

打开 exampleudf/pom.xml,用下面的 XML 替换现有的 <dependencies> 条目:

```xml
<dependencies>
    <dependency>
        <groupId> org.apache.hadoop </groupId>
        <artifactId> hadoop-client </artifactId>
        <version> 2.7.3 </version>
        <scope> provided </scope>
    </dependency>
    <dependency>
        <groupId> org.apache.hive </groupId>
        <artifactId> hive-exec </artifactId>
        <version> 1.2.1 </version>
        <scope> provided </scope>
    </dependency>
</dependencies>
```

这些条目指定 HDInsight 3.6 中包含的 Hadoop 和 HIVE 版本。可以从 HDInsight 组件版本控制文档中找到关于随 HDInsight 提供的 Hadoop 和 Hive 版本的信息。

在文件末尾的</Project>行之前添加一个<Build>节。本节应包含以下 XML：

```xml
<build>
  <plugins>
    <!-- build for Java 1.8. This is required by HDInsight 3.6 -->
    <plugin>
        <groupId> org.apache.maven.plugins </groupId>
        <artifactId> maven-compiler-plugin </artifactId>
        <version> 3.3 </version>
        <configuration>
      <source> 1.8 </source>
      <target> 1.8 </target>
    </configuration>
  </plugin>
  <!-- build an uber jar -->
  <plugin>
  <groupId> org.apache.maven.plugins </groupId>
  <artifactId> maven-shade-plugin </artifactId>
  <version> 2.3 </version>
  <configuration>
    <!-- Keep us from getting a can't overwrite file error -->
    <transformers>
      <transformer
implementation = "org.apache.maven.plugins.shade.resource.ApacheLicenseResourceTransformer">
      </transformer>
      <transformer
implementation = "org.apache.maven.plugins.shade.resource.ServicesResourceTransformer">
        </transformer>
      </transformers>
    <!-- Keep us from getting a bad signature error -->
    <filters>
      <filter>
        <artifact> *:* </artifact>
        <excludes>
          <exclude> META-INF/*.SF </exclude>
          <exclude> META-INF/*.DSA </exclude>
          <exclude> META-INF/*.RSA </exclude>
        </excludes>
      </filter>
```

```xml
        </filters>
      </configuration>
      <executions>
        <execution>
          <phase> package </phase>
          <goals>
            <goal> shade </goal>
          </goals>
        </execution>
      </executions>
    </plugin>
  </plugins>
</build>
```

这些条目定义了如何构建项目。具体地说,项目使用的 Java 版本以及如何构建 Uberjar 以部署到集群。

一旦修改完成,就保存文件。

将 exampleudf/src/main/java/com/microsoft/examples/App.java 改名为 ExampleUDF.java,然后在编辑器中打开文件。

用以下方法替换 ExampleUDF.java 文件的内容,然后保存文件。

```java
package com.microsoft.examples;

import org.apache.hadoop.hive.ql.exec.Description;
import org.apache.hadoop.hive.ql.exec.UDF;
import org.apache.hadoop.io.*;

// Description of the UDF
@Description(
    name = "ExampleUDF",
    value = "returns a lower case version of the input string.",
    extended = "select ExampleUDF(deviceplatform) from hivesampletable limit 10;"
)
public class ExampleUDF extends UDF {
    // Accept a string input
    public String evaluate(String input) {
        // If the value is null, return a null
        if(input == null)
            return null;
        // Lowercase the input string and return it
```

```
        return input.toLowerCase();
    }
}
```

编译并安装这个 UDF。

使用以下命令编译和打包 UDF：

```
mvn compile package
```

此命令将 UDF 构建并封装到 exampleudf/target/ExampleUDF-1.0-SNAPSHOT.jar 文件中。

使用 SCP 命令将文件复制到 HDInsight 集群：

```
scp ./target/ExampleUDF-1.0-SNAPSHOT.jar myuser@mycluster-ssh.azurehdinsight
```

用集群的 SSH 用户账户替换 myuser。用集群名称替换 mycluster。如果使用密码来保护 SSH 账户，则提示您输入密码。如果使用证书，则可能需要使用-i 参数来指定私钥文件。

使用 SSH 连接到集群：

```
ssh myuser@mycluster-ssh.azurehdinsight.net
```

从 SSH 会话，将 JAR 文件复制到 HDInsight 存储：

```
hdfs dfs -put ExampleUDF-1.0-SNAPSHOT.jar /example/jars
```

5. 使用 Hive 中的 UDF

使用以下内容从 SSH 会话启动 Beeline 客户机：

```
beeline -u 'jdbc:hive2://localhost:10001/;transportMode=http'
```

此命令假定为集群的登录账户使用默认的 ADMIN。

一旦到达 jdbc:hive2://localhost:10001/>，提示输入以下内容将 UDF 添加到蜂箱中，并将其作为一个函数公开：

```
ADD JAR wasb:///example/jars/ExampleUDF-1.0-SNAPSHOT.jar;
CREATE TEMPORARY FUNCTION tolower as 'com.microsoft.examples.ExampleUDF';
```

注：此示例假定 Azure 存储是集群的默认存储。如果集群使用数据湖存储，则将 wasb:/// 值更改为 adl:///。

使用 UDF 将从表中检索的值转换为小写字符串：

```
SELECT tolower(deviceplatform) FROM hivesampletable LIMIT 10;
```

此查询从表中选择设备平台（Android、Windows、iOS 等），将字符串转换为小写，然后显示它们。输出类似于下面的文本：

```
+ ------ + -- +
|   _c0  |
+ ------ + -- +
| android |
| android |
| android |
| android |
| android |
| android |
| android |
| android |
| android |
| android |
+ ------ + -- +
```

6. 实例数据

HDInsight 上的 Hive 预装了一个内部表,名为 hivesampletable。HDInsight 还提供可以与 Hive 一起使用的示例数据集。这些数据集存储在/example/data 和/HdiSamples 目录中。这些目录存在于集群的默认存储区中。

7. Hive 查询示例

下面的 HiveQL 语句将项目列到/example/data/sample.log 文件中:

```
set hive.execution.engine = tez;
DROP TABLE log4jLogs;
CREATE EXTERNAL TABLE log4jLogs (
    t1 string,
    t2 string,
    t3 string,
    t4 string,
    t5 string,
    t6 string,
    t7 string)
ROW FORMAT DELIMITED FIELDS TERMINATED BY ' '
STORED AS TEXTFILE LOCATION '/example/data/';
SELECT t4 AS sev, COUNT(*) AS count FROM log4jLogs
    WHERE t4 = '[ERROR]' AND INPUT__FILE__NAME LIKE '%.log'
    GROUP BY t4;
```

在前面的示例中,HiveQL 语句执行以下操作:

```
set hive.execution.engine = tez;
```

将执行引擎设置为使用 Tez。使用 Tez 可以使查询性能提高。

注:只有在使用基于 Windows 的 HDInsight 洞集集群时才需要使用此语句。Tez 是基于 Linux 的 HDInsight 的默认执行引擎。

- DROP TABLE:如果表已经存在,请删除它。
- CREATE EXTERNAL TABLE:在 Hive 中设置一个新的外部工作台。外部表只将表定义存储在 Hive 中。数据保留在原始位置和原始格式。
- ROW FORMAT:告诉 Hive 如何格式化数据。在这种情况下,每个日志中的字段由一个空间分隔。
- STORED AS TEXTFILE LOCATION:告诉 HIve 数据存储在哪里(example/data 目录),并将其存储为文本。数据可以在一个文件中或在目录内的多个文件上传播。
- SELECT:选择字段 T4 包含值[ERROR]的所有行的计数。此语句返回值为 3,因为有三行包含此值。
- INPUT__FILE__NAME LIKE '%.log':HIVE 尝试将架构应用到目录中的所有文件。在这种情况下,目录包含与模式不匹配的文件。为了防止结果中的垃圾数据,该语句告诉 HIVE,我们应该只返回从.log 中结束的文件中的数据。

注:当期望基础数据由外部源更新时,应使用外部表。例如,自动数据上传过程或 MapReduce 操作。

删除外部表时不删除数据,它只删除表定义。

若要创建内部表而不是外部表,请使用以下 HiveQL:

```
set hive.execution.engine = tez;
CREATE TABLE IF NOT EXISTS errorLogs (
    t1 string,
    t2 string,
    t3 string,
    t4 string,
    t5 string,
    t6 string,
    t7 string)
STORED AS ORC;
INSERT OVERWRITE TABLE errorLogs
SELECT t1, t2, t3, t4, t5, t6, t7
    FROM log4jLogs WHERE t4 = '[ERROR]';
```

这些语句执行以下操作:

- CREATE TABLE IF NOT EXISTS:如果该表不存在,则创建它。因为不使用外部关键字,所以该语句创建内部表。该表存储在蜂箱数据仓库中,完全由蜂箱管理。
- STORED AS ORC：将数据存储在优化行列（ORC）格式中。ORC 用一种高度优化和高效的格式存储 Hive 数据。
- INSERT OVERWRITE...SELECT：从包含错误的 log4jLogs 表中选择行，然后将数据插入到 errorLogs 表中。

注：与外部表不同，删除内部表也会删除基础数据。

8. Hive 查询性能的提高

Apache Tez 是一种允许数据密集型应用程序（如蜂箱）在规模上运行效率更高的框架。默认情况下，Tez 是基于 Linux 的 HDInsight 集群启用的。

注：Tez 目前默认为基于 Windows 的 HDInsight 集群，必须启用它。为了利用 Tez，必须为蜂箱查询设置以下值：

set hive.execution.engine = tez;

Tez 是基于 Linux 的 HDInsight 集群的默认引擎。

为了帮助使用 Tez 运行调试任务，HDInsight 提供了下列 Web UI，允许查看 Tez jobs 的详细信息。

9. 使用 Hive View

使用以下步骤运行使用 Tez 的 Hive 查询：

在 Web 浏览器中，导航到 HTTPS://Clustername.AZUREHDINSIGHT.NET，其中 Clustername 是 HDInsight 集群的名称。

从页面顶部的菜单中选择视图图标。这个图标看起来像一连串的正方形。在出现的下拉菜单中，选择 Hive View，如图 9-16 所示。

图 9-16 Hive View

当 Hive 视图加载时，将下面的查询粘贴到查询编辑器中，然后单击执行按钮。

select market, state, country from hivesampletable where deviceplatform = 'Android' group by market, country, state;

一旦作业完成，就应该看到查询过程结果段中显示的输出。结果应类似于以下文本：

```
market    state       country
en-GB     Hessen      Germany
```

en – GB Kingston　　Jamaica

10. 使用 Tez View

从页面顶部的菜单中选择视图图标。在出现的下拉菜单中,选择 Tez View,如图 9 – 17 所示。

图 9 – 17　Tez View

当 Tez 视图加载时,会看到当前正在运行或已经在集群上运行的蜂箱查询列表,如图 9 – 18 所示。

图 9 – 18　Tez 视图中的查询列表

如果只有一个条目,则是在上一节中运行的查询。如果有多个条目,则可以使用页面顶部的字段进行搜索。

为 Hive 查询选择查询 ID。显示有关查询的信息,如图 9 – 19 所示。
此页上的选项卡允许查看以下信息:
- Query Detail(查询详情):关于 Hive 查询细节。
- Time Line(时间线):处理每个阶段的时间的信息。
- Configurations(配置):用于此查询的配置。

从查询详细信息中,可以使用链接来查找有关此查询的应用程序或 DAG 的信息。

应用程序链接显示关于该查询的纱线应用程序的信息。从这里可以访问纱线应用日志。

DAG 链接显示关于该查询的有向无环图的信息。从这里可以查看 DAG 的图形表示,也可以在 DAG 内找到有关顶点的信息。

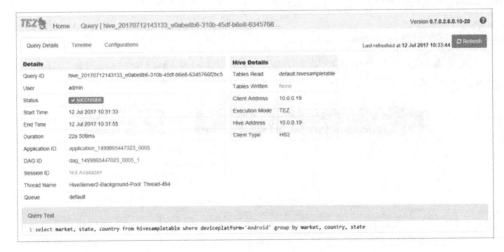

图 9-19 查询的详细信息

11. 在 HDInsight 中为 Hadoop 作业上传数据

Azure HDInsight 在 Azure 存储和 Azure 数据湖存储中提供了一个全功能的 Hadoop 分布式文件系统（HDFS）。Azure 存储和数据湖存储被设计为一个 HDFS 扩展，为客户提供无缝体验。它们使 Hadoop 生态系统中的全部组件能够直接运行在其管理的数据上。Azure 存储和数据湖存储是不同的文件系统，它们被优化，以存储数据和对该数据进行计算。

12. 向 Azure 存储上传数据

（1）命令行实用程序

表 9-11 列出了微软为 Azure 存储提供的实用工具。

表 9-11 实用工具

Tool	Linux	OS X	Windows
Azure Command-Line Interface	√	√	√
Azure PowerShell			√
AzCopy		√	√
Hadoop command	√	√	√

① Azure CLI。Azure CLI 是一个跨平台的工具，允许管理 Azure 服务。使用以下步骤将数据上传到 Azure 存储。

为 Mac、Linux 和 Windows 安装和配置 Azure CLI。

打开命令提示符、BASH 或其他 Shell，并使用以下方法对 Azure 订阅进行身份验证。

② Azure login。当提示时,输入订阅的用户名和密码。
输入以下命令,列出订阅的存储账户:

```
azure storage account list
```

选择包含要使用的 Blob 的存储账户,然后使用以下命令检索该账户的密钥:

```
azure storage account keys list <storage-account-name>
```

此命令返回主密钥和次密钥。复制主键值,因为它将在下一步中使用。
使用以下命令检索存储账户内的 Blb 容器列表:

```
azure storage container list -a <storage-account-name> -k <primary-key>
```

(2) 上载和下载文件

使用以下命令将文件上载并下载到 Blub:
上载文件:

```
azure storage blob upload -a <storage-account-name> -k <primary-key> <source-file> <container-name> <blob-name>
```

下载文件:

```
azure storage blob download -a <storage-account-name> -k <primary-key> <container-name> <blob-name> <destination-file>
```

注:如果始终使用相同的存储账户,则可以设置以下环境变量,而不是为每个命令指定账户和密钥:
- 存储账户名;
- 存储账户密钥。

13. Azure PowerShell

Azure PowerShell 是一个脚本环境,可用于控制和自动化 Azure 中工作负载的部署和管理。

Azure PowerShell 支持使用 Azure 服务管理器管理 HD 洞察力资源,并于 2017 年 1 月 1 日被删除。本书中的步骤,使用新的 HDInsight cmdlet 与 Azure 资源管理器。请保证安装最新版本的 PowerShell。

14. 将本地文件上传到 Azure 存储

打开 Azure PowerShell 控制台,在下面的脚本中设置前五个变量的值:

```
$ resourceGroupName = "<AzureResourceGroupName>"
$ storageAccountName = "<StorageAccountName>"
$ containerName = "<ContainerName>"

$ fileName = "<LocalFileName>"
```

```
$ blobName = " <BlobName> "

# Get the storage account key
$ storageAccountKey = (Get - AzureRmStorageAccountKey - ResourceGroupName $ resource-GroupName - Name $ storageAccountName)[0].Value
# Create the storage context object
$ destContext = New - AzureStorageContext - StorageAccountName $ storageAccountName - StorageAccountKey $ storageaccountkey

# Copy the file from local workstation to the Blob container
Set - AzureStorageBlobContent - File $ fileName - Container $ containerName - Blob $ blobName - context $ destContext
```

粘贴脚本到 Azure PowerShell 控制台运行复制文件。

15. AzCopy

AzCopy 是一个命令行工具,用于简化将数据传送进出 Azure Storage 账户的任务。可以将其用作独立的工具,或者将该工具合并到现有的应用程序中。下载 AzCopy:https://docs.microsoft.com/en-us/azure/storage/common/storage-use-azcopy。

AzCopy 命令行是:

AzCopy <Source> <Destination> [filePattern [filePattern...]] [Options]

AzCopy 命令的基本语法是:

AzCopy /Source:<source> /Dest:<destination> [Options]

(1) 下载单个 Blob

AzCopy /Source:https://myaccount.blob.core.windows.net/mycontainer /Dest:C:\myfolder /SourceKey:key /Pattern:"abc.txt"

(2) 从次要区域下载单个 Blob

AzCopy /Source:https://myaccount - secondary.blob.core.windows.net/mynewcontainer /Dest:C:\myfolder /SourceKey:key /Pattern:"abc.txt"

(3) 在容器中下载所有 Blob

AzCopy /Source:https://myaccount.blob.core.windows.net/mycontainer /Dest:C:\myfolder/ SourceKey:key /S

假设下列 Blob 位于指定容器中:abc.txt、abc1.txt、abc2.txt、vd1\a.txt、vd1\abcd.txt。

下载运行后,C:\myfolder 包括下列文件:

C:\myfolder\abc.txt;

C:\myfolder\abc1.txt;

C:\myfolder\abc2.txt;

C:\myfolder\vd1\a.txt;

C:\myfolder\vd1\abcd.txt。

(4) 下载带有特定前缀的 Blob

AzCopy /Source:https://myaccount.blob.core.windows.net/mycontainer /Dest:C:\myfolder /SourceKey:key /Pattern:a /S

所有从前缀为 A 开始的 Blob 都被下载了。

(5) 将导出文件的最后修改时间设置为与源块相同

AzCopy /Source:https://myaccount.blob.core.windows.net/mycontainer /Dest:C:\myfolder /SourceKey:key /MT

(6) 上传到一个 Blob

AzCopy /Source:C:\myfolder /Dest:https://myaccount.blob.core.windows.net/mycontainer /DestKey:key /Pattern:"abc.txt"

如果指定的目标容器不存在,则 AzCopy 创建它并将文件上载到该容器中。

(7) 将单个 Blob 上传到虚拟目录

AzCopy /Source:C:\myfolder /Dest:https://myaccount.blob.core.windows.net/mycontainer/vd /DestKey:key /Pattern:abc.txt

(8) 将所有 Blob 上传到文件夹中

AzCopy /Source:C:\myfolder /Dest:https://myaccount.blob.core.windows.net/mycontainer /DestKey:key /S

指定选项/S 递归地将指定目录的内容上传到 Blob 存储,这意味着所有子文件夹及其文件也被上传。

(9) 上传与特定模式匹配的 Blob

AzCopy /Source:C:\myfolder /Dest:https://myaccount.blob.core.windows.net/mycontainer /DestKey:key /Pattern:a* /S

(10) 在 Blob 存储中复制 Blob

将单个 Blob 从同一个存储容器中复制到另一个容器中。

AzCopy /Source:https://myaccount.blob.core.windows.net/mycontainer1 /Dest:https://myaccount.blob.core.windows.net/mycontainer2 /SourceKey:key /DestKey:key /Pattern:abc.txt

(11) 将单个 Blob 从一个存储账户复制到另一个存储账户

AzCopy /Source: https://sourceaccount.blob.core.windows.net/mycontainer1 /Dest: https://destaccount.blob.core.windows.net/mycontainer2 /SourceKey:key1 /DestKey:key2 /

Pattern:abc.txt

(12) 将单个 Blob 从次要区域复制到主区域

AzCopy /Source:https://myaccount1-secondary.blob.core.windows.net/mynewcontainer1 /Dest:https://myaccount2.blob.core.windows.net/mynewcontainer2 /SourceKey:key1 /DestKey:key2 /Pattern:abc.txt

(13) 将容器中的所有 Blob 复制到另一个存储账户

AzCopy /Source:https://sourceaccount.blob.core.windows.net/mycontainer1 /Dest:https://destaccount.blob.core.windows.net/mycontainer2 /SourceKey:key1 /DestKey:key2 /S

16. Hadoop 命令行

Hadoop 命令行仅在数据已经存在于集群头节点上时用于将数据存储到 Azure 存储块中。

为了使用 Hadoop 命令，必须首先使用以下方法之一连接到头节点：

① SSH 或者远程桌面；

② 使用以下命令将文件上载到存储区：

hadoop -copyFromLocal <localFilePath> <storageFilePath>

HDInsight 的默认文件系统是在 Azure 存储中：/example/data.txt。也可以将文件引用为 wasb:///example/data/data.txt，或者 wasb://<ContainerName>@<StorageAccountName>.blob.core.windows.net/example/data/davinci.txt。

9.4 使用 Hadoop 的 Pig 和 MapReduce

在 Hadoop 上使用 MapReduce 进行 HDInsight。

了解如何在 HDISCT 集群上运行 MapReduce 作业。

表 9-12 列出了查找 MapReduce 与 HDInsight 可以使用的各种方法。

表 9-12　可以使用的方法及目标

方　　法	达成目标	Cluster 操作系统	客户端操作系统
SSH	使用 Hadoop 命令通过 SSH	Linux	Linux、Unix、Mac OS X 或 Windows
REST	使用 REST 远程提交作业 Job（例如：use cURL）	Linux 或 Windows	Linux、Unix、Mac OS X 或 Windows
Windows PowerShell	使用 Windows PowerShell 远程提交作业	Linux 或 Windows	Windows

1. 什么是 MapReduce

Hadoop MapReduce 是用于编写处理大量数据作业的软件框架。输入数据被分割成独立的块。每个块在集群中的节点上并行处理。MapReduce 作业由两个功能组成：

① Mapper：消耗输入数据，分析它（通常使用筛选和排序操作），并发出元组（键/值对）。

② Reducer：消耗 Mapper 发出的元组，并执行汇总操作。该操作从 Mapper 数据创建更小的组合结果。

基本单词计数 MapReduce 作业示例如图 9-20 所示。

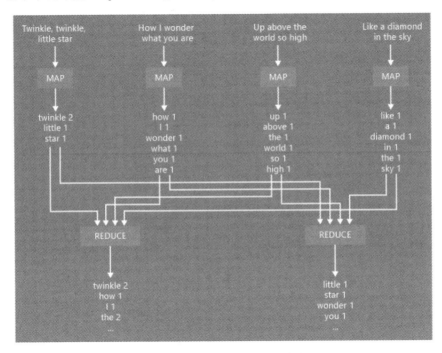

图 9-20 MapReduce 的原理图

这个作业的输出是文本中每个单词出现次数的计数。

映射器 Mapper 将输入文本中的每一行作为输入，并将其分解成单词。每当单词出现时，它就会发出键/值对，后面跟着一个 1。在将输出发送到减速器之前将其排序。

Reducer 对每个单词的这些单个计数进行求和，并发出单个键/值对，该对包含的单词的后面跟着其出现次数的总和。

MapReduce 可以用多种语言来实现。Java 是最常见的实现，并用于本文档中的演示目的。

2. 开发语言

基于 Java 和 Java 虚拟机的语言或框架可以直接作为 MapReduce 作业运行。本文档中使用的示例是 JavaMapReduce 应用程序。非 Java 语言，如 C#、Python 或独立的可执行文件，必须使用 Hadoop 流。

Hadoop 流通过 STDIN 和 STDOUT 与映射器和减速机通信。映射器和减速机从 STDIN 一次读取一行数据，并将输出写入 STDOUT。映射器和减速机读取或发射的每一行必须是键/值对的格式，由 Tab 字符分隔：

[key]/t[value]

(1) 实例数据

HDInsight 提供各种示例数据集，这些示例数据集存储在/example/data 和/HdiSamples 目录中。这些目录位于集群的默认存储区中。在本文中，使用/example/data/gutenberg/davinci.txt 文件。这个文件包含了达·芬奇的笔记本。

(2) 实例 MapReduce

一个示例 MapReduce 单词计数应用程序包含在 HDLISCT 集群中。此示例位于集群的默认存储区的/example/jars/hadoop-mapreduce-examples.jar 中。

下面的 Java 代码是 hadoop-mapreduce-examples.jar 文件中包含的 MapReduce 应用程序的源代码：

```java
package org.apache.hadoop.examples;

import java.io.IOException;
import java.util.StringTokenizer;

import org.apache.hadoop.conf.Configuration;
import org.apache.hadoop.fs.Path;
import org.apache.hadoop.io.IntWritable;
import org.apache.hadoop.io.Text;
import org.apache.hadoop.mapreduce.Job;
import org.apache.hadoop.mapreduce.Mapper;
import org.apache.hadoop.mapreduce.Reducer;
import org.apache.hadoop.mapreduce.lib.input.FileInputFormat;
import org.apache.hadoop.mapreduce.lib.output.FileOutputFormat;
import org.apache.hadoop.util.GenericOptionsParser;

public class WordCount {

    public static class TokenizerMapper
        extends Mapper <Object, Text, Text, IntWritable> {
```

```java
private final static IntWritable one = new IntWritable(1);
private Text word = new Text();

public void map(Object key, Text value, Context context
        ) throws IOException, InterruptedException {
    StringTokenizer itr = new StringTokenizer(value.toString());
    while (itr.hasMoreTokens()) {
    word.set(itr.nextToken());
    context.write(word, one);
    }
}
}


public static class IntSumReducer
        extends Reducer <Text,IntWritable,Text,IntWritable> {
private IntWritable result = new IntWritable();

public void reduce(Text key, Iterable <IntWritable> values,
        Context context
        ) throws IOException, InterruptedException {
    int sum = 0;
    for (IntWritable val : values) {
    sum += val.get();
    }
    result.set(sum);
    context.write(key, result);
}
}


public static void main(String[] args) throws Exception {
Configuration conf = new Configuration();
String[] otherArgs = new GenericOptionsParser(conf, args).getRemainingArgs();
if (otherArgs.length != 2) {
    System.err.println("Usage: wordcount <in> <out> ");
    System.exit(2);
}
Job job = new Job(conf, "word count");
job.setJarByClass(WordCount.class);
job.setMapperClass(TokenizerMapper.class);
job.setCombinerClass(IntSumReducer.class);
job.setReducerClass(IntSumReducer.class);
job.setOutputKeyClass(Text.class);
```

```
            job.setOutputValueClass(IntWritable.class);
            FileInputFormat.addInputPath(job, new Path(otherArgs[0]));
            FileOutputFormat.setOutputPath(job, new Path(otherArgs[1]));
            System.exit(job.waitForCompletion(true) ? 0 : 1);
        }
    }
```

(3) 运行这个 MapReduce

HDInsight 可以通过使用各种方法来运行 HIVEQL 作业。表 9-13 所列可决定哪种方法适合你，可从中选择一种来用。

表 9-13 运行 HIVEQL 作业的不同方法及完成目标

方法	完成目标	集群操作系统	客户端操作系统
SSH	使用 Hadoop command 通过 SSH	Linux	Linux、Unix、Mac OS X 或 Windows
Curl	远程提交作业 job remotely 通过使用 REST 接口	Linux 或 Windows	Linux、Unix、Mac OS X 和 Windows
Windows PowerShell	远程提交作业 job remotely 通过使用 Windows PowerShell	Linux 或 Windows	Windows

因为 SSH 使用比较多，故这里介绍如何用 SSH 运行 MapReduce 作业。

① 用 SSH 连接：

ssh sshuser@myhdinsight-ssh.azurehdinsight.net

如果使用 SSH 身份验证的密码，则需要在提示时提供密码。

如果使用证书密钥进行 SSH 身份验证，则可能需要指定客户端系统上私钥的位置，例如：

ssh -i ~/mykey.key sshuser@myhdinsight-ssh.azurehdinsight.net

② 使用 Hadoop 命令来运行作业：

yarn jar /usr/hdp/current/hadoop-mapreduce-client/hadoop-mapreduce-examples.jar wordcount /example/data/gutenberg/davinci.txt /example/data/WordCountOutput

这个命令启动 wordcount 类，它包含在 hadoop-mapreduce-examples.jar 文件中。它使用 /example/data/gutenberg/davinci.txt 文档作为输入，输出存储在 /example/data/WordCountOutput 中。

作业在处理过程中发出详细信息，并且在作业完成时返回与下面的文本类似的信息：

```
File Input Format Counters
    Bytes Read = 1395666
File Output Format Counters
```

```
Bytes Written = 337623
```

当作业完成时,使用以下命令列出输出文件:

```
hdfs dfs -ls /example/data/WordCountOutput
```

此命令显示两个文件:_SUCCESS 和 part-r-00000。part-r-00000 文件包含此作业的输出。

注:一些 MapReduce 作业可以将结果分割成多个 part-r-♯♯♯♯♯文件。如果是,请使用♯♯♯♯♯后缀来指示文件的顺序。

若要查看输出,请使用以下命令:

```
hdfs dfs -cat /example/data/WordCountOutput/part-r-00000
```

此命令显示 wasb://example/data/gutenberg/davinci.txt 文件中包含的单词列表以及每个单词出现的次数。下面的文本是文件中包含的数据的一个示例:

```
wreathed      3
wreathing     1
wreaths       1
wrecked       3
wrenching     1
wretched      6
wriggling     1
```

③ 使用 Hadoop 的 Pig 作业:

Pig 是一个平台,用于创建程序的 Hadoop,使用的程序语言称为 PigLatin。Pig 是 Java 创建 MapReduce 解决方案的替代品,它包含了 Azure HDInsight。使用表 9-14 查找 Pig 可以与 HDInsight 一起使用的各种方式。

表 9-14 Pig 可以与 HDInsight 一起使用的各种方式

方 法	交互式操作	批处理	集群操作系统	客户端操作系统
SSH	√	√	Linux	Linux、Unix、Mac OS X 或 Windows
REST API		√	Linux 或 Windows	Linux、Unix、Mac OS X 或 Windows
.NET SDK for Hadoop		√	Linux 或 Windows	Windows (for now)
Windows PowerShell		√	Linux 或 Windows	Windows

3. 为什么要使用 Pig

在 Hadoop 中使用 MapReduce 处理数据的挑战之一是通过只使用 map 和 reduce 函数来实现处理逻辑。对于复杂的处理,必须将处理分解为多个 MapReduce

操作,这些操作被链接在一起以实现期望的结果。

Pig 允许将处理定义为数据流过的一系列转换,以产生期望的输出。

Pig Latin 拉丁语允许描述来自原始输入的数据流,通过一个或多个转换,以产生所需的输出。Pig Latin 拉丁语程序遵循一般模式如下:

- 加载:从文件系统读取要操作的数据;
- 变换:操纵数据;
- 转储或存储:输出数据到屏幕或存储它用于处理。

(1) 自定义函数

Pig Latin 还支持用户定义函数(UDF),它允许调用实现难以用 Pig Latin 建模的逻辑的外部组件。

(2) 实例数据

HDInsight 提供各种示例数据集,这些示例数据集存储在/example/data 和/HdiSamples 目录中。这些目录位于集群的默认存储区中。本文档中的 Pig 示例使用了来自/example/data/sample.log 的 Log4J 文件。

文件中的每个日志都由一行字段组成,其中包含[LOG LEVEL]字段以显示类型和严重性,例如:

2012-02-03 20:26:41 SampleClass3 [ERROR] verbose detail for id 1527353937

在前面的示例中,日志级别是 ERROR。

注:还可以使用 Apache Log4J 日志工具生成 Log4J 文件,然后将该文件上传到 Blob 中。请参阅上载数据到 HDInsight 指令。有关 Azure 存储中的斑点如何使用 HDinsight 的更多信息,请参见使用 Azure Blob 存储与 HDInsight。

(3) 实例作业

下面的 Pig Latin 作业从 HDInsight 集群的默认存储中加载 sample.log 文件。然后,执行一系列转换,计算每个日志级别在输入数据中发生的次数。结果写入 STDUT。

```
LOGS = LOAD 'wasb:///example/data/sample.log';
LEVELS = foreach LOGS generate REGEX_EXTRACT( $ 0, '(TRACE|DEBUG|INFO|WARN|ERROR|FATAL)', 1) as LOGLEVEL;
FILTEREDLEVELS = FILTER LEVELS by LOGLEVEL is not null;
GROUPEDLEVELS = GROUP FILTEREDLEVELS by LOGLEVEL;
FREQUENCIES = foreach GROUPEDLEVELS generate group as LOGLEVEL, COUNT(FILTEREDLEVELS.LOGLEVEL) as COUNT;
RESULT = order FREQUENCIES by COUNT desc;
DUMP RESULT;
```

图 9-21 显示了每一个数据转换的总结。

(4) 经营 Pig Latin 工作

HDInsight 可以通过使用各种方法来运行 Pig Latin 作业。表 9-15 决定哪种

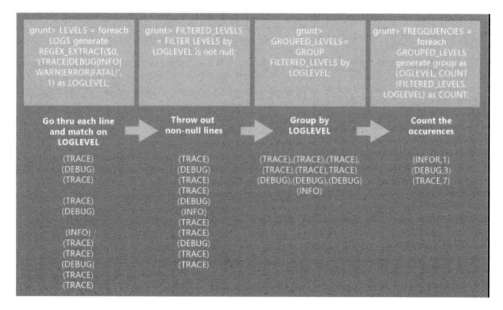

图 9-21 每一个数据转换的总结

方法适合你,然后按照其方法进行操作。

表 9-15 Pig Latin 可以运行的几种方式

方法	交互式	批处理	集群操作系统	客户端操作系统
SSH	√	√	Linux	Linux、Unix、Mac OS X 或 Windows
Curl		√	Linux 或 Windows	Linux、Unix、Mac OS X 或 Windows
.NET SDK for Hadoop		√	Linux 或 Windows	Windows(for now)
Windows PowerShell		√	Linux 或 Windows	Windows

4. 用 Pig 命令(SSH)在基于 Linux 的集群上运行 Pig 作业

了解如何交互式地从 SSH 连接到 HDInsight 集群运行 Pig 作业。Pig 拉丁编程语言允许描述应用于输入数据,以产生所需输出的转换。

(1) 与 SSH 连接

使用 SSH 连接到 HDInishgt 集群。下面的示例连接名为 myhdinsight 的集群,称为 SSHUSER 账户:

ssh sshuser@myhdinsight-ssh.azurehdinsight.nets

(2) 使用 Pig 命令

一旦连接,使用以下命令启动 Pig 命令行接口(CLI):

pig

输入以下语句:

LOGS = LOAD '/example/data/sample.log';

此命令将 SAMPL.E.log 文件的内容加载到日志中。可以使用以下语句查看文件的内容:

DUMP LOGS;

接下来,通过应用正则表达式来转换数据,以使用以下语句仅从每个记录中提取日志记录级别:

LEVELS = foreach LOGS generate REGEX_EXTRACT($ 0, '(TRACE|DEBUG|INFO|WARN|ERROR|FATAL)', 1) as LOGLEVEL;

转换后可以使用 DUP 查看数据。在这种情况下,使用:

DUMP LEVELS;

使用下面的语句继续应用转换:

FILTEREDLEVELS = FILTER LEVELS by LOGLEVEL is not null;
GROUPEDLEVELS = GROUP FILTEREDLEVELS by LOGLEVEL;
FREQUENCIES = foreach GROUPEDLEVELS generate group as LOGLEVEL, COUNT(FILTEREDLEVELS.LOGLEVEL) as COUNT;
RESULT = order FREQUENCIES by COUNT desc;

第一句是移除日志级别中包含空值的行,并将结果存储到筛选级别。

第二句是按日志级别对行进行分组,并将结果存储到 GROUPEDLEVELS 中。

第三句是创建一组数据,该数据集包含每个唯一日志级别值以及发生的次数。数据集被存储到 FREQUENCIES 中。

第四句通过计数(降序)命令日志级别并存储到 RESULT。

还可以使用 Store 语句保存转换的结果。例如,以下语句将 RESULT 保存到集群默认存储上的/example/data/pigout 目录:

STORE RESULT into '/example/data/pigout';

若要退出 GRUNT 提示,请输入以下语句:

QUIT;

(3) Pig latin 批处理文件

还可以使用 Pig 命令来运行包含在文件中的 Pig latin。

退出 GRUNT 提示符后,使用以下命令创建名为 pigbatch.pig 的文件。

nano ~/pigbatch.pig

键入或粘贴以下几行：

LOGS = LOAD '/example/data/sample.log';

LEVELS = foreach LOGS generate REGEX_EXTRACT($ 0 , '(TRACE|DEBUG|INFO|WARN|ERROR|FATAL)', 1) as LOGLEVEL;

FILTEREDLEVELS = FILTER LEVELS by LOGLEVEL is not null;

GROUPEDLEVELS = GROUP FILTEREDLEVELS by LOGLEVEL;

FREQUENCIES = foreach GROUPEDLEVELS generate group as LOGLEVEL, COUNT(FILTEREDLEVELS.LOGLEVEL) as COUNT;

RESULT = order FREQUENCIES by COUNT desc;

DUMP RESULT;

完成后，使用 Ctrl＋X, Y, 然后进入保存文件。

使用 Pig 命令运行 pigbatch.pig 文件。

pig ~/pigbatch.pig

批量作业完成后，会看到以下输出：

(TRACE,816)
(DEBUG,434)
(INFO,96)
(WARN,11)
(ERROR,6)
(FATAL,2)

9.5 用 HDInsight 创建 Spark

1. 使用 Azure 门户创建 HDInsight 中的集群

（1）创建高性能 Spark

在 Azure 门户中，选择 Create a resource→Data＋Analytics→HDInsight，如图 9-22 所示。

在 Basics 页面下，提供以下输入，如表 9-16 所列。

表 9-16 创建 HDInsight 集群的参数

属　性	描　述
集群名	给 HDInsight Spark 集群命名。用于此 Spark 的集群名称是 myspark20180403
订阅	从下拉菜单中选择用于此集群的 Azure 订阅

续表 9-16

属 性	描 述
集群类型	展开该项，然后选择 Spark 作为集群类型，并指定 Spark 集群版本 
集群登录名	输入集群登录用户名，默认名称是 Admin。可以使用这个账户在 Jupyter 笔记本中登录，然后再快速入门
集群登录密码	请输入集群登录密码
SSH 登录名	输入 SSH 用户名。用于这个快速启动的 SSH 用户名是 SSHUSER。默认情况下，此账户共享与集群登录用户名账户相同的密码
资源组	指定是否要创建新的资源组或使用现有的资源组。资源组是为 Azure 解决方案保存相关资源的容器。此处是 myspark20180403rg
地点	选择一个比较近的数据中心，需要注意的是，有的数据中心可能不支持 HDInsight 的某些版本之类的信息

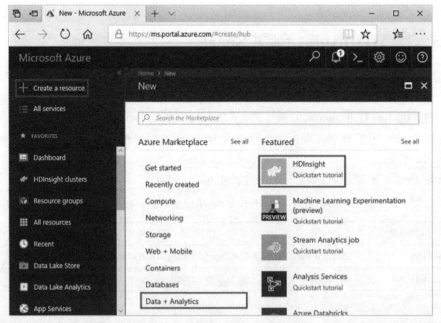

图 9-22 创建一个 HDInsight

将以上参数输入图 9-23。

图 9-23 输入创建 Spark cluster 的参数

单击 Next 按钮继续存储页面。

在存储条件下,提供以下值:

选择一个存储账户:选择创建新的,然后给新的存储账户命名。用于此快速启动的存储账户名称是 myspark20180403store,如图 9-24 所示。

默认容器具有默认名称。如果需要,可以更改名称。

单击 Next 按钮继续到摘要页。

在 Summary,选择 Create。创建集群需要大约 20 分钟。必须先创建集群,然后才能进行下一个会话。

(2) 创建一个 Jupyter 笔记本

Jupyter 笔记本电脑是一种支持各种编程语言的交互式笔记本环境。笔记本允许与数据交互,将代码与标记文本结合起来,并执行简单的可视化。

① 打开 Azure Portal。

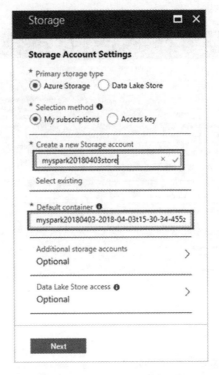

图 9-24 输入 Spark 需要的存储账号

② 选择 HDInsight Cluster，然后选择所创建的集群，如图 9-25 所示。

图 9-25 选择集群

③ 从门户中选择集群仪表板，然后选择 Jupyter 笔记本。如果提示，则请输入集群的登录凭据，如图 9-26 所示。

④ 选择 New→PySpark 创建笔记本，如图 9-27 所示。

⑤ 一个新的笔记本被创建和打开，名字没有命名（Untitled.pynb）。

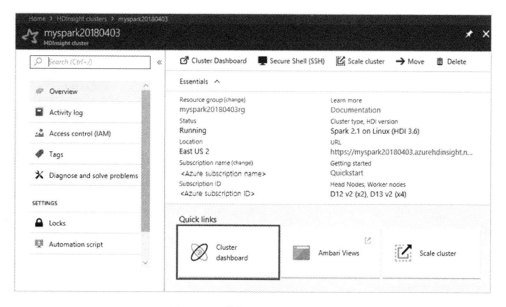

图 9-26　选择 Cluster dashboard

图 9-27　创建笔记本

2. 运行 Spark SQL 语句

SQL(结构化查询语言)是查询和定义数据最常用、最广泛使用的语言。使用熟悉的 SQL 语法,Spark SQL 作为 Apache Spark 的扩展,用于处理结构化数据。

验证内核是否已准备就绪。当您在笔记本中看到内核名称旁边的空心圆时(见图 9-28),内核就准备好了。实心圆表示内核忙。

当第一次启动笔记本时,内核在后台执行一些任务,等待内核准备就绪。

将下列代码粘贴到空单元格中,然后按 Shift+Enter 键来运行代码。命令列出集群上的 Hive 表:

图 9-28　内核准备好了

```
%%sql
SHOW TABLES
```

当在 HDInsight Spark 集群中使用 Jupyter Notebook 时,将获得一个预设的 sqlContext,可以使用它来使 Spark SQL 运行 Hive 查询。%%sql 告诉 Jupyter 笔记本使用预置的 sqlContext 来运行 Hive 查询。查询从 Hive 表(hivesampletable)中检索前 10 行,默认情况下所有的 HDInsight 集群都是这样的。获得结果需要大约 30 秒。输出看起来如图 9-29 所示。

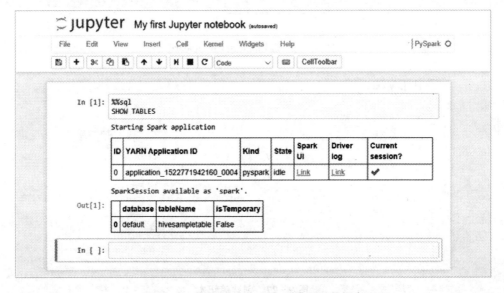

图 9-29　在 Jupyter notebook 中显示数据表

每次在 Jupyter 运行查询时,Web 浏览器窗口标题都会显示一个(繁忙)状态和笔记本标题。在右上角的 PySpark 文本旁边还可以看到一个实心圆。

运行另一个查询来查看 hivesampletable 中的数据。

```
%%sql
SELECT * FROM hivesampletable LIMIT 10
```

屏幕应刷新以显示查询输出,如图 9-30 所示。

从笔记本上的文件菜单中选择关闭并停止。关闭笔记本会释放集群资源。

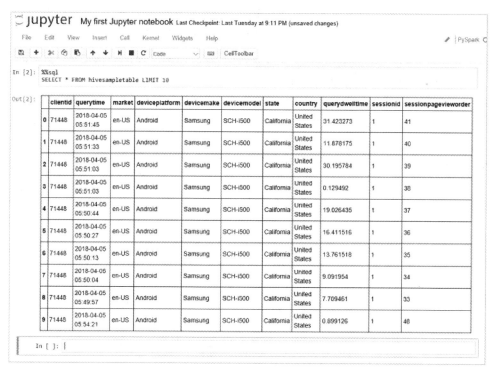

图 9-30　显示的 hivesampletable 的数据

3. 清理资源

HDInsight 在 Azure 存储或 Azure Data Lake 存储中保存数据,因此当不使用时,可以安全地删除集群。还需要为 HDInsight 集群付费,即使它不在使用中。由于集群的费用是存储费用的许多倍,所以当集群不使用时删除集群在经济上是有意义的。如果计划立即处理下一步中列出的教程,则可能需要保留集群。

切换回 Azure 门户,并选择 Delete,如图 9-31 所示。

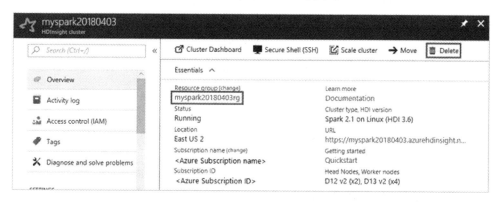

图 9-31　清理资源时可以单击 Delete 按钮将资源删除

还可以选择资源组名称来打开资源组页,然后选择删除资源组。通过删除资源组 resource group,可以删除 HDInsight Spark 集群和默认存储账户。

4. 在 Azure HDInsight 中加载数据并在 Apache Spark 集群上运行查询

(1) 从 CSV 文件创建数据文件

应用程序可以从现有的弹性分布式数据集(RDD)、Hive 表或使用 SQLContext 对象的数据源创建数据帧。下面的屏幕截图显示了本教程中使用的 HVAC.CSV 文件的快照。CSV 文件带有所有 HDInsight Spark。数据捕获了一些建筑物的温度变化,如图 9-32 所示。

Date	Time	TargetTemp	ActualTemp	System	SystemAge	BuildingID
6/1/2013	0:00:01	66	58	13	20	4
6/2/2013	1:00:01	69	68	3	20	17
6/3/2013	2:00:01	70	73	17	20	18
6/4/2013	3:00:01	67	63	2	23	15
6/5/2013	4:00:01	68	74	16	9	3
6/6/2013	5:00:01	67	56	13	28	4
6/7/2013	6:00:01	70	58	12	24	2
6/8/2013	7:00:01	70	73	20	26	16
6/9/2013	8:00:01	66	69	16	9	9

图 9-32 样本数据

打开在前提条件下创建的 Jupyter 笔记本。

将下面的代码粘贴到笔记本的空单元格中,然后按 Shift+Enter 键来运行代码。代码导入此场景所需的类型:

```
from pyspark.sql import *
from pyspark.sql.types import *
```

当在 Jupyter 中运行交互式查询时,Web 浏览器窗口或选项卡标题将显示(繁忙)状态以及笔记本标题。在右上角的 PySpark 文本旁边还可以看到一个实心圆。作业完成后,它变成一个空心圆圈,如图 9-33 所示。

通过运行以下代码创建数据文件和临时表(HVAC):

```
# Create an RDD from sample data
csvFile = spark.read.csv('wasb:///HdiSamples/HdiSamples/SensorSampleData/hvac/HVAC.csv', header = True, inferSchema = True)
csvFile.write.saveAsTable("hvac")
```

注:通过使用 PySpark 内核创建笔记本,当运行第一个代码单元时,SQL 上下文会自动为您创建。不需要显式创建任何上下文。

(2) 对数据文件运行查询

一旦创建表,就可以对数据运行交互式查询。

在笔记本的空单元格中运行以下代码:

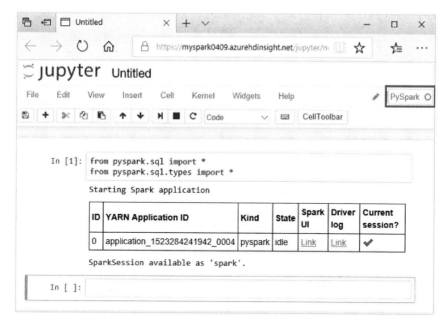

图 9-33　导入所需库

```
%%sql
SELECT buildingID, (targettemp - actualtemp) AS temp_diff, date FROM hvac WHERE date = \"6/1/13\"
```

因为 PySpark 内核在笔记本中使用，所以现在可以直接在临时表 HVAC 上运行交互式 SQL 查询。

显示以下表格输出，如图 9-34 所示。

图 9-34　导入的 CSV 文件数据

也可以在其他可视化中看到结果。若要查看相同输出的区域图，请选择区域，然后设置其他值,如图 9-35 所示。

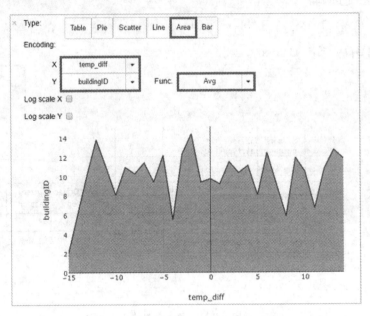

图 9-35 导入的数据可视化所成图形

从笔记本上的文件菜单中选择保存和检查点。

如果现在开始下一个教程,则把笔记本打开。否则,关闭笔记本以释放集群资源:从笔记本上的文件菜单中,选择关闭并停止。

第 10 章

Azure Automation

10.1 简 介

Azure Automation 提供基于云的自动化和配置服务，在 Azure 和非 Azure 环境中提供一致的管理。它由过程自动化、更新管理和配置特征组成。Azure Automation 在部署、操作以及工作负载和资源退役期间提供完全控制。本章简要介绍了 Azure Automation，并回答了一些常见的问题。有关不同功能的更多信息，请访问贯穿本概述的链接。

1. 过程自动化

Azure Automation 提供了对于自动化频繁、耗时和容易出错的云管理任务的能力。这种自动化有助于您专注于增加业务价值的工作。通过减少错误和提高效率，也有助于降低运营成本。您可以集成部署、配置和管理端到端流程中需要的 Azure 服务和其他公共系统。该服务允许以图形方式在 PowerShell 或 Python 中编写运行簿。通过使用混合 Author RunBooks，可以在跨平台环境中编排统一管理。Webbooks 通过触发来自 ITSM、DevOps 和监视系统的自动化，提供了一种满足请求并确保连续交付和操作的方法。

2. 配置管理

Azure Automation 期望的状态配置是基于云的 PowerShell DSC 解决方案，它提供企业环境所需的服务。在 Azure Automation 中管理 DSC 资源，并将配置从 Azure 云中的 DSC Pull Server 应用到虚拟机或物理机。它提供丰富的报告，通知重要事件，例如当节点偏离指定的配置时，可以监视并自动更新跨云中或内部物理和虚拟机（Windows 或 Linux）的机器配置。

可以在客户资源中获取有关安装应用程序和其他配置项的可见性的清单。丰富的报告和搜索功能可用于快速查找详细信息，以帮助理解操作系统内配置了什么。可以跟踪跨服务、守护进程、软件、注册表和文件的更改，以快速识别可能引起问题的内容。此外，DSC 可以帮助诊断和警告在环境中发生的不必要的更改。

3. 更新管理

在 Azure Automation 的混合环境中更新 Windows 和 Linux 系统。可以在 Azure、Office 和其他云上获得更新符合性的可见性。可以创建时间表部署，以便在定义的维护窗口中协调更新的安装。如果不应在机器上安装更新，则可以从部署中排除这些更新。

4. 共享能力

Azure Automation 包括一组共享资源，使其更易于自动化和配置环境。

① 基于角色的访问控制——使用自动化操作员角色控制对账户的访问，该角色允许在不提供创作能力的情况下运行任务。

② 变量——提供一种方法来保存可用于运行手册和配置的内容。可以更改值，而不必修改引用它们的任何运行库和配置。

③ 凭据——安全地存储敏感信息，可以在运行时使用 RunBooad 和配置。

④ 证书——存储和运行时可用，因此它可以用于验证和保护部署的资源。

⑤ 连接——存储一个名称/值对信息，这些信息在连接到连接资源中的系统时包含公共信息。连接由模块作者定义，用于在运行库和配置中使用。

⑥ 调度——用于服务在预定时间触发自动化。

⑦ 与源代码控制的集成——将配置作为代码，运行簿或配置可以签入源代码控制系统。

⑧ 电源模块——用于管理 Azure 和其他系统。导入微软、第三方、社区或自定义的 CMDLet 和 DSC 资源的自动化账户。

5. Windows 与 Linux

Azure Automation 自动化设计用于跨越混合云环境，也适用于 Windows 和 Linux。它提供了一种一致的方式来自动化配置部署的工作负载和正在运行的操作系统。

6. 社区画廊

浏览 Automation 图库以获得运行簿和模块，以便快速开始从 PowerShell 图库和 Microsoft Script Center 集成和创作流程。

7. 自动化的常见场景

Azure Automation 管理基础设施和应用程序的整个生命周期。将知识转移到系统上，组织如何交付和维护工作负载。在通用语言，如 PuthBek、期望状态配置、Python 和图形 RunBook 中，获取目标、报告和遵从的部署资源的完整清单，识别可能导致的配置错误并提高操作符合性的更改。

- 构建/部署资源——使用 RunBooad 和 Azure 资源管理器模板在混合环境中部署 VMS，集成到开发工具，如詹金斯和 VisualStudio 团队服务中。

- 配置 VM——评估并配置 Windows 和 Linux 机器所需的基础设施和应用程序配置。
- 监控器——识别引起问题的机器上的变化并补救或升级到管理系统。
- 保护——Quarantine VM 可引发安全警告,可设置客人要求。
- 管理——为团队建立基于角色的访问控制,恢复未使用的资源。

8. Azure 自动化的价格

每个 Region 区域的价格可能略有不同。需要查询相应区域的价格。这里就以 East US 2 为例说一下价格,East US 2 价格是以美元为单位的。

定价细节如下。

(1) 过程自动化

过程自动化包括 RunBook 和监视器。作业的计费基于该月使用的作业运行时的分数,监视器的计费基于一个月使用的小时数。每当作业或观察者运行时,都会发生过程自动化的费用。表 10-1 列出了过程自动化收费标准。

表 10-1 过程自动化收费标准

类 别	免费单元(每月)	价 格
作业运行	500 分钟	0.002 美元/分钟
监视器运行	744 小时	0.002 美元/小时

(2) 配置管理

配置管理包括配置拉动服务和更改跟踪功能。计费是基于向服务注册的节点的数量和存储在 Azure Log Analytics 服务中的日志数据。

配置管理的费用在节点向服务注册时开始,在节点从服务注销时停止。节点是由配置管理来管理配置的任何机器。这可能是一个 Azure 虚拟机(VM)、驻地虚拟机、物理主机,或在另一个公共云中的 VM。节点计费按小时计费。表 10-2 列出了配置管理收费标准。

表 10-2 配置管理收费标准

类 别	免费单元(每月)	价 格
Azure 节点	不适用	免费
非 Azure 节点	5 个节点	6 美元/节点

(3) 更新管理

更新管理包括对环境中的更新的可视性和部署。该服务没有收费,只需支付存储在 Azure 日志分析服务中的日志数据的存储费。表 10-3 列出了更新管理的收费标准。

表 10-3 更新管理的收费标准

类　别	免费单元(每月)	价　格
任何节点	不适用	免费

10.2　创建 Azure Automation 账号

1. 创建自动化账户

单击 Azure 左上角的 Create a resource 按钮。

选择管理工具,然后选择自动化。

输入账户信息。若要创建 Azure Run As Account,请选择 Yes,以便简化对 Azure 进行身份验证的工件自动启用。需要注意的是,在创建自动化账户时,在选择了该名称之后不能更改该名称。完成后,单击 Create 按钮,启动自动化账户部署,如图 10-1 所示。

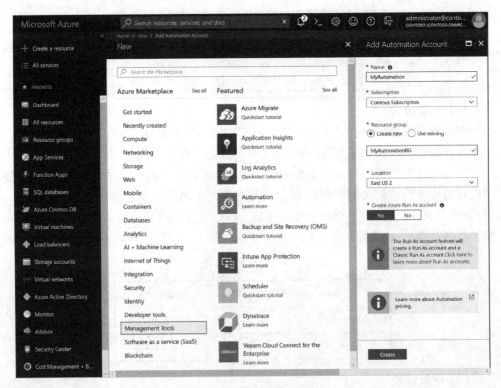

图 10-1　创建 Azure Automation 账号

当部署完成后,单击"All Services,选择 Automation Accounts 并选择所创建的自动化 Account。查看此账号,如图 10-2 所示。

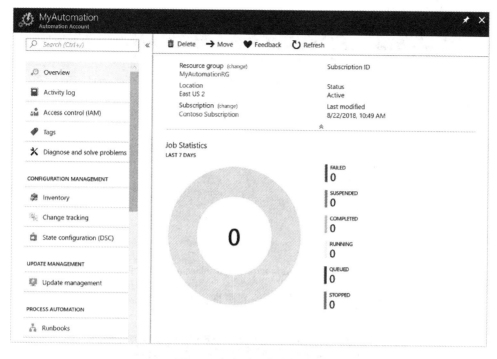

图 10 - 2　查看此自动化账号

2. 运行 Runbooks

运行教程中的一个 Runbooks。

单击进程自动化下的 RunBooks，显示 Runbooks 列表。默认情况下，在账户中启用了几个教程运行手册，如图 10 - 3 所示。

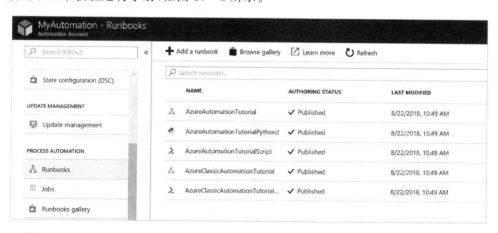

图 10 - 3　Runbooks 菜单里列出的已有 Runbook

选择 AzureAutomationTutorialScript Runbook，此操作将打开 Runbook 概述

页，如图 10-4 所示。

图 10-4　AzureAutomationTutorialScript 的详情页

单击 Start，然后在 Start 中的 Runbook 页面上单击确定按钮以启动 Runbook，如图 10-5 所示。

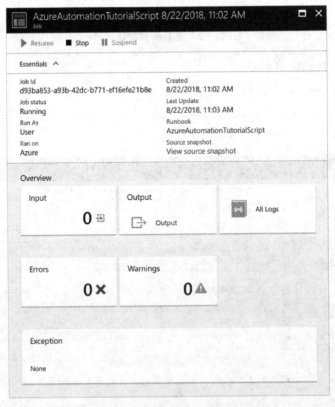

图 10-5　此 Runbook 运行后的窗口

在作业状态运行之后,单击 Output 或 All Logs 查看运行库作业的输出。对于本教程 Runbook,输出是 Azure 资源的列表。

3. 清理资源

当不再需要时,删除资源组、自动化账户和所有相关资源。为此,请选择自动化账户的资源组,然后单击 Delete 按钮。

以上运行的是教程 Runbook。让我们创建一些自己的 Runbook。

10.3 创建 Azure Automation Runbook

1. 创建 Runbook

① 创建一个 Runbook。在这个快速启动中创建的示例 Runbook 默认输出 Hello World。

② 打开自动化账户。

③ 单击进程自动化下的 Runbook,显示 Runbook 列表。

④ 单击列表顶部的 Add Runbook 按钮。在添加 Runbook 页上,选择 Quick Create。

⑤ 为 Runbook 名称输入 Hello-World,并选择 PowerShell 作为 RealBook 类型。单击 Create 按钮,如图 10-6 所示。

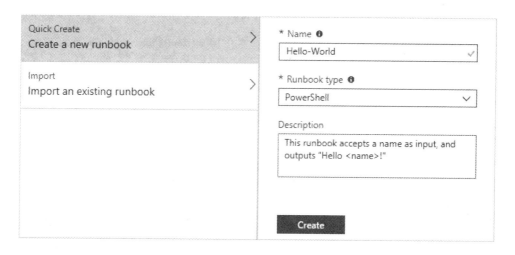

图 10-6　快速创建 Runbook

⑥ 创建 Runbook 并且 Edit PowerShell Runbook 页面打开了,如图 10-7 所示。

⑦ 键入或复制下面的代码并将其粘贴到编辑窗格中。它创建一个名为 Name 的可选输入参数,默认值为 World,并输出使用该输入值的字符串:

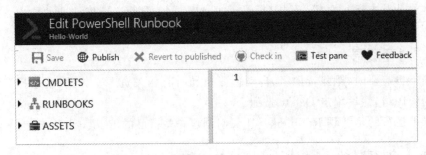

图 10-7 编辑 PowerShell Runbook 页面

```
param
(
    [Parameter(Mandatory = $false)]
    [String] $Name = "World"
)

"Hello $Name!"
```

单击 Save 按钮，保存 Runbook 的草稿副本，如图 10-8 所示。

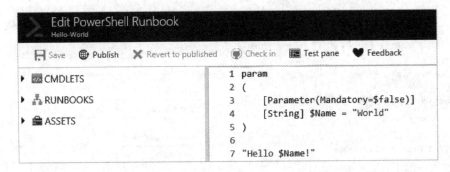

图 10-8 输入 PowerShell 脚本

2. 测试 Runbook

一旦创建了 Runbook，就可以测试 Runbook 来验证它是否有效。

① 单击测试窗格打开测试页。

② 输入名称的值，然后单击 Start 按钮。测试作业开始，作业状态和输出显示如图 10-9 所示。

③ 单击右上角的×，关闭测试页。在弹出窗口中单击 OK 按钮。

④ 在编辑 PowerShell Runbook 页面中，单击 Publish 按钮将 Runbook 作为账户中的 Runbook 的正式版本发布，如图 10-10 所示。

3. 运行 Runbook

一旦发布了 Runbook，将显示概述页。

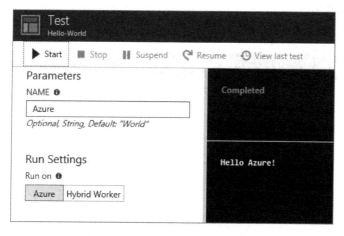

图 10 – 9　测试刚建好的脚本

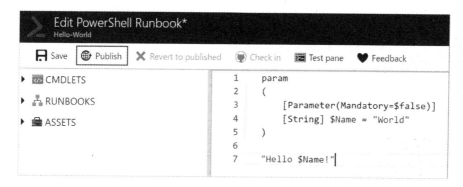

图 10 – 10　部署改好的脚本，作为正式的脚本运行

在 Runbook 概览页面中，单击 Start 打开该 Runbook 的 Start Runbook 配置页，如图 10 – 11 所示。

图 10 – 11　运行发布的正式脚本

将名称保留为空白,以便使用默认值,然后单击确定按钮。提交 Runbook 作业,并显示作业页,如图 10-12 所示。

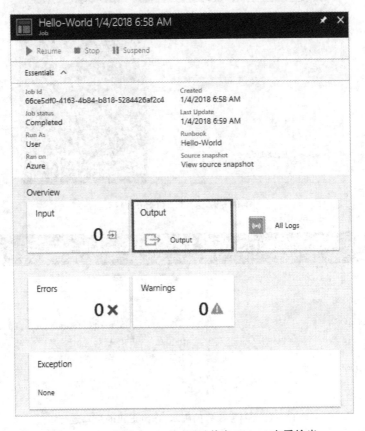

图 10-12 运行 Runbook,可以单击 Output 查看输出

当作业状态正在运行或完成时,单击 Output 打开输出窗格并查看 Runbook 输出,如图 10-13 所示。

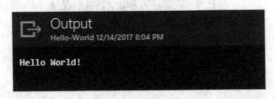

图 10-13 以上脚本的输出

10.4 Azure Automation Runbook 的类型

Azure 自动化支持几种类型的 Runbook,它们将在表 10-4 中简要描述。下面

的部分提供关于每种类型的进一步信息,包括何时使用每种类型的考虑。

表 10 - 4 Runbook 的类型

类型	描述
Graphical	基于 Windows PowerShell,在 Azure 门户的图形编辑器中进行完整的创建和编辑
Graphical PowerShell Workflow	基于 Windows PowerShell 的工作流,在 Azure 门户的图形编辑器中进行完整的创建和编辑
PowerShell	基于 Windows PowerShell 脚本的文本 Runbook
PowerShell Workflow	基于 Windows PowerShell 工作流的文本 Runbook based
Python	基于 Python 的文本 Runbook

1. 图形 Runbook

图形化和图形化的 PowerShell 工作流 Runbook 是在 Azure 门户中用图形编辑器创建和编辑的。可以将它们导出到一个文件中,然后将它们导入另一个自动化账户中,但是不能用其他工具创建或编辑它们。图形 Runbook 生成 PowerShell 代码,但不能直接查看或修改代码。图形 Runbook 不能转换为文本格式之一,也不能将文本运行库转换为图形格式。图形 Runbook 可以在导入过程中转换为图形 PowerShell 工作流 Runbook,反之亦然。

(1) 优　势
- 可视化插入链接配置创作模型。
- 关注数据如何通过流程。
- 直观地表示管理过程。
- 将其他 Runbooks 作为子运行簿来创建高级工作流。
- 鼓励模块化编程。

(2) 局限性
- 无法编辑 Azure 门户之外的 Runbook。
- 可能需要包含 PowerShell 代码的代码活动来执行复杂的逻辑。
- 无法查看或直接编辑由图形工作流创建的 PowerShell 代码。注意,可以查看在任何代码活动中创建的代码。

2. PowerShell Runbook

PowerShell Runbook 是基于 Windows PowerShell 的。可以使用 Azure 门户中的文本编辑器直接编辑 Runbook 的代码。还可以使用任何脱机文本编辑器,并将 Runbook 导入 Azure Automation。

(1) 优　势
- 用 PowerShell 代码实现所有复杂的逻辑,而避开 PowerShell 工作流的额外复杂性。
- Runbook 比 PowerShell 工作流 Runbook 更快地启动,因为它不需要在运行之前编译。

(2) 局限性
- 必须熟悉 PowerShell 脚本。
- 不能使用并行处理并行执行多个动作。
- 无法在错误的情况下使用检查点恢复 Runbook。
- 通过使用 Start-AzureAutomationRunbook cmdlet,PowerShell 工作流运行库和图形运行库只能作为子运行库包含该 cmdlet 创建新作业。

(3) 已知问题

下面是 PowerShell Runbook 的当前已知问题。
- PowerShell Runbook 无法检索具有空值的未加密的可变资产。
- PowerShell Runbook 不能用名称检索变量资产。
- 在 PowerShell Runbook 中的循环过程可能在大约 80 次迭代后崩溃。

如果试图将非常大量的数据一次性写入输出流,则 PowerShell Runbook 可能会失败。通常可以通过在处理大型对象时输出所需的信息来解决此问题。例如,可以使用 Get-Process | Select ProcessName,CPU 仅输出所需的字段,而不是输出 Get-Process 之类的内容。

3. PowerShell 工作流 Runbook

PowerShell 工作流 Runbook 是基于 Windows PowerShell 工作流的文本运行手册。可以使用 Azure 门户中的文本编辑器直接编辑 Runbook 的代码。还可以使用任何脱机文本编辑器,并将 Runbook 导入 Azure Automation。

(1) 优　势
- 用 PowerShell 工作流代码实现所有复杂的逻辑。
- 如果出现错误,请使用检查点恢复 Runbook。
- 使用并行处理并行执行多个动作。
- 可以包括其他图形运行手册和 PowerShell 工作流手册作为子运行簿来创建高级工作流。

(2) 局限性
- 作者必须熟悉 PowerShell 工作流。
- Runbook 必须处理 PowerShell 工作流的额外复杂性,如反序列化对象。
- PowerShell 工作流 Runbook 比 PowerShell Runbook 启动时间要长,因为它需要在运行之前编译。
- PowerShell 运行簿只能通过使用 Start-AzureAutomationRunbook cmdlet 作为子运行簿包含,该 cmdlet 创建新的作业。

4. Python Runbook

Python 运行手册在 Python 2 下编译。可以使用 Azure 门户中的文本编辑器直接编辑 Runbook 的代码,也可以使用任何脱机文本编辑器将 Runbook 导入 Azure Automation。

(1) 优　势

可以利用 Python 的健壮标准库。

(2) 局限性

- 必须熟悉 Python 脚本。
- 此时只支持 Python 2,这意味着 Python 3 特定功能将失败。

(3) 已知问题

以下是 Python Runbook 当前已知的问题。

为了利用第三方库,运行簿必须在 Windows Hybrid Runbook.r 或 Linux Hybrid Runbook.r 上运行,在运行簿启动之前,这些库已经安装在机器上。

5. 考虑事项

在确定特定的 Runbook 使用哪种类型时,应考虑以下附加因素。

不能将 Runbook 从图形转换为文本类型,反之亦然。

使用不同类型的 Runbook 作为子 Runbook 是有局限性的。有关 Azure AutoCAD 中的子手册,请参阅更多信息。

10.5　创建图形化 Runbook

1. 创建图形 Runbook

① 创建一个简单的 Runbook,输出文本 Hello World。

② 在 Azure 门户中,打开自动化账户。

在自动化账户页面可以快速查看此账户中的资源。此时应该已经拥有了一些资产。这些资产中的大部分是自动包含在一个新的自动化账户中的模块;也应该有在前提条件中提到的凭证资产。

③ 在流程管理下选择 Runbook 以打开 Runbook 列表。

④ 通过选择＋添加一个 Runbook 创建一个新的 Runbook,然后单击创建一个新的 Runbook。

⑤ 给 Runbook 一个名称:MyFirstRunbook-Graphical。

在这种情况下,将创建一个图形 Runbook。在 Runbook 类型选择图形,如图 10-14 所示。

⑥ 单击 Create 按钮创建 Runbook 并打开图形编辑器。

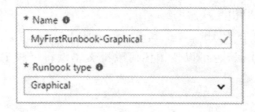

图 10-14 创建 Runbook 时选择图形

2. 添加活动

① 编辑器左侧的库控件允许选择要添加到 Runbook 中的活动。此时将添加一个写输出 CMDLET，从 Runbook 输出文本。

② 在库控件中，单击搜索文本框并键入 Write-Output。

③ 向下滚动到列表的底部。可以右击 Write-Output 并选择 Add to canvas（加到画布上），或者单击 cmdlet 旁边的省略号，然后选择 Add to canvas。

④ 单击画布上的写输出活动。此操作打开配置控制页，允许配置活动。

⑤ 标签默认为 CMDLET 的名称，但可以将其更改为更友好的内容。更改它来写入 Hello World 到输出。

⑥ 单击参数以提供 CMDLET 参数的值。

有些 CMDLET 具有多个参数集，并且需要选择要使用的参数集。在这种情况下，写输出只有一个参数集，因此不需要选择一个参数集。

⑦ 选择输入对象参数。这是指定要发送到输出流的文本的参数。

⑧ 在数据源下拉中，选择 PowerShell 表达式。数据源下拉提供了用于填充参数值的不同源。

可以使用这些源的输出，如另一个活动、自动化资产或 PowerShell 表达式。在这种情况下，输出只是 Hello World。可以使用 PowerShell 表达式并指定字符串。

⑨ 在表达式框中，键入 Hello World，然后单击确定按钮两次，返回到画布。

⑩ 单击保存按钮，保存 Runbook。

3. 测试 Runbook

在发布 Runbook 以使其在生产中可用之前，需要测试它以确保其正常工作。当测试一个 Runbook 时，应运行它的草稿版本并交互式地查看它的输出。

① 选择测试窗格打开测试页。

② 单击 Start 开始测试。这应该是唯一启用的选项。

③ 创建一个 Runbook 作业，并在窗格中显示其状态。

④ 作业状态以队列的形式开始，表示它正在等待云中的 Runbook 工作者使用。当工人提取作业执行，然后运行时，Runbook 开始运行。

⑤ 当 Runbook 完成时，将显示其输出。在这种情况下，可以看到 Hello World，

如图 10-15 所示。

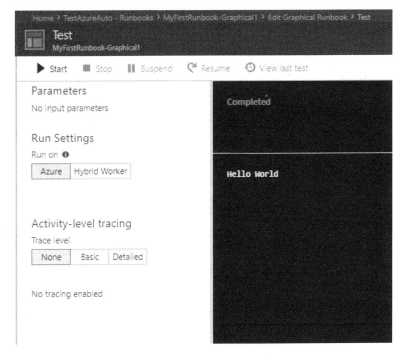

图 10-15　Runbook 运行完之后的输出

⑥ 关闭测试页以返回画布。

4. 发布并启动 Runbook

至此，创建的 Runbook 仍然处于草稿模式，它需要发布，然后才能在生产中运行。当发布 Runbook 时，用草稿版本覆盖现有发布的版本。在这种情况下，还没有发布版本，因为刚刚创建了 Runbook。

选择发布，以发布 Runbook，然后在提示时发布 Yes。

如果向左滚动以查看 Runbook 页面中的 Runbook，则显示已发布的创作状态。

滚动回到右边查看 MyFirstRunbook-Graphical。

上面的选项允许我们启动运行簿，将其安排在未来某个时间启动，或者创建一个 Web hook，以便可以通过 HTTP 调用启动它。

当启动 Runbook 时，选择 Start，然后选择 Yes。

为创建的 Runbook 作业打开作业页。验证作业状态是否已完成。

一旦运行库状态显示完成，单击 Output。输出页面打开，可以在窗格中看到 Hello World。

关闭输出页。

单击所有日志打开 Runbook 作业的流 Stream 页。应该只在输出流中看到 Hello World,但是这里也可以显示 Runbook 作业的其他流,例如 Verbose 和 Error(如果 Runbook 写到这些流)。

关闭所有日志页和作业页以返回 MyFirstRunbook-Graphical。

若要查看 Runbook 的所有作业,请关闭作业页并在资源下选择作业。这列出了该 Runbook 创建的所有作业。你只应该看到一个 Job,因为你只做过一次工作。

可以单击此作业打开在启动 Runbook 时查看的相同作业窗格。这允许及时返回并查看为特定的 Runbook 创建的任何作业的详细信息。

5. 创造变量资产

至此已经测试并发布了 Runbook,但到目前为止,它并没有做任何有用的事情。如果想让它管理 Azure 的资源,则在配置运行簿以进行身份验证之前,应创建一个变量来保存订阅 ID,并在进行身份验证的活动之后对其进行引用,包括对订阅上下文的引用,这样就可以轻松地在多个订阅之间工作。在继续之前,请从导航窗格中的订阅选项中复制订阅 ID。

① 在自动化账户页中,选择共享资源下的变量。

② 选择添加变量。

③ 在新建变量页的名称框中,输入 Azure Subscription Id,在值框中输入订阅 ID。为类型保留字符串,为加密保留默认值。

④ 单击 Create 按钮创建变量。

6. 添加认证

现在,已经有了一个变量来保存订阅 ID,可以将运行簿配置为使用在先决条件中引用的 Run As 凭证进行身份验证。通过将 Azure 作为连接资产和 Connect-AzureRmAccount cmdlet 添加到画布中来实现这一点。

① 导航回到 Runbook,并选择 MyFirstRunbook-Graphical 页上的编辑。

② 如果不再需要 Write Hello World to output,则单击省略号(…)并选择 Delete。

③ 在库控件中,选择添加到画布中,扩展资产、连接,并将 AzureRunAsConnection 添加到画布中。

④ 在库控件中,在搜索文本框中键入 Connect-AzureRmAccount。

⑤ 将 Connect-AzureRmAccount 添加到画布中。

⑥ 悬停在 Get Run As Connection,直到一个圆圈出现在底部。单击圆圈并拖动箭头至 Connect-AzureRmAccount。您创建的箭头是一个链接。Runbook 以 Get Run As Connection 开始,然后运行 Connect-AzureRmAccount,如图 10-16 所示。

⑦ 在画布上,在标签文本框中选择 Connect-AzureRmAccount,在配置控制页输入 Login to Azure 到 Label 文本框。

⑧ 单击参数并显示活动参数配置页。

Connect-AzureRmAccount 有多个参数集，所以在提供参数值之前需要选择一个参数集。单击参数集，然后选择 ServicePrincipalCertificate 证书参数集。

一旦选择了参数集，这些参数就会显示在活动参数配置页中。单击 APPLICATIONID，如图 10 – 17 所示。

⑨ 在 Parameter Value 页中，选择数据源的活动输出，并从列表中选择 Get Run As Connection，在字段路径文本框类型中选择 ApplicationId，然后单击确定按钮。您正在指定字段路径的属性名称，因为该活动输出具有多个属性的对象。

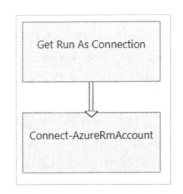

图 10 – 16　两个步骤

图 10 – 17　选择参数

⑩ 单击 CERTIFICATETHUMBPRINT，在参数值页中，从数据源中选择 Activity output 活动输出。选择 Get Run As Connection，在字段路径文本框类型 CertificateThumbprint 中，单击 OK 按钮。

⑪ 单击 SERVICEPRINCIPAL，在参数值页中，从数据源中选择 Constant value,单击选项 TRUE,然后单击 OK 按钮。

⑫ 单击 TENANTID,在参数值页中,从数据源中选择 Activity output。选择 Get Run As Connection,在字段路径文本框类型 Tenant Id 中,单击 OK 按钮两次。

⑬ 在库控件中,在搜索文本框中键入 Set-AzureRmContext。

⑭ 在画布中添加 Set-AzureRmContext。

⑮ 在画布上,选择 Set-AzureRmContext,并在配置控制页中指定标签文本框中输入 Specify Subscription Id。

⑯ 单击 Parameters,然后 Activity Parameter Configuration 页显示。

⑰ Set-AzureRmContext 有多个参数集,因此需要在提供参数值之前选择一个参数集。单击 Parameter Set,然后选择 SubscriptionId 参数集。

当你选择参数集后,这些参数显示在 Activity Parameter Configuration 页。点击 Subscription ID。

⑱ 在参数值页面,在 Data Source 中选择变量资产(Variable Asset),从列表中选择 Azure Subscription Id,并单击 OK 按钮两次。

⑲ 在 Login to Azure 链接上移动鼠标,直到一个圆圈出现在底部,点击此圆圈并拖动此箭头到 Specify Subscription Id。

此时,Runbook 应该如图 10-18 所示。

7. 加入活动去启动一个 VM

在这里加入一个 Start-AzureRmVM 活动区启动一个 VM。可以选择任何一个订阅中的 VM 并且现在得将 VM 名字写入 cmdlet。

① 在库控件中,在 Search 文本框中输入 Start-AzureRm。

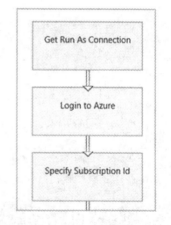

图 10-18 若干步骤的 Runbook

② 加入 Start-AzureRmVM 到画布并拖动它到 Specify Subscription Id 之下。

③ 在 Specify Subscription Id 上盘旋鼠标直到一个小圆圈出现,点击此圆圈并拖动箭头到 Start-AzureRmVM。

④ 选择 Start-AzureRmVM,点击 Parameters 并点击 Parameter Set 来查看 Start-AzureRmVM。选择 ResourceGroupNameParameterSetName 参数集。

⑤ ResourceGroupName 和 Name 有 have 感叹符在旁边,这表示这些是必须输入的字段。请注意此两个字段需要输入字符串。

⑥ 选择 Name,选择 PowerShell 表达式作为 Data source 并且输入 VM 的名字,并用双引号括起来,单击 OK 按钮。

⑦ 选择 ResourceGroupName,用 PowerShell 表达式作为 Data source 并且输入资源组的名字,用双引号括起来,单击 OK 按钮。

⑧ 点击 Test 面板,可以测试此 Runbook。

⑨ 点击 Start 来启动测试,一旦它完成了,检查此 VM 是否启动了。

此时你的 Runbook 应该如图 10-19 所示。

8. 加入额外的输入参数

Runbook 目前启动此 VM,在 Start-AzureRmVM cmdlet 和资源组中指定 VMOur。如果指定资源组名和 VM 名并输入参数到此 Runbook 中,则此 Runbook 将会变得更有用。

① 点击 MyFirstRunbook-Graphical 面板,点击 Edit 打开图形编辑器。

② 选择 Input and output 并且进入此 Runbook 的 Input Parameter 面板。

③ 在 Name 下输入 VMName 。请输入字符串。要求必须输入(Mandatory 设置成 Yes)。单击 OK 按钮。

④ 创建第二个必需的输入参数,叫做 ResourceGroupName 并单击 OK 按钮来关闭 Input 和 Output 面板,如图 10-20 所示。

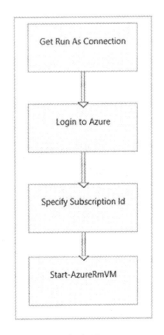

图 10-19　加入了 Start-Azure-RmVM 动作的脚本

⑤ 选择这个 Start-AzureRmVM 活动并点击 Parameters。

⑥ 在 Name 下选择 VMName 作为 Runbook input。

⑦ 在 RESOURCEGROUPNAME 下选择 ResourceGroupName 作为 Runbook input,如图 10-21 所示。

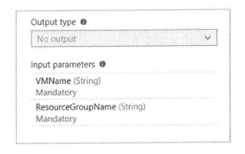

图 10-20　Input 和 Output 面板

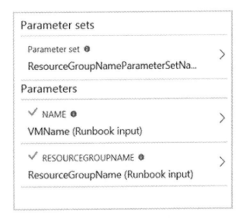

图 10-21　参数选择

⑧ 保存此 Runbook，打开 Test 面板，可以提供此两参数的值。

⑨ 打开 Test 面板。

⑩ 点击 Publish，这是 Runbook 的新版本。

⑪ 停止此前启动的 VM。

⑫ 点击 Start 来启动此 Runbook，输入 VM Name 和 ResourceGroup Name，作为将要启动 VM 的参数。

⑬ 当 Runbook 完成后，检查此启动的 VM。

9. 创建有条件的链接

现在修改此 Runbook，所以只有在 VM 没有启动的情况下，此 Runbook 试图启动此 VM。首先将 Get-AzureRmVM cmdlet 加到此 Runbook，此 Runbook 取得此 VM 的实例 Level 状态。然后加入一个 PowerShell Workflow code 模块，将其命名为 Get Status，输入一小段 PowerShell 代码来检测此 VM 是否运行或者停止。Get Status 模块的一个条件链接只有在当前状态是停止的时候才运行 Start-AzureRm-VM 命令。最后，输出一个消息告知 VM 成功地启动了（如果没有用 PowerShell Write-Output cmdlet）。

① 在图形编辑器里打开 MyFirstRunbook-Graphical。

② 移除 Specify Subscription Id 和 Start-AzureRmVM 之间的连接：点击它并按 Delete 键。在库控件，在 Search 文本框中输入 Get AzureRm。

③ 加入 Get-AzureRmVM 到画布。

④ 选择 Get-AzureRmVM 并且选择 Parameter Set 来查看 Get-AzureRmVM 集合。选择 GetVirtualMachineInResourceGroupNameParamSet 参数集。ResourceGroupName 和 Name 旁有感叹号，这表示必须输入字符串。

⑤ 在数据源节 Name 下，选择 Runbook 输入并且选择 VMName，然后单击 OK 按钮。

⑥ 在数据源节 ResourceGroupName 下，选择 Runbook 输入并且选择 ResourceGroupName，然后单击 OK 按钮。

⑦ 在数据源节 Status 下，选择 Constant value 并且选择 True，然后单击 OK 按钮。

⑧ 创建一个链接从 Specify Subscription Id 到 Get-AzureRmVM。

⑨ 在库控件中，展开 Runbook 控制并加入 Code 到画布。

⑩ 创建一个链接从 Get-AzureRmVM 到 Code。

⑪ 点击 Code，并且在配置面板改变 Label 为 Get Status。

⑫ 选择 Code 的参数，并且 Code 编辑器页出现了。

⑬ 在代码编辑器里，粘贴下列代码片段：

```
$ StatusesJson = $ ActivityOutput['Get - AzureRmVM'].StatusesText
$ Statuses = ConvertFrom - Json $ StatusesJson
$ StatusOut = ""
foreach ( $ Status in $ Statuses){
if( $ Status.Code - eq "Powerstate/running"){ $ StatusOut = "running"}
elseif ( $ Status.Code - eq "Powerstate/deallocated" ) { $ StatusOut = "stopped"}
}
$ StatusOut
```

⑭ 创建一个链接从 Get Status 到 Start-AzureRmVM，如图 10-22 所示。

⑮ 选择此链接，并且在配置页改变 Apply 条件为 Yes。注意此链接变成了打点的线，这表示目标活动只在条件为真时才会运行。

⑯ 对于条件表达式，输入 $ActivityOutput['Get Status'] -eq "Stopped"，则 Start-AzureRmVM 只在 VM 停止状态下才会运行。

⑰ 在库控件中，展开 Cmdlets 并且 Microsoft.PowerShell.Utility.加入 Write-Output 到 canvas 两次。

⑱ 在第一个 Write-Output 控件，点击 Parameters 并改变 Label 值为 Notify VM Started。

⑲ 对于 InputObject，改变 Data source 为 PowerShell 表达式，并输入表达式 $VMName successfully started。注：$VMName 是一个变量名，表示 VM 的名字。

⑳ 在第二个 Write-Output 控件，点击 Parameters 并改变 Label 值为 Notify VM Start Failed。

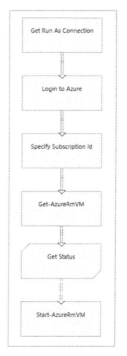

图 10-22 从 Get Status 到 Start-AzureRmVM 建立链接

㉑ 对于 InputObject，改变 Data source 为 PowerShell 表达式，并输入表达式 $VMName could not start。

㉒ 创建一个链接从 Start-AzureRmVM 到 Notify VM Started 和 Notify VM Start Failed。

㉓ 选择此链接到 Notify VM Started 并改变 Apply condition 为 True。

㉔ 选择链接到 Notify VM Start Failed 并改变 Apply condition 为 True。

㉕ 对于条件表达式（Condition expression），输入 $ActivityOutput['Start-AzureRmVM'].IsSuccessStatusCode-ne $true，此 Write-Output 控件只在 VM 没有成功启动时才运行。此时 Runbook 应该看起来如图 10-23 所示。

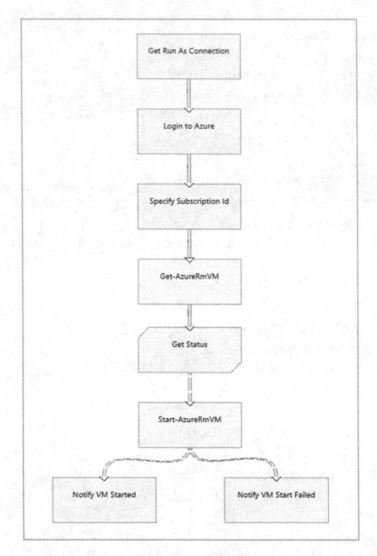

图 10-23 条件表达式

㉖ 保存 Runbook 并打开 Test 面板。
㉗ 启动此 Runbook（当 VM 停止时），随后 VM 就启动了。

10.6 创建 PowerShell Runbook

1. 创建新的 Runbook

① 启动创建一个简单的（如输出文本 Hello World）Runbook。
② 在 Azure Portal，打开你的自动化账户。

此自动化账户页面给出一个快速的资源预览。你应该已经有一些资产,其中大部分是一些模块,被自动地包含在一个新的自动化账户中。你应该也有一些登录账户密码的资产。

③ 点击 Process Automation 下面的 Runbooks 来打开 Runbook 列表。

④ 点击这个"＋",加一个 Runbook 按钮来创建一个新的 Runbook。

给此 Runbook 命名为 MyFirstRunbook-PowerShell。

在此例中,你将创建一个 PowerShellRunbook,所以选择了 Powershell 的 Runbook 类型。

⑤ 点击 Create 按钮,创建此 Runbook 并打开文本编辑器。

2. 加入代码到 Runbook

① 可以要么直接输入代码到此 Runbook,要么从库控件选择 cmdlet、runbooks 及资产,并将它们加入到此 Runbook,输入参数。在此处,可以直接输入代码到此 Runbook。

② 你的 Runbook 当前是空的,输入：Write-Output "Hello World"的脚本语句,如图 10－24 所示。

图 10－24 加入 PowerShell 脚本

③ 单击 Save 按钮来保存。

3. 测试此 Runbook

在你发布此 Runbook 之前,需要测试此 Runbook 来保证它正常工作。当你测试此 Runbook 时,运行它的草稿版本,然后即可查看它的交互式输出。

① 点击 Test 面板,打开此 Test 面板。

② 单击 Start 按钮启动测试。一个 Runbook 作业启动了并且它的状态也显示了。

此作业的状态如果为 Queued,则表示此作业等待 Runbook 工作者调度可用。当它变成 Starting 状态时,表示有一个 Runbook 工作者开始运行此作业。最后是 Running 状态,则表示此作业正在运行。

当此 Runbook 作业完成后,它的输出将被显示,如 Hello World,如图 10－25 所示。

图10-25 脚本输出的信息

③ 关闭 Test 面板,回到画布。

4. 发布并启动此 Runbook

至此,所创建的 Runbook 还是在草稿模式,需要发布然后才可以在生产环境中看到它。当发布一个 Runbook 时,复写已有的发布版本,如果还没有发布过,则此草稿版本将变成一个发布版本。

点击发布 Publish 来发布此 Runbook,然后单击 Yes 按钮确认发布。

如果滚动到左边查看此 Runbook 在 Runbooks 面板中的样子,则它会显示此 Runbook 是已经发布的(Published)。

再滚回来(往右滚)查看 MyFirstRunbook-PowerShell 的面板。

顶边的那些选项允许启动此 Runbook、查看此 Runbook、安排一个时间启动,或者创建一个新的根据 http 调用就可以启动的 Runbook。

若想启动此 Runbook,则单击 Start 按钮,然后单击 OK 按钮确认启动。

一个 Job 页打开了,这是为此作业对应的作业。可以关闭此页,也可以让其保持打开,则可以看到此作业的进度。

此作业的状态显示在作业汇总并且这些状态与测试此 Runbook 时得到的状态是一致的,如图 10-26 所示。

当此 Runbook 状态显示为完成时,在概览之下,点击 Output,则 Output 面板打开了,可以看到输入是 Hello World,如图 10-27 所示。

关闭 Output 面板。

点击 All Logs 来打开流 Streams 面板,应该只能看到 Hello World。在 Output 流,可以显示 Runbook 其他流输出的内容,比如 Verbose 和 Error。如果 Runbook 写入到这些流输出,则点击 All Logs 来查看 log,如图 10-28 所示。

关闭流页和作业页以返回 MyFirstRunbook-PowerShell 页。

在详细信息下,单击作业打开此 Runbook 的作业窗格。这列出了本 Runbook

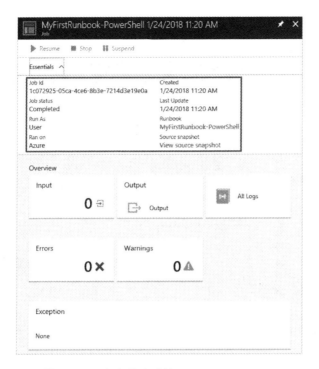

图 10-26　运行发布后的 PowerShell Runbook

图 10-27　点击 Output 查看其输出

图 10-28　查看 log

创建的所有作业。只应该看到一份 Job 作业,因为只运行过一次 Job 作业,如图 10-29 所示。

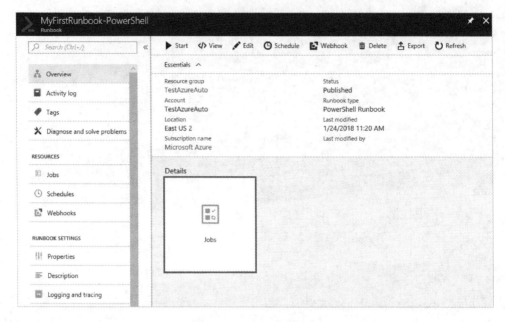

图 10-29　查看 Job 作业

可以单击此作业打开在启动 Runbook 时查看的相同作业窗格。这允许您及时返回并查看为特定的 Runbook 创建的任何作业的详细信息。

5. 添加身份验证管理 Azure 资源

现在,已经测试并发布了 Runbook,但到目前为止,它并没有做任何有用的事情。若想让它管理 Azure 的资源,它是无法做到这一点的,除非使用了在先决条件中引用的凭据进行身份验证。可以使用 Connect-AzureRmAccount cmdlet 进行此操作。

① 通过单击 MyFirstRunbook-PowerShell 页面上的编辑打开文本编辑器。

② 您不再需要写 Write-Output 行,所以继续删除它。

③ 键入或复制并粘贴以下代码,这些代码以您的自动化运行为账户处理身份验证:

```
$ Conn = Get-AutomationConnection -Name AzureRunAsConnection
Connect-AzureRmAccount -ServicePrincipal -Tenant $ Conn.TenantID'
  -ApplicationId $ Conn.ApplicationID -CertificateThumbprint $ Conn.CertificateThumbprint
```

④ 单击测试面板,以便测试 Runbook。

⑤ 单击 Start 按钮开始测试。一旦完成,应该接收与下面类似的输出,从账户中显示基本信息。这证实凭证是有效的,如图 10-30 所示。

```
Completed


Environments                                                                                          Context
------------                                                                                          -------
{[AzureCloud, AzureCloud], [AzureChinaCloud, AzureChinaCloud], [AzureUSGovernment, AzureUSGovernment]} Microsoft.Azur...
```

图 10-30 显示账户中的基本信息

6. 添加代码启动虚拟机

既然 Runbook 正在对 Azure 订阅进行身份验证，那么就可以管理资源。添加一个命令来启动虚拟机。可以在 Azure 订阅中选择任何虚拟机，然后在 Runbook 中硬编码该名称。

在 Connect-AzureRmAccount 之后，键入 Start-AzureRmVM -Name 'VMName' -ResourceGroupName 'NameofResourceGroup'，提供要启动的虚拟机的名称和资源组名称。

```
$ Conn = Get - AutomationConnection - Name AzureRunAsConnection
Connect - AzureRmAccount - ServicePrincipal - Tenant $ Conn.TenantID'
 - ApplicationID $ Conn.ApplicationID - CertificateThumbprint $ Conn.CertificateThumbprint
Start - AzureRmVM - Name 'VMName' - ResourceGroupName 'ResourceGroupName'
```

7. 向 RunBook 添加输入参数

Runbook 当前启动 Runbook 中硬编码的虚拟机，但如果在 Runbook 启动时指定虚拟机，则更有用。向 Runbook 添加输入参数以提供该功能。

将 VMName 和 ResourceGroupName 的参数添加到 Runbook 中，并将这些变量与 Start-AzureRmVM cmdlet 一起使用，如下面的示例所示。

```
Param(
 [string] $ VMName,
 [string] $ ResourceGroupName
)
$ Conn = Get - AutomationConnection - Name AzureRunAsConnection
Connect - AzureRmAccount - ServicePrincipal - Tenant $ Conn.TenantID'
 - ApplicationID $ Conn.ApplicationID - CertificateThumbprint $ Conn.CertificateThumbprint
Start - AzureRmVM - Name $ VMName - ResourceGroupName $ ResourceGroupName
```

保存 Runbook 并打开测试窗格。现在可以为测试中使用的两个输入变量提供值。

关闭测试面板。

单击发布，发布 RunBook 的新版本。

停止在上一步启动的虚拟机。

单击 OK 按钮，启动 Runbook。输入要启动的虚拟机的 VMName 和 ResourceGroupName，如图 10-31 所示。

图 10-31　运行 Runbook 前输入参数

当 Runbook 完成时，检查虚拟机是否已启动。

PowerShell Automation 与 PowerShell 工作流 Automation 的差异如下：

① PowerShell 运行库具有与 PowerShell 工作流运行库相同的生命周期、能力和管理方法，但也存在一些差异和限制。

② PopeScriptRunbook 与 PowerShell 工作流手册相比运行速度更快，因为它们没有编译步骤。

③ PowerShell 工作流 Runbook 支持检查点，使用检查点，PowerShell 工作流运行库可以从 Runbook 中的任何点恢复，而 PowerShell Runbook 只能从一开始恢复。

④ PowerShell 工作流 Runbook 支持并行和串行执行，而 PowerShell Runbook 只能连续执行命令。

⑤ 在 PowerShell 工作流 Runbook 中，活动、命令或脚本块可以具有自己的运行空间，而在 PowerShell Runbook 中，脚本中的所有内容都在单个运行空间中运行。在本机 PowerShell Runbook 和 PowerShell 工作流 Runbook 之间也存在一些句法差异。

10.7 创建 PowerShell 工作流 Runbook

1. 创建新的 Runbook

① 创建一个简单的 Runbook,输出文本 Hello World。

② 在 Azure 门户中,打开自动化账户。

在自动化账户页面可以快速查看此账户中的资源。你应该已经拥有了一些资产。其中大部分是自动包含在一个新的自动化账户中的模块。你也应该有在前提条件中提到的凭证资产。

③ 单击进程自动化下的 Runbook,打开 Runbook 列表。

④ 通过点击"+"添加一个 Runbook 按钮,创建一个新的 Runbook,然后创建一个新的 Runbook。给 RunBook 命名为 MyFirstRunbook-Workflow。

在这种情况下,您将创建一个 PowerShell 工作流手册,因此选择 Runbook 类型为 PowerShell 工作流。

⑤ 单击 Creat,创建 Runbook 并打开文本编辑器。

2. 向 Runbook 中添加代码

可以直接将代码输入到运行簿中,或者可以从 Library 控件中选择 cmdlet、Runbook 和资产,并使用任何相关参数将它们添加到 Runbook 中。对于此演练,可以直接键入代码到此 Runbook。

您的 Runbook 当前是空的,只有必需的工作流关键字、Runbook 的名称和包围整个工作流的大括号,如下:

```
Workflow MyFirstRunbook - Workflow
{
}
```

键入 Write-Output "Hello World"。

```
Workflow MyFirstRunbook - Workflow
{
Write-Output "Hello World"
}
```

单击保存按钮保存 Runbook。

3. 测试手册

在发布 Runbook 以使其在生产中可用之前,需要测试它以确保其正常工作。当测试一个 Runbook 时,应运行它的草稿版本并交互式地查看它的输出。

① 单击测试,打开测试面板。

② 单击 Start 开始测试。这应该是唯一启用的选项。

③ 创建一个 Runbook 并显示其状态。

作业状态将从队列开始,表示它正在等待云中的 Runbook 工作者使用。当一个工作者要求作业,然后在 Runbook 变成运行时,它开始运行。

当 Runbook 完成时,将显示其输出。在这种情况下,你应该看到 Hello World,如图 10-32 所示。

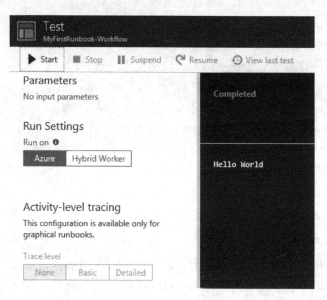

图 10-32 运行 PowerShell 工作流 Runbook

④ 关闭测试窗格以返回画布。

4. 发布并启动 Runbook

至此创建的 Runbook 仍然处于草稿模式。需要在生产之前运行它。当发布 Runbook 时,用草稿版本覆盖现有发布的版本。在这种情况下,还没有发布版本,因为刚刚创建了 Runbook。

单击发布按钮,发布 Runbook,然后在提示时单击 Yes 按钮。

如果现在滚动左边以查看 Runbook 窗格中的 Runbook,则显示已发布的创作状态。

滚动回到右边查看 MyFirstRunbook-Workflow 窗格。

上面的选项允许启动运行簿,将其安排在未来某个时间启动,或者创建一个 Webhook,以便可以通过 HTTP 调用启动它。

若只想启动 Runbook,则单击 Start 按钮,然后在提示时单击 Yes 按钮,如图 10-33 所示。

为您创建的 Runbook 打开作业窗格。在这种情况下,您可以看到作业的进展。也可以关闭这个窗格。

图 10 - 33　Start 按钮

作业状态显示在作业摘要中,并与测试 Runbook 时所看到的状态相匹配,如图 10 - 34 所示。

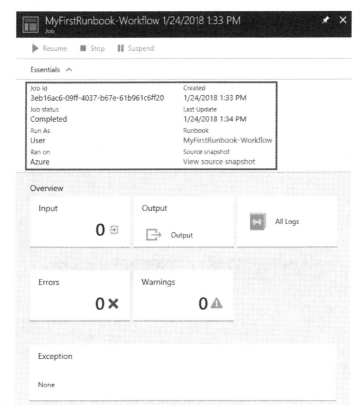

图 10 - 34　运行部署的 PowerShell 工作流 Runbook

一旦运行库状态显示完成,单击 Output。输出窗格打开,可以看到您的 Hello World,如图 10 - 35 所示。

关闭输出窗格。

单击所有日志以打开 Runbook 作业的流窗格。应该只在输出流中看到 Hello World,但是这可以显示 Runbook 作业的其他流,例如 Verbose 和 Error(如果 Runbook 写到这些流),如图 10 - 36 所示。

关闭流窗格和作业窗格,以返回 MyFirstRunbook 窗格。

单击作业打开此 Runbook 的作业窗格。这里列出了本 Runbook 创建的所有作业。你只应该看到一份作业,因为你只运行过一次作业,如图 10 - 37 所示。

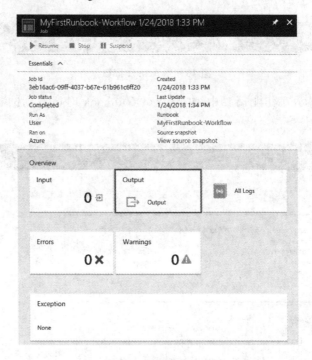

图 10-35 查看输出

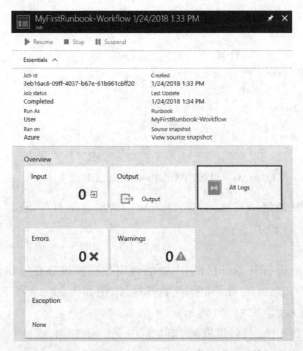

图 10-36 查看 PowerShell 工作流 Runbook 的 log

图 10-37　Jobs 菜单

5．添加身份验证管理 Azure 资源

至此，已经测试并发布了您的 Runbook，但到目前为止，它并没有做任何有用的事情。若想让它管理 Azure 的资源，除非使用在先决条件中引用的凭证对其进行身份验证，否则它不能这样做。可以使用 Connect-AzureRmAccount cmdlet 进行此操作。

① 通过单击 MyFirstRunbook-Workflow 窗格上的编辑，打开文本编辑器。

② 您不再需要写 Write-Output 行，所以继续删除它。

③ 将光标定位在括号之间的空白行上。

键入或复制并粘贴下面的代码，这些代码将以您的自动化运行为账户处理身份验证：

```
$Conn = Get-AutomationConnection -Name AzureRunAsConnection
Connect-AzureRmAccount -ServicePrincipal -Tenant $Conn.TenantID
-ApplicationId $Conn.ApplicationID -CertificateThumbprint $Conn.CertificateThumbprint
```

④ 单击测试窗格，以便测试 Runbook。

⑤ 单击 Start 按钮开始测试。一旦完成，您应该接收与下面类似的输出，从您的账户中显示基本信息。这证实凭证是有效的，如图 10-30 所示。

6．添加代码启动虚拟机

既然 Runbook 正在对您的 Azure 订阅进行身份验证，故可以管理资源。添加一个命令来启动虚拟机。可以在 Azure 订阅中选择任何虚拟机，现在要在 Runbook 中硬编码该名称。

① 在 Connect-AzureRmAccount 之后，键入 Start-AzureRmVM -Name 'VMName' -ResourceGroupName 'NameofResourceGroup'，提供要启动的虚拟机的名称和资源组名称。

```
workflow MyFirstRunbook-Workflow
{
$Conn = Get-AutomationConnection -Name AzureRunAsConnection
```

```
    Connect-AzureRmAccount -ServicePrincipal -Tenant $Conn.TenantID -ApplicationId
$Conn.ApplicationID -CertificateThumbprint $Conn.CertificateThumbprint
    Start-AzureRmVM -Name 'VMName' -ResourceGroupName 'ResourceGroupName'
}
```

② 保存 Runbook,然后单击测试窗格,以便测试它。

③ 单击 Start 按钮开始测试。一旦完成,检查虚拟机是否已启动。

7. 向 Runbook 添加输入参数

Runbook 当前启动其中硬编码的虚拟机,但是如果可以在 Runbook 启动时指定虚拟机,则更有用。向 Runbook 添加输入参数以提供该功能。

① 将 VMName 和 ResourceGroupName 的参数添加到 Runbook 中,并将这些变量与 Start-AzureRmVM Cmdlet 一起使用,如下面的示例所示:

```
workflow MyFirstRunbook-Workflow
{
Param(
[string]$VMName,
[string]$ResourceGroupName
)
    $Conn = Get-AutomationConnection -Name AzureRunAsConnection
    Connect-AzureRmAccount -ServicePrincipal -Tenant $Conn.TenantID -ApplicationId
$Conn.ApplicationID -CertificateThumbprint $Conn.CertificateThumbprint
    Start-AzureRmVM -Name $VMName -ResourceGroupName $ResourceGroupName
}
```

② 保存 Runbook 并打开测试窗格。现在可以为测试中的两个输入变量提供值。

③ 关闭测试窗格。

④ 单击发布按钮,发布 Runbook 的新版本。

⑤ 停止在上一步启动的虚拟机。

⑥ 单击 Start 按钮,启动 Runbook。输入要启动的虚拟机的 VMName 和 ResourceGroupName,如图 10-38 所示。

⑦ 当 Runbook 完成时,检查虚拟机是否已启动。

图 10-38 输入参数

第 11 章

Power BI

11.1 简 介

Power BI 是微软的一款集成的企业数据分析工具,可以通过 Excel、数据库、网页等数据导入或者连接,经过数据处理、绘制等操作,最后生成交互式报表。它支持多种数据源,集成了强大的 Power Query 进行数据分析,最后通过多种图表来显示数据分析的结果。这样就能监测业务数据,及时发现复杂业务背后的关键点,然后采取必要的行动。

通过 Power BI,你可以:

① 把多种数据源集合到一起,并转化成图、表格或者图表。
② 通过 Power Query 挖掘出数据模型。
③ 用生动的图表显示挖掘的结果、趋势和模式。

Power BI 由以下部分组成:

① 桌面应用 Power BI Desktop。
② 在线的 SAAS 服务 Power BI Service。
③ Power BI Apps,应用于 Windowsphones、Tables、iOS 以及 Android。

本章集中介绍 Power BI Desktop 和 Power BI Service。本章不介绍 Power BI Apps,想尝试的朋友可以在 Android 或者 iPhone 上安装试试。

11.2 Power BI Desktop

11.2.1 安装 Power BI Desktop

安装条件:

① Windows 8.1 或者 Windows 10。
② Power BI Desktop 安装包。

Power BI Desktop 安装包可以直接在官网上下载。从浏览器中打开官网:https://powerbi.microsoft.com/en-us/,然后打开 Products→Power BI Desktop,就

可以看到如图 11-1 所示的界面。

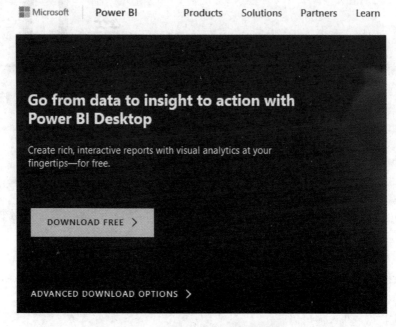

图 11-1 下载 Power BI 的界面

点击 Download Free，按照屏幕提示就可以直接下载安装包了。双击下载的安装包，按照屏幕提示安装即可。本书的 Power BI 是 June 2018 版本（2.59.5135.601），安装界面如图 11-2 所示。

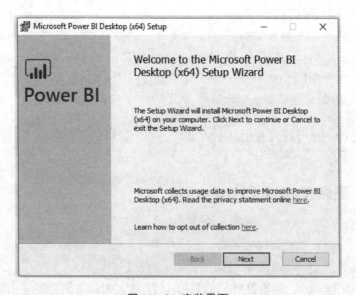

图 11-2 安装界面

11.2.2 Power BI Desktop 界面

在开始菜单中打开 Power BI,就能看到其主菜单,如图 11-3 所示。

图 11-3 Power BI 的主菜单

可以看到有如下的菜单页:
- File;
- Home;
- View;
- Modeling;
- Help。

在 Ribbon 下面左边的菜单,称为 Models Bar,可以切换不同的模式,如图 11-4 所示。

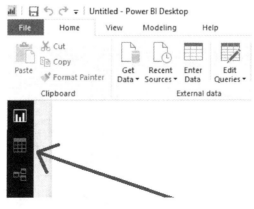

图 11-4 Models Bar

Models Bar 有三种模式:
- Report 模式;
- Data 模式;
- Relationship 模式。

(1) Report 模式

Report 模式是 Power BI 的关键功能,主要是用各种视图展示数据,使得数据可视化,能够一眼就看到问题的关键。

(2) Data 模式

Data 模式用来创建数据。当数据被导入到 Power BI 后,点击数据模式可以看到所导入的数据,如图 11-5 所示。在右边可以看到所有的数据表和每个表对应的列。

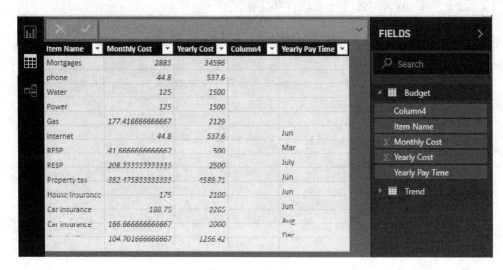

图 11-5 Data 模式

(3) Relationship 模式

Relationship 模式中可以看到所有的数据表格,可以将不同的表关联起来,与数据库的外键连接类似,如图 11-6 所示。

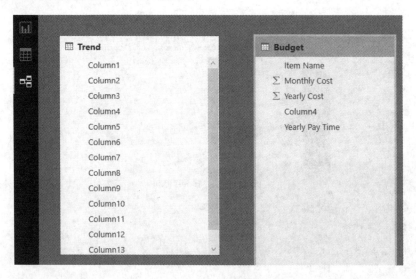

图 11-6 有关系的数据间建立连接

导入数据的方法如下:

Power BI 支持从多种数据源导入数据。在 Home 菜单上点击 Get Data 菜单,就可以选择不同的数据源,然后按照屏幕的提示进行数据导入即可。数据导入后可以在前面提到的 Data 模式中看到,如图 11-7 所示。

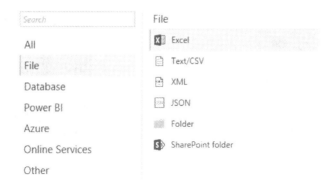

图 11-7 导入数据

如图 11-8 所示,当切换到 Data 模式时,就可以看到新导入的数据。

图 11-8 新导入的数据

有了这个数据之后,切换到 Report 模式,在 Visualizations 中选择饼图,然后从 Fields 中把相应的字段拖到相应的位置,Power BI 就可以画出饼图,如图 11-9 所示。

11.2.3 Query Editor

Query Editor 可以在导入数据前修改和预览数据。比如当导入数据后,可能会发现以下问题:

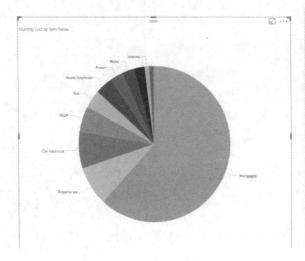

图 11-9　用导入数据画饼图

① 有些列没有列头；
② 有些行或者列是空的。

这时我们就需要由 Query Editor 来处理数据了。

在导入数据到 Power BI 后，就可以修改数据源的设置。在 Edit Queries 中选择下拉箭头，然后就可以看到 Data source settings 菜单，如图 11-10 所示。

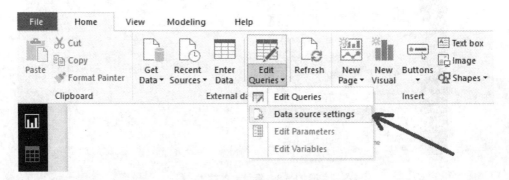

图 11-10　Data source settings 菜单

打开 Data source settings 后，就可以看到如图 11-11 所示窗口。

在此窗口上，可以修改 Data Source。如需修改权限或者清除权限，点击相应的按钮即可。

要修改 Query 内容，在 Home 的 Ribbon 中点击 Edit Queries 就可以打开 Query Editor，如图 11-12 所示。

打开 Query Editor 后，就会看到如图 11-13 所示的界面。

Power BI 11

[图 11-11 Data source settings 对话框]

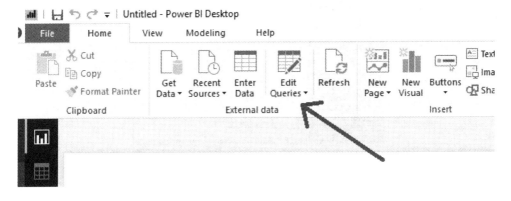

图 11-12 Edit Queries 菜单

图 11-13 Power Query Editor

左边列出了所有的 Queries，右边是对应的 Queries 的历史。在每个 Query 上点击右键，可以看到右键菜单，如图 11-14 所示。

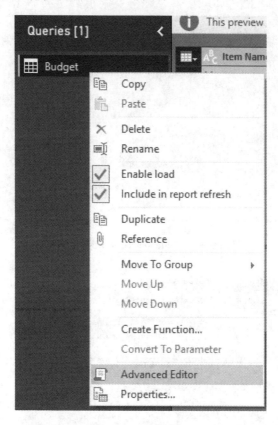

图 11-14　在 Query 上的 Advanced Editor

我们可以复制粘贴 Query。先在目标 Query 上单击 Copy，然后在空白处单击粘贴就可以了，如图 11-15 所示。

图 11-15　复制再粘贴一个 Query

Delete、Rename 和 Duplicate 菜单很明显,不需要过多解释。

如果点击 Advanced Editor,就可以手动直接编辑。Query Editor 有自己的语法,如图 11-16 所示。

图 11-16　高级编辑器查看查询语句源码

Power BI 的这种语法叫做 Power Query Formula Language,一般简称 Power Query M Language。在下面网址可以找到其官方网站:

https://msdn.microsoft.com/en-us/query-bi/m/power-query-m-reference。

最简单的 M Language 例子如下:

```
Let
    X = 1
In
    X
```

Let 和 In 是 M Language 的保留关键字。编辑完后的代码如图 11-17 所示。

Power BI M Language 是由 Let 和 In 组成的。"x=1"定义了一个变量名为"x"并将其值设置为 1。M Language 的变量名中可以有空格,但如果有空格,则要求变量名前面有♯,并要求将变量名放到引号内,比如 ♯"Variable Name"。

如果有多个语句,则每条语句要求用","分开。最后,M Language 的输出是在 In 语句里面。

M Language 的参考页面可以在下面的地址找到:

https://msdn.microsoft.com/en-us/query-bi/m/expressions-values-and-let-expression。

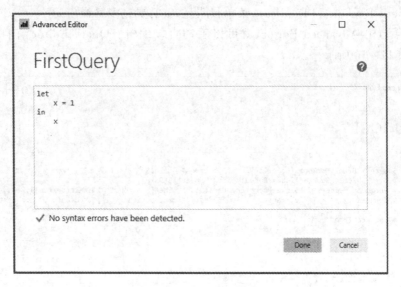

图 11-17 第一个查询的语句

M Language 支持的数据类型如表 11-1 所列。

表 11-1 内置类型

数据类型	例　子
二进制类型	00
日期型	6/28/2016
日期时间型	6/28/2016 10:40:50AM
日期时间加时区型	6/28/2016 10:40:50AM -08:00
时间段型	15:35:00
逻辑型	True 或者 False
Null	Null
数字型	0,1,1.5,2.3E5
文本型	"abcd"
时间类型	12:34:12PM

此外，M Language 中可以使用下列数据结构：

① List；

② Record；

③ Table；

④ 其他数据结构。

完成后单击 Done 按钮，在 Query 中就多了一个 Query，如图 11-18 所示。

最后,可以用 Card Chart 把结果输出来,如图 11-19 所示。

图 11-18 第一个查询新建好了

图 11-19 第一个查询显示的结果

如果在 Queries 中空白处点击右键,则会出现如下菜单,可以从这儿建立新的 Query,如图 11-20 所示。

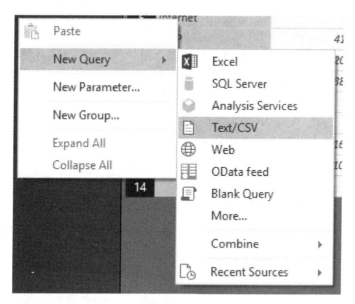

图 11-20 New Query 菜单

为了更好地规范化导入的数据,一些常见的做法如下:

① 将有意义的数据尽可能地细化。将一些列分解为有意义的多列,比如将姓名分解为姓和名两列。

② 将不必要的数据删除。将不必要的列和行全部删除以节约空间和处理的时间。

③ 清理错误数据。错误数据会导致错误的结论,所以要先清除错误数据。

④ 设置正确的数据类型,比如货币、日期、数据、文字等。

⑤ 表名列名要规范化,并尽量命名为有意义的名称。

⑥ 将数据标准化。比如将数据中的 $ 符号删除,这样数据就更容易处理。

以上介绍了 Power Query Editor 的右键菜单。现在我们来看看其主菜单。Power Query Editor 有以下几个 Tab:

① Home;

② Transform;

③ Add Column;

④ View;

⑤ Help。

Home Tab 如图 11-21 所示。

图 11-21 Home Tab

比如这里已有 Choose Columns,那么选定自己想要的列,然后单击 OK 按钮就可以了,如图 11-22 所示。

图 11-22 选择栏目 Choose Columns

比如这里选择几列后,再打开 Advance Editor,就可以看到相应的 Power BI M

Language,如图 11-23 所示。

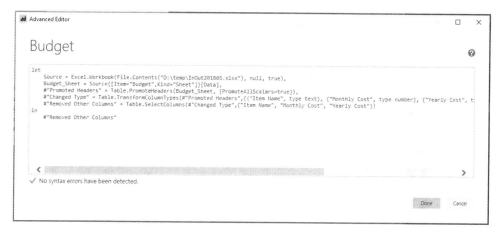

图 11-23 高级编辑器编辑其查询语句

相应的 Power BI M Language 还是比较简单的,如果有什么疑问,可以查找相应的介绍。

Transform Tab 如图 11-24 所示。

图 11-24 Transform Tab

这里有一些常用的功能,比如可以设置数据类型,将第一列设置为表头等。在这里也以用 Power BI 来调用 R 脚本。

Add Column Tab 如图 11-25 所示。

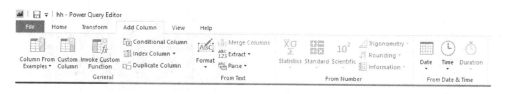

图 11-25 Add Column Tab

这里可以往 Query 中添加新的列,结果如图 11-26 所示。

在每一列的表头右击,可以看到如下菜单。可以根据需要选择相应的功能,如图 11-27 所示。

最后,在窗口的右边是 QUERY SETTINGS 窗口,这里我们可以对每一个步骤进行设置,如图 11-28 所示。

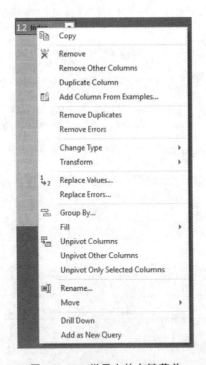

图 11-26 往 Query 中加入 Column

图 11-27 栏目上的右键菜单   图 11-28 Query settings

每一个步骤的前面有一个删除按钮,可以删除此步骤。如果某一步骤可以通过界面来修改,则该步骤的右边会有一修改按钮。

当然,也可以在前面介绍的 Advance Editor 中直接修改。所以,如果能够直接用 Power BI M Language 来编写导入数据,就肯定方便多了。

11.2.4 创建 Report 和数据可视化

前面演示了如何创建报告以使数据可视化,这里再细说一下,如图 11-29 所示。

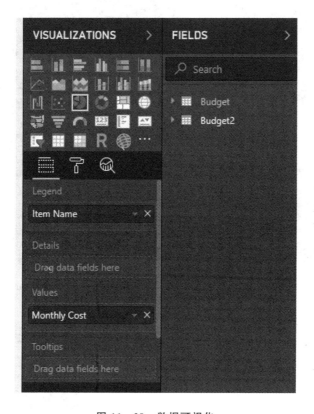

图 11-29 数据可视化

在 VISUALIZATIONS 中,先选择一个想要的图表,然后在右边的 FIELDS 中选择数据源,将对应的列拖到对应的位置即可。比如图 11-29,设置好图表的图例字段和数据字段,然后 Power BI 就自动画出图了。

在格式按钮中,可以设置图表的不同格式,比如图表的标题、边框的颜色等。这些设置可以让你的图表更加漂亮,如图 11-30 所示。

下面介绍几种常见的图表。

1. Stacked bar chart

先介绍 Stacked bar chart。单击 Stacked bar Chart 按钮,然后在 Report 中就多了一个空白的 Chart,如图 11-31 所示。

点击第一个 Stacked bar chart,将坐标和数值设置好后,就可以看到效果图了,如图 11-32 所示。

图 11-30　图表的格式

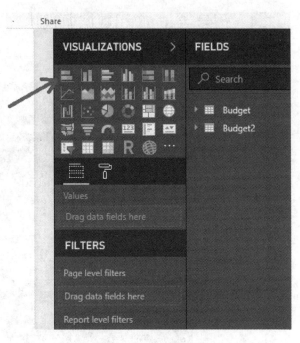

图 11-31　Stacked bar 图

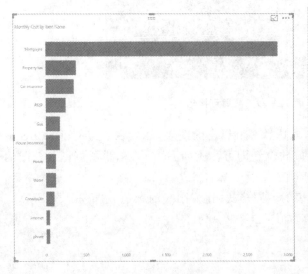

图 11-32　显示出的叠加起来的横条图

2. Stacked column chart

与 Stacked bar chart 类似,在其右边的那个就是 Stacked column chart。操作设置也类似,最后其结果如图 11-33 所示。

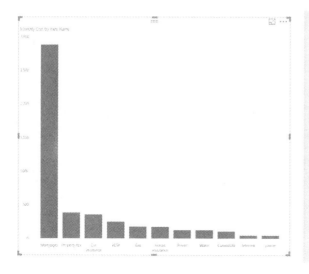 

图 11-33 柱状图

其实在完成 Stacked bar chart 后,只要单击 Stacked column chart,就可以直接将 Stacked bar chart 变成 Stacked column chart。

3. Line Chart

Line Chart 也是非常常见的图表。选择图标就可以插入 Line Chart,如图 11-34 所示。

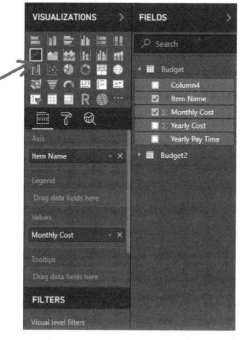

图 11-34 线图的图标

Line Chart 的效果如图 11-35 所示。

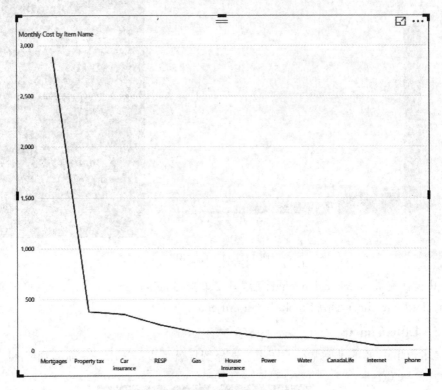

图 11-35 线 图

4. Waterfall chart

Waterfall chart 也是一种非常有意思的图表,可以清楚地看到数据是如何一步一步地叠加起来的。要使用 Waterfall chart,可以点击如图 11-36 所示的图标。

图 11-37 就是 Waterfall chart 的效果,可以看到数据是如何慢慢叠加的。

5. Donut chart

Donut chart 和 Pie chart 类似,其设置也类似。图 11-38 是 Donut chart 的效果。

6. Tree map

Tree map 是非常特殊的图表,使用图块大小来显示数据的关系。其操作方式与以上的图表类似。图 11-39 就是 Tree map 的效果图。

7. 新图表的导入

以上介绍了几种常见的图表。如果这些图表还不能满足工作需要,那么还可以通过导入图表定义到 Power BI,这样 Power BI 就可以使用新的图表了。单击最后的那个图标,就可以看到菜单,如图 11-40 所示。

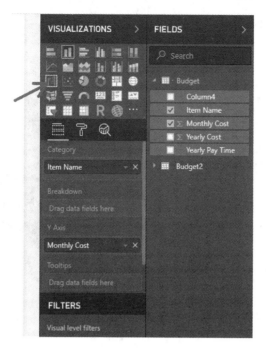

图 11-36 瀑布图图表

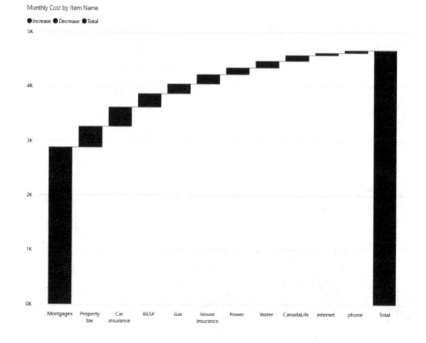

图 11-37 瀑布图实例

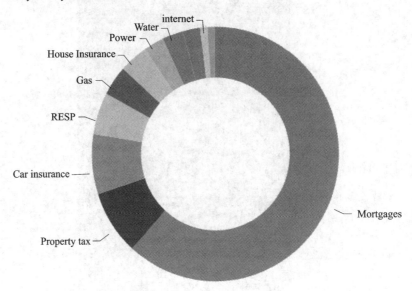

图 11-38　甜甜圈图

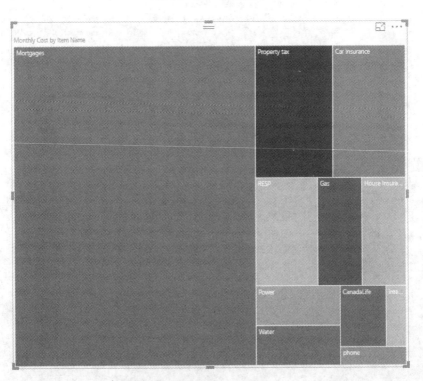

图 11-39　Tree map

图 11-40 Import from marketplace 菜单

比如可以选择从 Market Place 中导入新的图表。从 Market Place 导入后，就可以看到如图 11-41 所示的窗口。

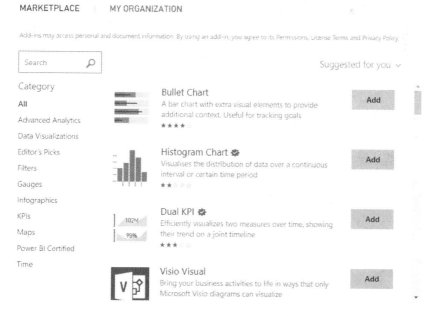

图 11-41 从 Market Place 导入一些图表

微软 Azure 实战参考

选择合适的图表,单击 Add,则可以将图表加入到 Power BI 中。如图 11-42 所示,新的图表就已经加入到 Power BI 中,然后就可以像别的图表那样使用了。

如果要删除一个图表,右击图表菜单后选择删除图表即可,如图 11-43 所示。

图 11-42 新导入的图表可以使用了

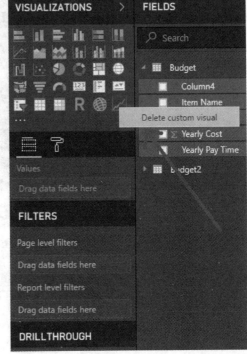

图 11-43 删除自定义图表

11.2.5 发布报表到 Power BI Service

当我们创建好 Power BI 报表后,就可以将报表发布到 Power BI Service。这样,你的同事或者客户就可以在网上直接看到你的报表。

发布报表很简单,在 Home 的 Ribbon 中点击 Publish 按钮,然后根据提示,登录到 Power BI Service,就可以轻松地发布到网上了,如图 11-44 所示。

图 11-44 发布图表到 Power BI Service

单击 Publish 按钮后,就会出现登录框,根据屏幕提示登录到 Power BI Service,
如图 11-45 所示。

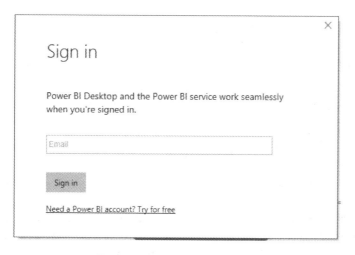

图 11-45　登录 Power BI Service

登录后选择 Workspace,就可以上传了。图 11-46 是发布成功的截屏。

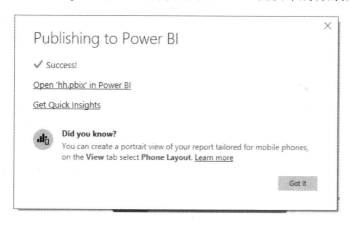

图 11-46　发布成功

登录到 Power BI Service,就可以在刚才选择的 Workspace 中找到上传的图表,
如图 11-47 所示。

11.2.6　Power BI M Language

现在用上面的那段例子来看看 M Language:

```
let
    Source = Excel.Workbook(File.Contents("D:\temp\InOut201805.xlsx"), null, true),
    Budget_Sheet = Source{[Item = "Budget", Kind = "Sheet"]}[Data],
```

图 11-47　查看 Power BI Service 中发布的图表

```
#"Promoted Headers" = Table.PromoteHeaders(Budget_Sheet, [PromoteAllScalars =
    true]),
#"Changed Type" = Table.TransformColumnTypes(#"Promoted Headers",{{"Item
    Name", type text}, {"Monthly Cost", type number}, {"Yearly Cost", type num-
    ber}, {"Column4", type any}, {"Yearly Pay Time", type text}})
in
#"Changed Type"
```

这里 Excel.Workbook 是 Power BI 的函数,将 Excel Workbook 导入 Source,取 Budget 的值给 Budget_Sheet,接着调用 Table.PromoteHeaders 函数将第一行设置为列名,然后调用 Table.TransformColumnTypes 函数改变列的类型。最后在 In 的语句中将结果输出。

可以看出 Power BI 的 M Language 其实就是不停地调用各种函数,将输入的数据进行各种处理,最后将结果输出。

我们可以在下面的链接中找到全部函数的列表:

https://msdn.microsoft.com/en-us/query-bi/m/power-query-m-function-reference。

下面介绍常用的函数。将这些常用函数分类如下。

1. 数据访问函数(见表 11-2)

这些函数用来访问数据并返回数据表值,大多数函数返回值称为 Navigation Table。Navigation Table 有两列,第一列是项目名,第二列是其对应的值。这个结构主要是被 Power Query 界面用来导航大规模的数据。

表 11-2　数据访问函数

函数名	描述
Access.Database	返回一个微软 Access 数据库的结构
ActiveDirectory.Domains	返回在当前域或可选的 Active Directory 可用的域信息表

续表 11-2

函数名	描述
AdoDotNet.DataSource	一个 ADO.NET 数据源返回的架构集合
AdoDotNet.Query	返回运行在 ADO.NET 数据源的本地查询结果
DB2.Database	返回有关指定 DB2 数据库中的表的数据
Excel.CurrentWorkbook	返回当前 Excel 工作簿中的表
Excel.Workbook	返回一个表,表示在给定的 Excel 工作簿中
File.Contents	返回该文件位于路径的二进制内容
Folder.Contents	返回一个表,内容包含属性的文件和文件夹,找到路径
Folder.Files	返回一个表,包含一个行的每个文件的文件夹路径和子文件夹,每行包含的文件夹或文件属性和链接的内容
Json.Document	返回一个 JSON 文档
Json.FromValue	产生一个给定的值与指定的文本编码,编码一个 JSON 表示
MySQL.Database	返回有关在指定的 MySQL 数据库中的表的数据表
Odbc.DataSource	返回从 ODBC 数据源的连接字符串指定一个表的 SQL 表和视图 ConnectionString
Odbc.Query	返回 ODBC 数据源的连接字符串指定一个表的 SQL 表和视图
OleDb.DataSource	返回 OleDb 数据源的连接字符串指定一个表的 SQL 表和视图 ConnectionString
Sql.Database	返回有关在指定的 SQL Server 数据库中的表的数据表
Web.Contents	返回的内容下载的 Web URL 作为一个二进制值
Web.Page	作为一个表返回一个 HTML 网页的内容
Xml.Document	作为一个分层表返回一个 XML 文档的内容(记录)

2. 日期函数(见表 11-3)

表 11-3 日期访问函数

函数名	描述
Date.AddDays	将日期增加指定的天数,并返回一个日期或者时间
Date.AddMonths	将日期增加指定的月数,并返回一个日期或者时间
Date.AddQuarters	将日期增加指定的季节数,并返回一个日期或者时间
Date.AddWeeks	将日期增加指定的星期数,并返回一个日期或者时间
Date.AddYears	将日期增加指定的年数,并返回一个日期或者时间
Date.Day	返回日期中的天数部分

续表 11-3

函数名	描述
Date.DayOfWeek	返回一个 0 和 6，代表从一个 DateTime 值返回这一周之间的数
Date.DayOfWeekName	返回星期的名字
Date.DayOfYear	返回日期中的年部分
Date.DaysInMonth	返回日期中的月部分
Date.EndOfDay	返回当天结束的时间
Date.EndOfMonth	返回当月结束的时间
Date.EndOfQuarter	返回当季结束的时间
Date.EndOfWeek	返回当星期结束的时间
Date.EndOfYear	返回当年结束的时间
Date.From	从一个数值中返回一个日期
Date.FromText	从一个文本中返回一个日期
Date.IsInCurrentDay	判断日期是不是在当天中
Date.IsLeapYear	判断是不是闰年

3. 时间函数(见表 11-4)

表 11-4 时间函数

函数名	描述
DateTime.From	将数值转化为时间
DateTime.FromText	将文本转化为时间
DateTime.LocalNow	返回当前当地时间
DateTime.ToText	将时间转化为文本

4. 数字函数(见表 11-5)

表 11-5 数字函数

函数名	描述
Number.IsEven	判断一个数是不是偶数
Number.IsOdd	判断一个数是不是奇数
Number.IsNaN	判断一个数是不是 NaN
Int32.From	将一个值转为 Int32
Number.Abs	返回一个数的绝对值
Number.Ln	返回一个数的自然底数的对数值

续表 11-5

函数名	描述
Number.Log10	返回一个以 10 为底数的对数值
Number.Random	返回一个 0 到 1 之间的随机值
Number.E	返回 2.718 281 828 459 045 1，即数学上的 E
Number.NaN	返回 NaN
Number.PI	返回 3.141 592 653 589 793 1，即数学上的 π

5. 分割函数(见表 11-6)

表 11-6 分割函数

函数名	描述
Splitter.SplitTextByDelimiter	根据分割符分割字符串
Splitter.SplitTextByLengths	根据长度分割字符串
Splitter.SplitTextByWhitespace	根据空白字符分割字符串
Splitter.SplitTextByEachDelimiter	根据每个分割符分割字符串
Splitter.SplitTextByPositions	根据位置分割字符串

6. 表函数(见表 11-7)

表 11-7 表函数

函数名	描述
Table.FromList	将一个列表转换成一个表
Table.FromRecords	将一个记录集合转换成一个表
Table.FromRows	将多行转换成一个表
Table.View	返回一个表的视图
Table.ToList	将一个表转换成一个列表
Table.ColumnCount	返回列的总数
Table.First	返回表的第一行
Table.Last	返回表的最后一行
Table.SelectRows	根据条件选择行
Table.ColumnNames	返回列的名字集合
Table.PromoteHeaders	将表的第一行提升为列名
Table.AddColumn	将一列添加到表中
Table.FillDown	对表进行向下填充

续表 11-7

函数名	描述
Table.Contains	检测表中是否包含某一行
Table.Distinct	将重复行移除
Table.Sort	根据条件对表进行排序

7. 文本函数(见表 11-8)

表 11-8 文本函数

函数名	描述
Text.Length	返回文本的长度
Guid.From	将给定的值转换为 Guid
Text.FromBinary	将二进制值编码为文本
Text.ToBinary	将文本编码为二进制
Text.At	返回某个位置的字符
Text.Insert	返回新插入的文本
Text.Replace	替换指定的子串并返回新字符串
Text.Contains	判断字符串中是否包含特定的子串
Text.Lower	将字符串转换为小写
Text.Reverse	将字符串反转
Text.Trim	将空白字符串移除后返回

8. URI 函数(见表 11-9)

表 11-9 URI 函数

函数名	描述
Uri.BuildQueryString	根据输入生成一个 URI 连接串
Uri.Combine	根据相对路径的部分生成一个 URI
Uri.EscapeDataString	对 URI 中的特殊字符进行编码
Uri.Parts	将 URI 部分分成不同部分并返回

9. 记录函数(见表 11-10)

表 11-10 记录函数

函数名	描述
Record.FieldCount	返回记录的字段数

续表 11-10

函数名	描述
Record.AddField	将新的字段名和值返回到记录中
Record.Combine	将列表中的值合并到记录中
Record.RemoveFields	将记录中的字段删除后返回
Record.SelectFields	将记录中的字段根据条件返回,原来的字段值不变
Record.FromList	根据列表值生成新的记录
Record.ToList	将记录转化为一个列表
Record.FieldNames	根据记录字段的顺序返回字段名称列表
Record.HasFields	判断一个字段或者多个字段是否存在记录中
Record.RenameFields	将记录的字段重新命名后返回

11.3 Power BI Service

可以从 Power BI Desktop 中直接发布到 Power BI Service。登录到 https://powerbi.microsoft.com,点击 Sign in 链接就可以登录到 Power BI Service,如图 11-48 所示。

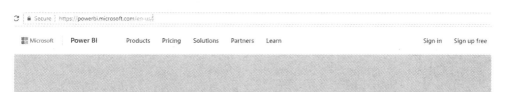

图 11-48 Power BI Service 门户页面

登录后就可以看到如图 11-49 所示的界面。

打开 My Workspace 菜单,注意到其下面有四个子菜单:

① DASHBOARDS。Dashboard 是一个专门用来显示图表的集合。可以将多个 Report 组合起来显示。也可以将 Excel Workbook 中的某部分数据特别显示出来。

② REPORTS。Power BI 的 Report 就是一个或者多个数据的可视化图表。

③ WORKBOOKS。Power BI Service 支持 Excel Workbook。可以将 Excel Workbook 直接上传到 Power BI Service。根据需要可以将 Excel Workbook 的数据直接显示到 Dashboard。

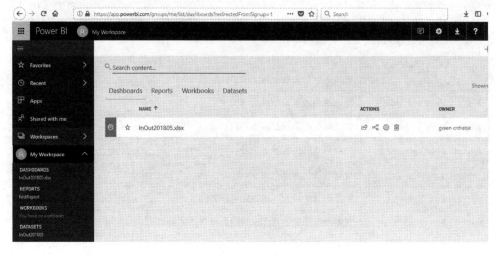

图 11-49 登录后的面板

④ DATASETS。Dataset 是一系列的导入或者连接的数据集合。Power BI 可以把不同的数据集合到一起。

11.3.1 导入 Dataset

打开 My Workspace 后,选择 Create→Dataset,就可以导入 Dataset,如图 11-50 所示。

图 11-50 导入 Dataset

点击 Dataset 后,就可以看到如图 11-51 所示的界面。可以选择从本公司的数据源导入,也可以从文件或者数据库创建新的数据源。

比如选择从文件中重新创建 Dataset,则从 Files 中单击 Get 按钮,就可以看到如图 11-52 所示的界面。

可以选择从本地导入文件,或者从云服务中导入数据。注意,如果要导入 Excel 文件,则 Excel 的数据必须在 Table 中。

在 Excel 中选中目标数据,然后按 Ctrl+T 键就可以插入一个 Table;也可以点击 Insert→Table 来插入一个 Table。插入 Table 后的效果如图 11-53 所示。

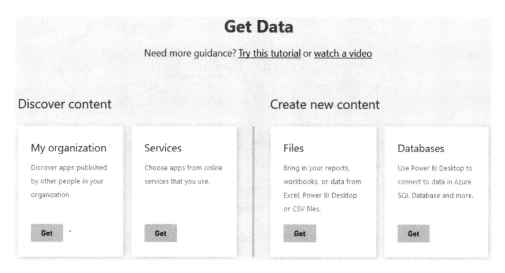

图 11-51　导入新数据源

图 11-52　从文件中导入数据

11.3.2　创建 Report

在上面提到的 Create 菜单中选择 Report，就可以看到如图 11-54 所示的界面。

在此界面中选择已经导入的 Dataset，然后单击 Create 按钮，就回到了熟悉的创建 Report 界面。这个界面跟 Power BI Desktop 基本类似，操作类似，如图 11-55 所示。

图 11-53　插入 Table 后的效果

图 11-54　创建报表

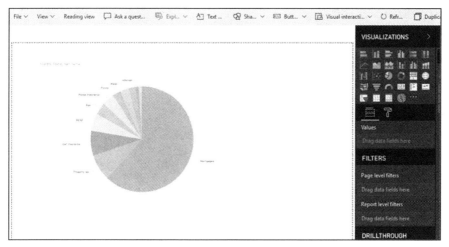

图 11 – 55　插入图表

11.3.3　创建 Dashboard

类似地，在 Create 菜单中选择 Dashboard，就会出现如图 11 – 56 所示的界面。

图 11 – 56　Greate dashboard 面板

输入 Dashboard 的名字，单击 Create 按钮就可以创建一个新的 Dashboard，如图 11 – 57 所示。

图 11 – 57　新建的面板

如何将 Report 放到 Dashboard 中呢？打开你的目标 Report，选择右上角的 Pin 按钮，如图 11-58 所示。

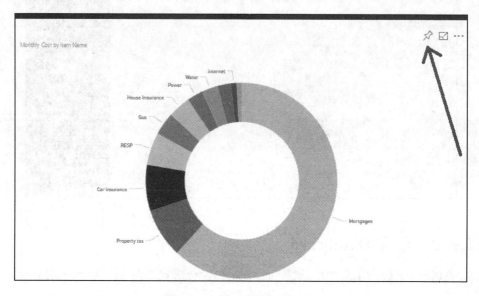

图 11-58　将报表钉到面板上

然后就可以看到如图 11-59 所示的界面，可以选择已经存在的 Dashboard 或者新创建一个 Dashboard。

图 11-59　Pin to dashboard 对话框

这里选择已经存在的 Dashboard，点击 Pin 按钮后，就可以把 Report 放到 Dashboard 中。这时，如果打开 Dashboard，就可以看到 Dashboard 中已经包含 Report 了，如图 11-60 所示。

图 11-60　Dashboard 中包含了 Report

如此一来,我们就完成了导入 Dataset、生成 Report 并导入到 Dashboard 中的全部过程。

11.3.4　创建 Workbook

Power BI Service 可以直接导入 Excel Workbook,导入的 Workbook 就会存放到 Power BI Service,然后可以插入到 Dashboard 中直接引用。

导入 Workbook 的步骤与导入 Dataset 类似,选择 Create 按钮→选中 Dataset→选择文件→本地文件→点击上传 Excel 文件到达 Power BI。如图 11-61 所示,选择 Upload 按钮即可。

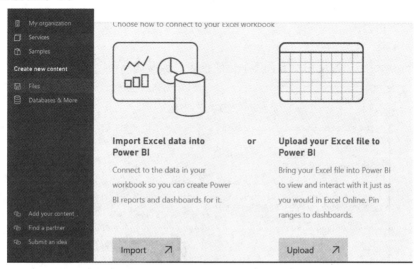

图 11-61　导入 Excel 以创建 Workbook

然后就可以在 Workbook 菜单中看到刚刚上传的文件,如图 11-62 所示。

图 11-62　查看创建的 Workbook

11.3.5　Data Gateway

把 Dataset 导入到 Power BI Service 后,可以用 Data Gateway 来更新 dataset 里的数据,这样就可同步更新 Report 和 Dashboard。

在 Power BI Service 的 Download 菜单里选择 Data Gateway,先下载 Data Gateway 的安装包,如图 11-63 所示。

图 11-63　选择 Data Gateway

然后双击打开下载的安装包,就可以看到如图 11-64 所示的界面,根据屏幕提示安装完 Data Gateway。

Data Gateway 的使用比较简单,按照屏幕提示操作即可。

图 11-64　Data Gateway 安装

11.4 Azure 中的 Power BI Embedded

Power BI Embedded 旨在简化 ISV 并提高开发人员使用功率 BI 的能力。Power BI Embedded 通过帮助您快速向应用程序添加令人惊叹的视图、报告和仪表板来简化 Power BI 功能。类似于微软 Azure 公司使用的应用程序，如机器学习和物联网。通过在他们的应用程序中启用易于导航的数据探索，ISV 允许他们的客户在上下文中快速做出明智的决定。

在 2017 年 5 月，微软宣布了 Power BI 和 Power BI Embedded 服务的融合。这种融合提供了一个 API 表面、一组一致的功能和跨两个服务访问的最新特性。此外，我们引入了基于容量的定价模型，简化了 BI 的功耗。

随着 Power BI Embedded 的出现，ISV 和开发人员在使用 Power BI API 将智能嵌入应用程序中就增加了灵活性。ISV 和开发人员可以利用最小化的开发工作，通过将微软世界级的分析引擎注入自己的应用程序中，来获得更快的市场营销时间。同样，开发人员也可以花时间关注自己的解决方案以满足客户的需求，而不是开发可视化分析特性。此外，Power BI Embedded 可以在已经使用的熟悉的开发环境 Visual Studio 和 Azure 中工作。

你是否有一个现有的应用程序与嵌入式 Power BI Content 使用 Power BI Premium？如果你是一个 ISV 或开发者交付应用程序或组织使用它们，不需要任何行动。你和你的客户可以不间断地继续使用这些应用程序。如果你有一个建立在 Power BI Workspace Collections 上的现有应用程序，并且有兴趣利用聚合的 API 表面和新的基于容量的 Azure SKU，请访问迁移指南的文档。

(1) Power BI 与 Power BI Premium 的比较

BI Embedded 软件是针对独立软件供应商(ISV)和为客户构建应用程序的开发人员的。它可以用作第三方商业智能服务，可以可视化应用程序数据，而不是自己构建该服务。Power BI Embedded 是一个平台即服务（Platform-as-a-Service，PAAS）分析解决方案，其中开发人员可以将报表和仪表板嵌入到客户的应用程序中。Power BI Premium 是一个软件即服务（Software-as-a-Service，SAAS）分析解决方案，它提供了最关键的业务数据的单个视图。

Power BI Embedded 是用多少功能付多少费，而 Power BI Premium 需要按月付费。

(2) 易于使用的工具

Power BI Embedded 可以让你专注于你做得最好的事情：建立实用的应用程序。可以使用你已经拥有的工具和技能来管理和开发 Power BI Embedded。

- Azure Portal：一个管理所有 Azure 服务的基于 Web 的应用程序。
- Visual Studio 代码：支持扩展的 Windows、MACOS 和 Linux 的免费的、可下

载的、开源的代码编辑器。
- Power BI 桌面：一个免费的、可下载的工具，用可视化分析创建丰富、交互式的报表使用 REST API。Power BI Embedded 允许使用任何语言开发。

11.4.1 在 Azure Portal 中创建 Power BI Embedded 容量

创建容量的步骤如下：
① 登录 Azure 门户网站。
② 选择创建一个资源→数据+分析。
③ 在搜索框中，搜索 Power BI Embedded。
④ 在 Power BI Embedded 中，选择创建。
⑤ 填写所需信息，然后选择创建，如图 11-65 所示。

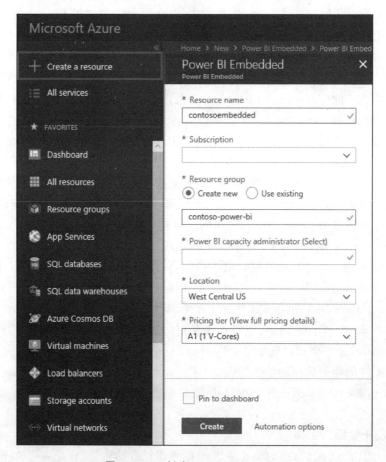

图 11-65 创建 Power BI Embeded

所需信息如表 11-11 所列。
⑥ 选择创建。

表 11-11 所需信息

设 置	描 述
Resource name（资源名）	识别容量的名称。除了 Azure 门户之外,在资源 BI 管理门户中显示资源名称
Subscription（订阅）	使用的订阅名,用来创建任何资源
Resource group（资源组）	用来管理零散资源,将它们放到一个组里管理
Power BI capacity administrator（Power BI 容量管理员）	此 Power BI capacity 的管理员,一般是在 Azure AD 里创建的账号。Power BI 容量管理员可以在 Power BI 管理门户中查看容量并向其他用户授予分配权限。默认情况下,容量管理员是你的账户。容量管理员必须在你的 Power BI 租户中
Location	选择数据中心的位置
Pricing tier	选择不同 tier,价格不同,应根据需要来选择。在价格允许的情况下,选择满足您需要的 SKU（V 核心计数和内存大小）。详情请参阅 Power BI Embedded 嵌入式定价页面

创作通常需要一分钟,有时只有几秒钟。如果选择 PIN 到仪表板,则可以导航到仪表板上,以查看新的容量。或者,可以导航到所有服务→Power BI Embedded,以查看容量是否已准备就绪,如图 11-66 所示。

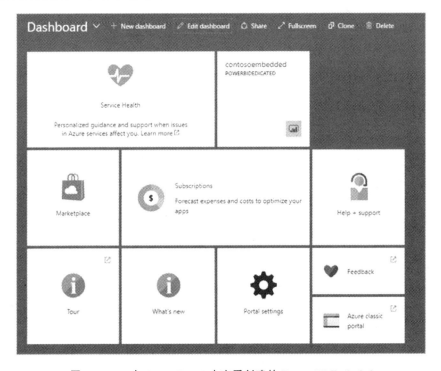

图 11-66 在 Azure Portal 中查看创建的 Power BI Embedded

11.4.2 将一个 Power BI 报告、仪表板或瓦片嵌入到客户的应用程序中

使用 Azure 中 Power BI Embedded，可以将报表、仪表板或瓦片嵌入到应用程序中。应用程序拥有的数据是关于使用 Power BI 作为其嵌入式分析平台的应用程序。使用应用程序拥有的数据通常是 ISV 开发人员的工作。作为 ISV 开发人员，可以创建 Power BI 内容，在完全集成和交互的应用程序中显示报告、仪表板或平铺，而不需要应用程序的用户具有 Power BI 许可证。本教程演示了如何使用 Power BI.NET SDK 以及 Power BI JavaScript API 将报表集成到应用程序中，这时使用应用程序拥有的数据来为客户使用 Power BI Embedded in Azure 服务。

1. 先决条件

开始，你需要一个 Power BI-Pro 账户（这个账户是你的主账户）和一个微软 Azure 订阅。

如果你没有注册 Power Bi-Pro，请在开始之前注册一个免费试用版。

如果你没有 Azure 订阅，请在开始之前创建一个免费账户。

你需要拥有自己的 Azure Active Directory 租户设置。

你需要安装 Visual Studio（版本 2013 或更高版本）。

2. 建立嵌入式 Embedded 分析开发环境

在开始将报告、仪表板或瓦片嵌入到应用程序中之前，需要确保环境允许嵌入 Power BI。

可以通过 Embedding 安装工具快速启动并下载示例应用程序，该应用程序帮助完成创建环境和嵌入报告的过程。

如果选择手动设置环境，则可以继续进行以下操作。

① 用 Azure Active Directory 注册应用程序，让你的应用程序访问电源 BIEST API。重新注册应用程序时允许为应用程序建立身份并指定 Power BI REST 资源的权限。

② 接受微软 Power BI API 条款。

③ 登录 Azure 门户网站，如图 11-67 所示。

④ 在左侧导航窗格中，选择所有服务，选择应用程序注册，然后选择新的应用程序注册，如图 11-68 所示。

⑤ 新建一个 App registrations，如图 11-69 所示。

⑥ 按照提示创建一个新的应用程序。对于应用程序拥有的数据，需要使用本地的应用程序类型，还需要提供一个重定向 URI，Azure AD 使用该 URI 返回令牌响应。输入一个特定于应用程序的值（例如：http://localhost:13526/Redirect），如图 11-70 所示。

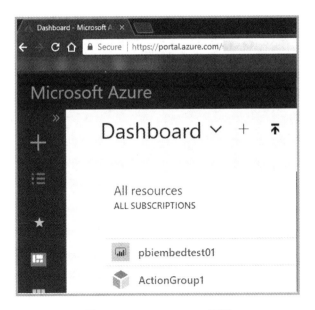

图 11-67　Azure Portal 面板

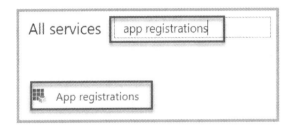

图 11-68　选择 App registrations

图 11-69　单击 New application registration 按钮

图 11-70 为 Power BI Embeded 创建 App registrations

3. 在 Azure Active Directory 中对应用程序设置应用权限

除了应用程序注册页面上提供的应用程序之外，还可以为应用程序启用附加权限。用您正在嵌入的主账户登录。主账户需是一个全局管理账户。

4. 使用 Azure Active Directory 门户

① 浏览 Azure 门户网站内的应用程序注册，并选择用于嵌入的应用程序，如图 11-71 所示。

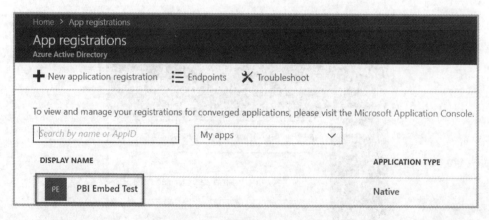

图 11-71 建好的 App registrations

② 选择设置，然后在 API 访问下选择所需的权限，如图 11-72 所示。

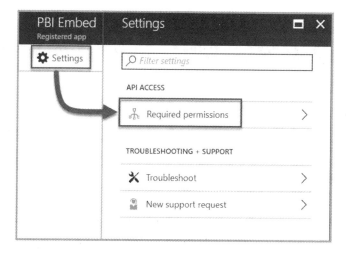

图 11-72　进入设置页面

③ 选择 Windows Azure Active Directory，然后确保选择了已签入的用户访问目录。选择保存，如图 11-73 所示。

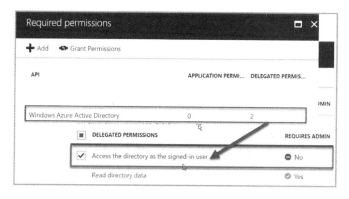

图 11-73　勾选 Access the directory as the signed-in user

④ 选择"＋Add"，如图 11-74 所示。
⑤ 选择一个 API，如图 11-75 所示。

图 11-74　选择 Add 按钮

图 11-75　选择一个 API

⑥ 选择 Power BI Service，然后选择 Select，如图 11-76 所示。

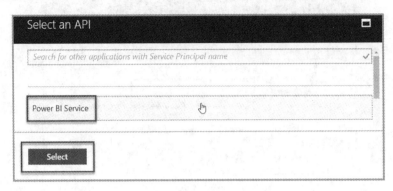

图 11-76　选择 Power BI Service

⑦ 选择委派权限下的所有权限。完成后选择保存，如图 11-77 所示。

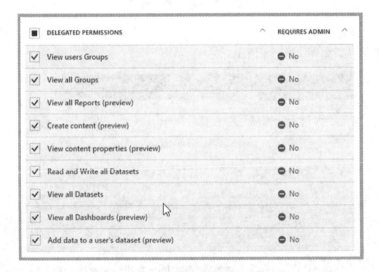

图 11-77　授予权限

⑧ 在所需权限中，选择授予权限。授予权限操作需要主账户以避免被 Azure AD 提示获得同意。如果执行此操作的账户是全局管理员，则需要向组织内的所有用户授予此应用程序的权限。如果执行此操作的账户是主账户，并且不是全局管理员，则只需向此应用程序的主账户授予权限，如图 11-78 所示。

5. 创建应用程序工作区

如果为客户嵌入报告、仪表板或平铺，则必须将内容放在应用程序工作区中。主账户必须是应用程序工作区的管理员。

① 创建工作区。选择 WorkStudio→创建应用程序工作区。在创建应用程序工作区中，放置应用程序需要访问的内容，如图 11-79 所示。

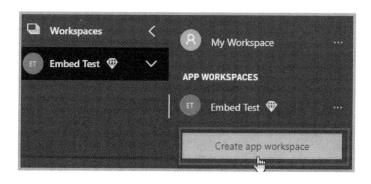

图 11-78 授权之后的列表

图 11-79 Create app workspace 菜单

② 给工作区一个名称。如果相应的工作区 ID 不可用，则编辑它以生成唯一的 ID，如图 11-80 所示。

图 11-80 Create an app workspace 界面

③ 有几个选项可以设置。如果选择公有，则组织中的任何人都可以看到工作区中的内容；反之，私有意味着只有工作区成员才能看到它的内容，如图 11-81 所示。

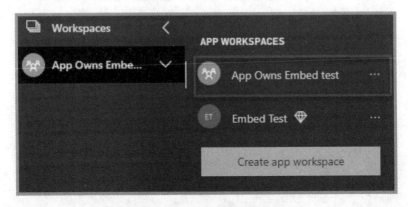

图 11-81 选择私有或者公有

在创建群组之后,不能更改公有/私有设置。

还可以选择成员是否可以编辑或只查看视图。

④ 添加要访问工作区的人的电子邮件地址,然后选择添加。不能添加组别名,只能是个人。

⑤ 决定每个人是会员还是管理员。管理员可以编辑工作区本身,包括添加其他成员。成员可以编辑工作区中的内容,除非他们只做视图访问。管理员和成员都可以发布应用程序。

⑥ 现在您可以查看新的工作区。电源 BI 创建工作空间并打开它。它出现在作为成员的工作区列表中。因为你是管理员,所以可以选择省略号(…)返回并对其进行更改、添加新成员或更改其权限,如图 11-82 所示。

图 11-82 查看新建的 Workspaces

6. 创建和发布报告

可以使用 Power BI 桌面创建报表和数据集,然后将这些报表发布到应用程序工作区。发布报表的最终用户需要有一个 Power BI-Pro 许可证才能发布到 App WorkStudio。

① 从 Github 下载示例博客演示,如图 11-83 所示。

② 在 BI 桌面上打开示例 PBIX 报告,如图 11-84 所示。

③ 发布到应用程序工作区,如图 11-85 所示。

图 11-83　从 Github 下载样例代码

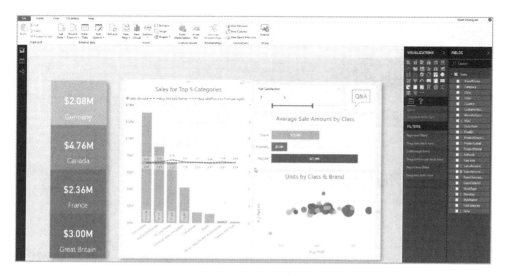

图 11-84　打开示例 PBIX 报告

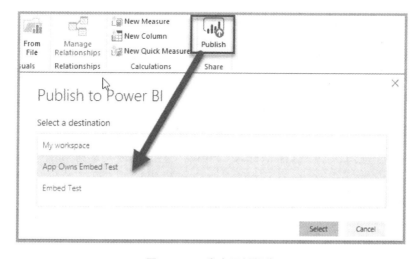

图 11-85　发布示例报告

④ 现在可以在网上的 Power BI Service 中查看报告，如图 11-86 所示。

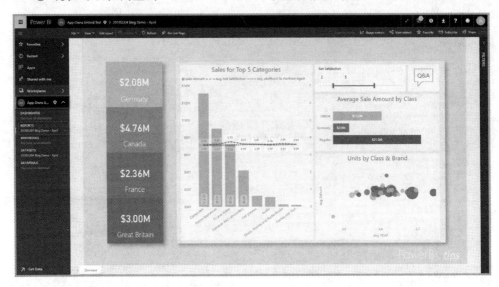

图 11-86　在 Power BI Service 中查看报告

7. 使用示例应用程序嵌入内容

按照以下步骤开始使用示例应用程序嵌入内容：

① 下载应用程序拥有 GITHUB 的数据样本以启动，如图 11-87 所示。

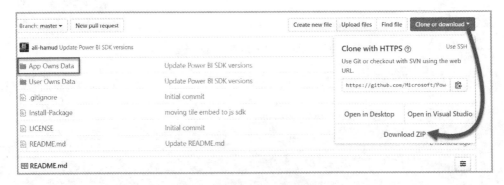

图 11-87　下载数据样本

② 在样例应用程序中打开 Web.CONFIG 文件。要成功运行应用程序，需要填写五个字段：applicationId、workspaceId、reportId、pbiUsername 和 pbiPassword，如图 11-88 所示。

③ 用 Azure 中的应用程序 ID 填写应用程序 ID 信息。应用程序 ID 用于应用程序向请求权限的用户标识自己。要获得应用程序 ID，请遵循以下步骤：

　　a. 登录 Azure 门户网站，如图 11-89 所示。

　　b. 在左侧导航窗格中，选择所有服务，并选择应用程序注册，如图 11-90 所示。

图 11-88 填写五个必需参数

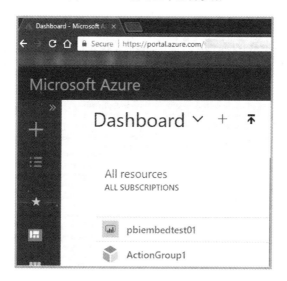

图 11-89 Azure Dashboard 界面

图 11-90 选择 App registrations

c. 选择要获取 Application ID 的应用程序,如图 11-91 所示。

图 11-91 查找之前注册的 Application

d. 应该可以看到列为 GUID 的应用程序 ID。使用此应用程序 ID 作为应用程序的 ID,如图 11-92 所示。

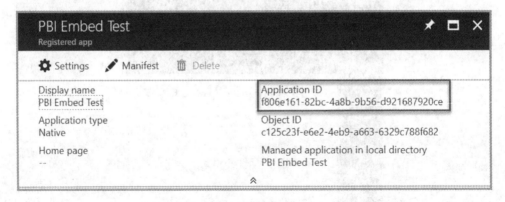

图 11-92 得到 Application ID,填写到 web.config 文件中

e. 用 Power BI 的 App Workspace GUID 填充 Workspace ID 信息,如图 11-93 所示。

图 11-93 得到 Workspace ID

f. 用 Power BI 的报告 GUID 填写报告信息,如图 11-94 所示。

图 11-94 获得报告的 GUID

g. 用 Power BI 主用户账户填写你的账号。用 Power BI 主用户账户的密码填写你的密码。

8. 运行应用程序

① 在 Visual Studio 中选择运行,如图 11-95 所示。

图 11-95　运行程序

② 选择嵌入报表。根据你选择测试的内容→报表、仪表板或瓦片,在应用程序中选择该选项,如图 11-96 所示。

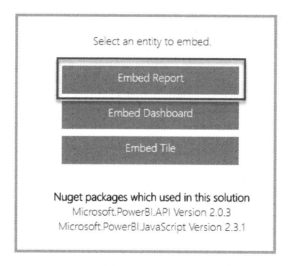

图 11-96　选择嵌入式报表 Embed Report

③ 在示例应用程序中查看报表,如图 11-97 所示。

9. 在应用程序中嵌入内容

即使嵌入内容的步骤可以使用 Power BI REST API 完成,本文中描述的示例代码也是用 .NET SDK 编写的。

在应用程序中嵌入客户,要求你从 Azure AD 获得主账户的访问令牌。在调用 Power BI REST API 之前,需要使用应用程序拥有的数据来获得 Power BI 应用程序的 Azure AD 访问令牌。

要使用访问令牌创建 Power BI 客户端,需要创建 Power BI 客户端对象,该对象允许你与 Power BI REST API 交互。这是通过用一个 Microsoft.Rest.TokenCredentials 对象包装 Access Token 来完成的。

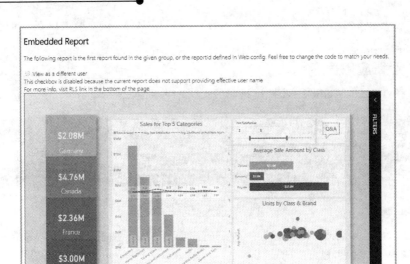

图11-97 在程序中查看报表

```
using Microsoft.IdentityModel.Clients.ActiveDirectory;
using Microsoft.Rest;
using Microsoft.PowerBI.Api.V2;

var tokenCredentials = new TokenCredentials(authenticationResult.AccessToken, "Bearer");

// Create a Power BI Client object. It is used to call Power BI APIs.
using (var client = new PowerBIClient(new Uri(ApiUrl), tokenCredentials))
{
    // Your code to embed items.
}
```

10. 获取要嵌入的内容项

可以使用Power BI客户端对象检索对要嵌入的项的引用。

下面是如何从给定工作区检索第一个报表的代码示例。

在示例应用程序中的Controllers\HomeController.cs文件中,可以获得内容项的示例,无论它是要嵌入的报告、仪表板还是瓦片。

```
using Microsoft.PowerBI.Api.V2;
using Microsoft.PowerBI.Api.V2.Models;

// You need to provide the workspaceId where the dashboard resides.
ODataResponseListReport reports = client.Reports.GetReportsInGroupAsync( workspa-
```

ceId);

```
// Get the first report in the group.
Report report = reports.Value.FirstOrDefault();
```

11. 创建嵌入令牌

需要生成嵌入令牌,可以从 JavaScript API 中引用。嵌入的令牌对嵌入的项是特定的。因此,在任何时候嵌入一块电源 BI 内容,都需要为它创建一个新的嵌入令牌。有关更多信息,包括使用哪个 Access 级别,请参见 GeaToToEngAPI。

这是一个为应用程序添加报表的嵌入令牌的示例。在示例应用程序中的 Controllers\HomeController.cs 文件中提供了为报告、仪表板或平铺创建嵌入令牌的示例。

```
using Microsoft.PowerBI.Api.V2;
using Microsoft.PowerBI.Api.V2.Models;

// Generate Embed Token.
var generateTokenRequestParameters = new GenerateTokenRequest(accessLevel: "view");
EmbedToken tokenResponse = client.Reports.GenerateTokenInGroup(workspaceId, report.Id, generateTokenRequestParameters);

// Generate Embed Configuration.
var embedConfig = new EmbedConfig()
{
    EmbedToken = tokenResponse,
    EmbedUrl = report.EmbedUrl,
    Id = report.Id
};
```

这假设为 EmbedConfig 和 TileEmbedConfig 创建了一个类。这些模型在 Models\EmbedConfig.cs 文件和 Models\TileEmbedConfig.cs 文件中可用。

12. 使用 JavaScript 加载一个项目

可以使用 JavaScript 将报表加载到网页上的 div 元素中。

对于使用 JavaScript API 的完整示例,可以使用游乐场工具。这是一种快速播放不同类型的 BI 嵌入样本的方法。还可以通过访问 PowerBI-JavaScript 维基页面获取更多关于 JavaScript API 的信息。

下面是一个使用嵌入结构模型和 TileEmbedConfig 模型以及报表视图的示例。在示例应用程序中的 Views\Home\EmbedReport.cshtml、Views\Home\EmbedDashboard.cshtml 或 Views\Home\Embedtile.cshtml 文件中,可以添加报表、仪表板或瓦片视图的示例。

```html
<script src="~/scripts/powerbi.js"> </script>
<div id="reportContainer"> </div>
<script>
    // Read embed application token from Model
    var accessToken = "@Model.EmbedToken.Token";

    // Read embed URL from Model
    var embedUrl = "@Html.Raw(Model.EmbedUrl)";

    // Read report Id from Model
    var embedReportId = "@Model.Id";

    // Get models. models contains enums that can be used.
    var models = window['powerbi-client'].models;

    // Embed configuration used to describe the what and how to embed.
    // This object is used when calling powerbi.embed.
    // This also includes settings and options such as filters.
    // You can find more information at https://github.com/Microsoft/PowerBI-JavaS-
    // cript/wiki/Embed-Configuration-Details.
    var config = {
        type: 'report',
        tokenType: models.TokenType.Embed,
        accessToken: accessToken,
        embedUrl: embedUrl,
        id: embedReportId,
        permissions: models.Permissions.All,
        settings: {
            filterPaneEnabled: true,
            navContentPaneEnabled: true
        }
    };

    // Get a reference to the embedded report HTML element
    var reportContainer = $('#reportContainer')[0];

    // Embed the report and display it within the div container.
    var report = powerbi.embed(reportContainer, config);
</script>
```

13. 走向生产环境

现在，已经完成了开发应用程序，是时候用专用的容量来备份应用程序工作区了。需要专门的能力来用于生产环境。

14. 创建专用容量

通过创建一个专用的容量,来为你的客户提供专用资源。可以在微软 Azure 门户网站购买专用容量。有关如何创建电源 BI 嵌入容量的细节,请参阅 Azure 门户中如何创建 Power BI Embedded 容量。

使用表 11-12 确定哪种 Power BI 嵌入容量最适合你的需要。

表 11-12 Power BI 嵌入容量

容量节点	总共核心数 (前端和后端)	后端核心	前端核心	直接查询或 在线链接限制/s	顶峰时的 最大页面数
A1	1 v-core(s)	0.5 core(s), 3 GB RAM	0.5 core(s)	0.5	1~300
A2	2 v-core(s)	1 core(s), 5 GB RAM	1 core(s)	10	301~600
A3	4 v-core(s)	2 core(s), 10 GB RAM	2 core(s)	15	601~1 200
A4	8 v-core(s)	4 core(s), 25 GB RAM	4 core(s)	30	1 201~2 400
A5	16 v-core(s)	8 core(s), 50 GB RAM	8 core(s)	60	2 401~4 800
A6	32 v-core(s)	16 core(s), 100 GB RAM	16 core(s)	120	4 801~9 600

使用 SKU,不能用一个免费的 BI 许可访问 BI 内容。

使用带有 PRO 许可证的嵌入令牌是为了进行开发测试,因此 Power BI 主账户可以生成的嵌入令牌的数量是有限的。必须购买专用的嵌入生产环境的容量。使用专用容量可以生成多少个嵌入令牌没有限制。使用 API AvailableFeatures(https://api.powerbi.com/v1.0/myorg/availableFeatures)来检查专用容量的使用情况。专用容量使用情况是以主账户的使用量为基础的。

关于 Power BI 容量,可查阅下面的链接来了解更多信息:https://aka.ms/pbiewhitepaper。

15. 将应用程序工作区分配到专用容量

一旦创建了专用容量,就可以将应用程序工作区分配到专用容量。要完成这一点,请遵循以下步骤:

① 在 Power BI 服务中,展开工作区并选择用于嵌入内容的工作区的省略号。然后选择编辑工作区,如图 11-98 所示。

图 11-98 编辑工作区 workspace

② 展开高级,启用专用容量,并选择创建的专用容量。然后选择保存,如图 11-99 所示。

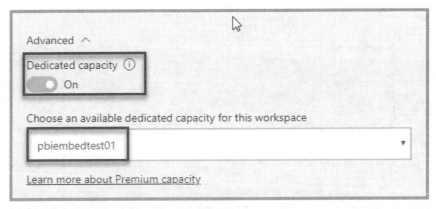

图 11-99 选择专用容量或者 Premium 容量

选择保存之后,应在应用程序工作区名称旁边看到一个菱形,如图 11-100 所示。

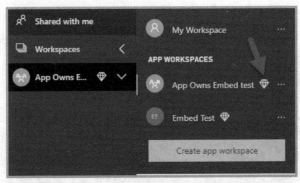

图 11-100 在工作区右边的一个菱形标志

第 12 章

部署 Linux 服务器到 Azure

12.1 虚拟机简介

虚拟机是 IAAS(基础设施即服务)的一部分。Azure IAAS 可以满足您独特的云基础设施需求。

用即时和可扩展的基础设施支持关键任务的应用程序和工作负载。使用 Azure IAAS,您可以将精力集中在您的业务需求上,而不必花费时间和金钱来计划、获取、保护和管理基础设施,从而满足您的应用程序需求。它还可以:

- 快速为所有工作负载提供基础设施;
- 部署与您的楼宇基础设施一致的混合环境;
- 用完整的安全和管理服务保护应用程序;
- 显著降低和优化基础设施成本。

图 12-1 为 IAAS、PAAS、SAAS 的对比。

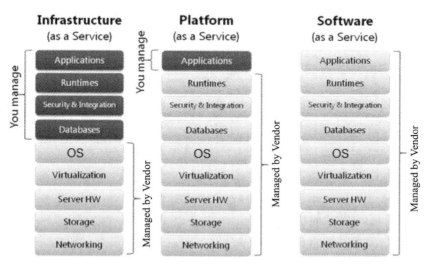

图 12-1 IAAS、PAAS、SAAS 对比

① IAAS：Infrastructure as a Service(基础设施即服务)。它将基础设施提供出来，供一些购买服务的用户使用。这些基础设施包括，操作系统(Operating System，简称OS)、虚拟化、服务器硬件(如CPU、RAM、主板、硬盘)等。还有共享存储空间(Storage)，以及网络基础设施，包括防火墙、网关路由器、交换机等。这些服务器硬件、操作系统、虚拟化技术、存储、网络等就构成了基础设施，然后云计算服务提供商将这些资源做成一个个的服务，可以按某些单位来计费，比如某个型号的服务器(虚拟出来的)，每个月的使用价格是200美元。这样给定一个价格以后，就可以给购买IAAS服务的用户使用，一个用户使用完成后，在释放虚拟服务器之前用户都已保存好数据，将资源释放，即从他的Azure账户中将此虚拟服务器资源删除，则对应的虚拟服务器资源被Azure平台回收，前一个用户留在服务器上的数据都会消失，之后虚拟服务器又可以给另外一个用户使用，不同用户之间不会看到其他用户的数据。

一些大的IAAS公司包括Amazon、Microsoft、VMWare、Rackspace、Google和Red Hat，都提供了IAAS服务。中国的厂商有阿里云、腾讯、百度等。用户可以按一定的方式购买虚拟服务器、网络等资源，然后开始使用。

② PaaS：Platform-as-a-Service(平台即服务)，简单地说是云计算提供商将一些中间件(包括安全、消息、集成、Xml Web Service或者RESTFul API等)、数据存储、运行时等提供出来供用户购买使用。用户可以按某种方式购买，比如存储容量、服务整体打包价格，还可以按调用次数等方式来计价。

③ Google提供了如Cloud BigTable的NoSql服务，还有诸如BigQuery的大数据分析服务等；Amazon提供了如RDS、Dynamo DB等数据库服务，以及MQ等消息服务、Amazon Machine Learning等。微软为Azure Storage提供了便宜的存储服务，Azure Service Bus提供了一个可靠的、高性能的消息服务，Azure Web Service提供了可自动进行伸缩的Web服务集群，可供开发者来开发不同的应用，比如Web应用，或者移动端应用加RESTFul API，同时也有Azure Machine Learning的人工智能的服务。阿里云也有对应的一些服务，如阿里RDS、阿里mongoDB、阿里MQ消息服务、阿里机器学习等。

其他的云计算服务提供商都提供了类似的服务。作为云计算领域的领先者，Amazon、微软、Google排在前面，中国的阿里云、腾讯、百度云等也在奋起直追，正在赶上排在前面的厂商。

④ SaaS：Software-as-a-Service(软件即服务)。它将整个应用软件作为一个服务提供给用户购买使用。这一层是和你的生活每天接触的一层，大多是通过网页浏览器来接入。任何一个远程服务器上的应用都可以通过网络来运行，就是SaaS了。你消费的服务完全是从网页如Netflix、MOG、Google Apps、Box.net、Dropbox或者苹果的iCloud那里进入这些分类。尽管这些网页服务是用作商务和娱乐或者两者都有，但这也算是云技术的一部分。一些用作商务的SaaS应用包括Citrix的Go To Meeting、Cisco的WebEx、Salesforce的CRM、ADP、Workday和SuccessFactors。

如图 12-1 所示,即所有层的服务(包括应用层)都由这个云服务提供商来提供。

各个云计算服务提供商在 IaaS 上提供的服务都差不多,可以支持不同的操作系统,比如 Windows 家族、Linux 家族的不同版本,其虚拟服务器是不同型号的,比如几个 CPU、若干 RAM、磁盘大小等。然后就是服务级别协议,保证可用率小数点后多少个 9 等。

在微软 Azure 中,虚拟机也有一些历史。有一部分创建得比较早的虚拟机,现在是作为虚拟机经典(classic)。这些旧模式的虚拟机调用了一些现在新虚拟机不再用的一些 SMAPI。现在新建立的虚拟机都用新式的。

12.2　用 Azure Portal 创建虚拟机

Azure 虚拟机(VMs)可以通过 Azure 门户创建。该方法提供基于浏览器的用户界面来创建 VMS 及其相关资源。这个快速入门向您展示了如何使用 Azure 门户在运行 Ubuntu 的 Azure 中部署 Linux 虚拟机(VM)。要查看您的 VM 在操作中,然后将 SSH 导入 VM 并安装 NGINX Web 服务器。

1. 创建 SSH 密钥对

您需要一个 SSH 密钥对来完成这个快速演练。如果现已有 SSH 密钥对,则可以跳过此步骤。

要创建 SSH 密钥对并登录到 Linux VM,请从 Bash shell 运行以下命令,并遵循屏幕上的提示。例如,可以使用 Azure 云 shell 或 Windows 子目录进行 Linux 操作。命令输出包括公钥文件的文件名:ssh-keygen -t rsa -b 2048。

将公钥文件的内容复制到剪贴板:cat ~/.ssh/id_rsa.pub。

登录 Azure Portal。

2. 创建虚拟机

① 选择在 Azure 门户的左上角创建一个资源。

② 在 Azure Marketplace 资源列表上方的搜索框中,搜索并选择 Ubuntu Server 16.04LTS by Canonical,然后选择 Create 按钮。

③ 在 Basics 选项卡的 PROJECT DETAILS 下,确保选择了正确的订阅,然后选择 Resource group。在弹出窗口中,为资源组的名称键入 myResourceGroup,然后选择 OK 按钮,如图 12-2 所示。

④ 在实例详细 INSTANCE DETAILS 信息下,为虚拟机名称键入 myVM,然后选择 East US 作为您的区域。保留其他默认值,如图 12-3 所示。

⑤ 在管理员账户 ADMINISTRATOR ACCOUNT 下,选择 SSH 公钥,键入用户名,然后将公钥粘贴到文本框中。删除公钥中的任何前导或尾随空白,如图 12-4 所示。

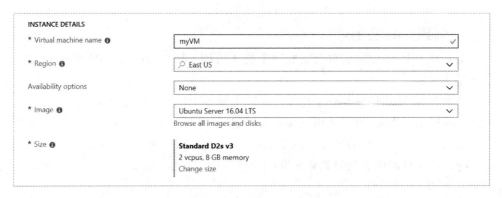

图 12-2　创建虚拟机

图 12-3　输入虚拟机的信息、选择型号等

图 12-4　输入登录用户名和密码

⑥ 在入站端口规则→Public 入站端口下，选择 Allow selected ports，然后从下拉列表中选择 SSH(22) 和 HTTP(80)，如图 12-5 所示。

部署 Linux 服务器到 Azure —12

图 12 - 5　输入允许的进入防火墙规则，默认只允许 SSH 和 HTTP

⑦ 剩余的都不用改，使用默认值，然后在页面底部选择 Review＋create 按钮。

3. 连接到虚拟机

① 用 VM 创建 SSH 连接。

② 选择 VM 概述页面上的连接按钮，如图 12 - 6 所示。

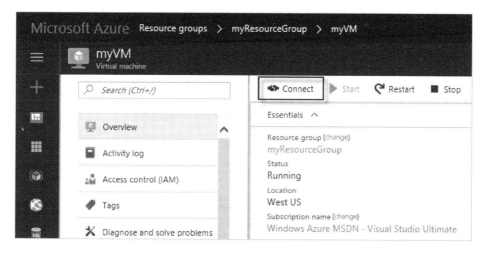

图 12 - 6　点击 Connect 准备连接到虚拟机

③ 在连接到虚拟机页面中，保持默认选项通过 DNS 名称在端口 22 上连接。在使用 VM 本地账户登录时，显示连接命令。单击按钮复制命令。下面的示例显示 SSH 连接命令的样子：

ssh azureuser@myvm-123abc.eastus.cloudapp.azure.com

④ 将 SSH 连接命令粘贴到 Shell 中，比如 Azure Cloud Shell 或 Bash on Ubuntu on Windows 以创建连接。

•473•

4. 安装 Web 服务器

要查看您的虚拟机在运行,请安装 NGINX Web 服务器。若要更新包源并安装最新的 NGINX 包,请从 SSH 会话中运行以下命令:

```
# update packages
sudo apt-get -y update

# install NGINX
sudo apt-get -y install nginx
```

完成后,退出 SSH 会话并返回到 Azure 门户中的 VM 属性。

5. 查看 Web 服务器的动作

使用 NGINX 安装和端口 80 打开到 VM,Web 服务器现在可以从因特网上访问。打开一个 Web 浏览器,并输入 VM 的公共 IP 地址。公共 IP 地址可以在 VM 概览页面上找到,也可以在添加入站端口规则的 Networking 页面的顶部找到,如图 12-7 所示。

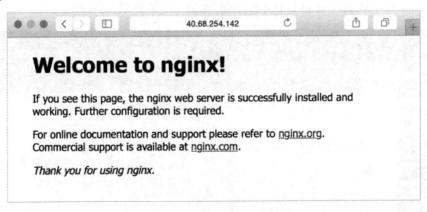

图 12-7 查看 nginx 默认页面

至此,此 Linux 虚拟机就创建好了。

12.3 用 Azure CLI 创建 Linux 虚拟机

Azure CLI 2 用于从命令行或脚本中创建和管理 Azure 资源。这个快速入门向您展示了如何使用 Azure CLI 2 在运行 Ubuntu 的 Azure 中部署 Linux 虚拟机(VM)。要查看您的 VM 在操作中,然后将 SSH 导入 VM 并安装 NGINX Web 服务器。

1. 打开 Azure Cloud Shell

Azure Cloud Shell 是一个免费的交互式 Shell,可以使用它来运行本文中的步骤。常见的 Azure 工具预先安装和配置在 Cloud Shell 中,供您使用您的账户。只需选择复制按钮复制代码,粘贴到 Cloud Shell,然后按 Enter 键运行它。有几种方法可以打开 Cloud Shell,表 12-1 列出了打开 Cloud Shell 的不同方式。

表 12-1 打开云 Shell 的不同方式

方　　式	图　　示
选择代码块右上角的 Try it	
在你的浏览器中打开 Cloud Shell	
选择 Azure 门户的右上角菜单上的 Cloud Shell 按钮	

2. 创建资源组

用 az group 创建命令创建资源组。Azure 资源组是 Azure 资源被部署和管理的逻辑容器。下面的示例在 EastUS 创建一个名为 myResourceGroup 的资源组:

az group create --name myResourceGroup --location eastus。

3. 创建虚拟机

用 AZ VM 创建命令创建一个 VM。

下面的示例创建一个名为 myVM 的 VM,添加一个名为 azureuser 的用户账户,如果在默认密钥位置(~/.ssh)中不存在 SSH 密钥,则生成 SSH 密钥。若要使用特定的密钥集,请使用--ssh-key-value 选项:

```
az vm create \
    --resource-group myResourceGroup \
    --name myVM \
    --image UbuntuLTS \
    --admin-username azureuser \
    --generate-ssh-keys
```

创建虚拟机和支持资源需要几分钟的时间。下面的示例输出显示 VM 创建操作成功。

```
{
  "fqdns": "",
  "id": "/subscriptions/< guid >/resourceGroups/myResourceGroup/providers/Microsoft.Compute/virtualMachines/myVM",
  "location": "eastus",
  "macAddress": "00-0D-3A-23-9A-49",
  "powerState": "VM running",
  "privateIpAddress": "10.0.0.4",
  "publicIpAddress": "40.68.254.142",
```

```
"resourceGroup": "myResourceGroup"
}
```

在 VM 的输出中注意自己的发布地址。此地址用于在下一步骤中访问虚拟机。

(1) 用于 Web 流量的开放端口 80

默认情况下,只有在 Azure 中创建 Linux VM 时才打开 SSH 连接。使用 AZ VM 开放端口打开 TCP 端口 80,以便与 NGINX Web 服务器一起使用:

```
az vm open-port --port 80 --resource-group myResourceGroup --name myVM。
```

(2) 连接到虚拟机

SSH 作为正常的 VM。用 VM 的公共 IP 地址替换 publicIpAddress,如前面的 VM 输出中所指出的:

ssh azureuser@publicIpAddress。

(3) 安装 Web 服务器

要查看您的虚拟机在运行,请安装 NGINX Web 服务器。若要更新包源并安装最新的 nginx 包,请从 SSH 会话中运行以下命令:

```
# update packages
sudo apt-get -y update

# install NGINX
sudo apt-get -y install nginx
```

(4) 查看 Web 服务器的动作

使用 nginx 安装,端口 80 现在从 Internet 上打开您的 VM,使用您选择的 Web 浏览器查看默认 nginx 欢迎页。使用前一步骤中获得的 VM 的公共 IP 地址。下面的示例显示默认的 nginx 网站,如图 12-8 所示。

图 12-8　使用公网 IP 访问此服务器

4. 清理资源

当不再需要时,可以使用 az group delete 命令来删除资源组、VM 和所有相关资

源。确保已退出 SSH 会话到 VM,然后删除资源:
az group delete --name myResourceGroup。

12.4　Automation Configuration

　　Automation Configuration 可以快速、一致地创建 VMs,自动化地将多个虚拟机用某一个模板进行配置,可以省去很多人工工作。
　　如何在第一次启动时使用 cloud-init 自定义 Azure 中的 Linux 虚拟机呢?
　　为了快速、一致地创建 VMs,通常需要某种形式的自动化。在第一次引导中定制 VM 的一种常见方法是使用 cloud-init。在本教程中,您将学习:
- 如何创建一个 cloud-init 配置文件;
- 如何创建一个使用 cloud-init 文件的虚拟机;
- 在创建 VM 后查看运行的 Node.js 应用程序;
- 使用密钥库来安全地存储证书;
- 用 cloud-init 实现 nginx 的安全部署。

1. 打开 Azure Cloud Shell

　　Azure Cloud Shell 是一个免费的交互式 Shell,您可以使用它来运行本文中的步骤。常见的 Azure 工具预先安装和配置在 Cloud Shell 中,供您使用您的账户。只需选择复制按钮复制代码,粘贴到 Cloud Shell,然后按 Enter 键运行它即可。有几种方法可以打开 Cloud Shell,参见表 12-1。
　　如果您选择在本地安装和使用 CLI,本教程要求您运行 Azure CLI 版本 2.0.30 或更高版本。运行 az --version 命令找到版本。

2. cloud-init 概述

　　cloud-init 是一种广泛使用的方法来定制 Linux VM。因为它是第一次启动,可以使用 cloud-init 来安装包和写入文件,或者配置用户和安全性。在初始引导过程中运行 cloud-init 时,没有额外的步骤或需要的代理来应用您的配置。
　　cloud-init 也适用于分布。例如,您不使用 apt-get install 或 yum install 来安装包;相反,您可以定义要安装的软件包列表。cloud-init 自动为您选择的发行版使用本机包管理工具。
　　我们正在与我们的合作伙伴合作,以获得 cloud-init 和工作的图像,它们提供给 Azure。表 12-2 概述了 Azure 平台图像上的当前 cloud-init 可用性。

3. 创建 cloud-init 配置文件

　　要在操作中看到 cloud-init,创建一个安装 nginx 的 VM,运行一个简单的"Hello World"Node.js 应用程序。下面的 cloud-init 配置安装所需的包,创建一个 Node.js 应用程序,然后初始化并启动应用程序。

表 12-2 cloud-init 可用性

别名	出版者	操作系统	SKU	Version
UbuntuLTS	Canonical	UbuntuServer	16.04-LTS	最新
UbuntuLTS	Canonical	UbuntuServer	14.04.5-LTS	最新
CoreOS	CoreOS	CoreOS	Stable	最新
	OpenLogic	CentOS	7-CI	最新
	RedHat	RHEL	7-RAW-CI	最新

在当前的 Shell 中,创建一个名为 cloud-init.txt 的文件,并粘贴以下配置。例如,在 Cloud Shell 中创建文件,而不是在本地机器上创建文件。你可以使用任何你想要的编辑器。输入 sensible-editor cloud-init.txt 以创建文件并查看可用编辑器的列表,确保整个 cloud-init 文件被正确复制,特别是第一行。

```
#cloud-config
package_upgrade: true
packages:
  - nginx
  - nodejs
  - npm
write_files:
  - owner: www-data:www-data
  - path: /etc/nginx/sites-available/default
    content: |
      server {
        listen 80;
        location / {
          proxy_pass http://localhost:3000;
          proxy_http_version 1.1;
          proxy_set_header Upgrade $http_upgrade;
          proxy_set_header Connection keep-alive;
          proxy_set_header Host $host;
          proxy_cache_bypass $http_upgrade;
        }
      }
  - owner: azureuser:azureuser
  - path: /home/azureuser/myapp/index.js
    content: |
      var express = require('express')
      var app = express()
      var os = require('os');
```

```
    app.get('/', function (req, res) {
      res.send('Hello World from host ' + os.hostname() + '!')
    })
    app.listen(3000, function () {
      console.log('Hello world app listening on port 3000!')
    })
runcmd:
  - service nginx restart
  - cd "/home/azureuser/myapp"
  - npm init
  - npm install express -y
  - nodejs index.js
```

4. 创建虚拟机

在创建 VM 之前,创建具有 az group 的资源组。下面的示例是在 East US 位置创建一个名为 myResourceGroupAutomate 的资源组:

az group create --name myResourceGroupAutomate --location eastus

现在用 az vm create 命令创建一个虚拟机。使用自定义数据参数传递到 cloud-init 配置文件中。如果您将文件保存在当前工作目录之外,则应提供 cloud-init.txt 配置文件的完整路径。下面的示例是创建一个名为 myAutomatedVM 的虚拟机。

```
az vm create \
    --resource-group myResourceGroupAutomate \
    --name myVM \
    --image UbuntuLTS \
    --admin-username azureuser \
    --generate-ssh-keys \
    --custom-data cloud-init.txt
```

要创建 VM 需要几分钟,要安装软件包和启动应用程序。在 Azure CLI 将您的任务返回到提示之后,还有后台任务继续运行。可能再过几分钟就可以访问应用程序了。当创建了 VM 时,注意 Azure CLI 显示的 publicIpAddress。这个地址用于通过 Web 浏览器访问 Node.js 应用程序。

为了允许 Web 流量到达您的虚拟机,使用 az vm open-port 从 Internet 打开端口 80:

az vm open-port --port 80 --resource-group myResourceGroupAutomate --name myVM。

5. 测试 Web 应用程序

现在您可以打开 Web 浏览器并在地址栏中输入 http://。在 VM 创建过程中提供您自己的公共 IP 地址。您的 Node.js 应用程序如图 12-9 所示。

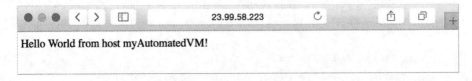

图 12-9 打开 Node.js 应用程序

6. 从关键库引入证书

这个可选的部分展示了如何在 Azure 密钥库中安全地存储证书并在 VM 部署期间引入它们。此过程确保在第一次引导时将最新的证书引入 VM，而不是使用包含已焙烧证书的自定义映像。在这个过程中，证书永远不会离开 Azure 平台，或者暴露在脚本、命令行历史或模板中。

Azure 密钥库保护密码密钥和机密，例如证书或密码。密钥库帮助简化密钥管理过程，并使您能够保持对访问和加密数据的密钥的控制。这个场景引入了一些 Key Vault 概念来创建和使用证书，但是并不是关于如何使用 Key Vault 的详尽概述。

下面的步骤说明了：

- 如何创造一个 Azure 的 Key Vault；
- 生成或上传证书到密钥库；
- 从证书中创建一个秘密以引入到虚拟机中；
- 创建一个虚拟机并引入证书。

7. 创建一个 Azure 的 Key Vault

首先，创建一个带有 az keyvault create 的密钥库，并在部署 VM 时启用它。每个密钥库需要一个唯一的名称，并且应该都是小写的。在下面的示例中用自己唯一的密钥库名称替换 mykeyvault：

```
keyvault_name=mykeyvault
az keyvault create \
    --resource-group myResourceGroupAutomate \
    --name $keyvault_name \
    --enabled-for-deployment
```

8. 在证书库中生成证书和存储

对于生产使用，您应该导入一个由 az keyvault certificate import 的可信提供者签署的有效证书。对于本教程，下面的示例演示如何用 az keyvault certificate create 生成使用默认证书策略的自签名证书：

```
az keyvault certificate create \
    --vault-name $keyvault_name \
```

```
--name mycert \
--policy "$(az keyvault certificate get-default-policy)"
```

9. 准备与 VM 一起使用的证书

若要在 VM 创建过程中使用证书,请使用 az keyvault secret list-versions 获取证书的 ID。VM 需要某种格式的证书来将其注入到引导中,因此用 az vm secret format 转换证书。下面的示例将这些命令的输出分配给变量以便于在下一步骤中使用:

```
secret=$(az keyvault secret list-versions \
    --vault-name $keyvault_name \
    --name mycert \
    --query "[?attributes.enabled].id" --output tsv)
vm_secret=$(az vm secret format --secret "$secret")
```

10. 创建 cloud-init 配置以确保 nginx 安全

创建 VM 时,证书和密钥存储在受保护的 /var/lib/waagent/ 目录中。为了自动将证书添加到 VM 并配置 nginx,可以使用前一个示例中更新的 cloud-init 配置。

创建一个名为 cloud-init-secured.txt 的文件,并粘贴以下配置。同样,如果使用 Cloud Shell,则在那里创建 cloud-init 配置文件,而不是在本地计算机上创建。使用 sensible-editor cloud-init-secured.txt 创建文件并查看可用编辑器的列表,确保整个 cloud-init 文件被正确复制,特别是第一行:

```
#cloud-config
package_upgrade: true
packages:
  - nginx
  - nodejs
  - npm
write_files:
  - owner: www-data:www-data
  - path: /etc/nginx/sites-available/default
    content: |
      server {
        listen 80;
        listen 443 ssl;
        ssl_certificate /etc/nginx/ssl/mycert.cert;
        ssl_certificate_key /etc/nginx/ssl/mycert.prv;
        location / {
          proxy_pass http://localhost:3000;
          proxy_http_version 1.1;
          proxy_set_header Upgrade $http_upgrade;
```

```
              proxy_set_header Connection keep-alive;
              proxy_set_header Host $host;
              proxy_cache_bypass $http_upgrade;
            }
          }
    - owner: azureuser:azureuser
    - path: /home/azureuser/myapp/index.js
      content: |
        var express = require('express')
        var app = express()
        var os = require('os');
        app.get('/', function (req, res) {
          res.send('Hello World from host ' + os.hostname() + '!')
        })
        app.listen(3000, function () {
          console.log('Hello world app listening on port 3000!')
        })
runcmd:
  - secretsname=$(find /var/lib/waagent/ -name "*.prv" | cut -c -57)
  - mkdir /etc/nginx/ssl
  - cp $secretsname.crt /etc/nginx/ssl/mycert.cert
  - cp $secretsname.prv /etc/nginx/ssl/mycert.prv
  - service nginx restart
  - cd "/home/azureuser/myapp"
  - npm init
  - npm install express -y
  - nodejs index.js
```

11. 创建安全虚拟机

现在创建一个带有 az vm create 的虚拟机。证书数据是从密钥库中用--secrets 参数注入的。与前一个示例一样，也通过"--custom-data"在 cloud-init 配置中传递：

```
az vm create \
    --resource-group myResourceGroupAutomate \
    --name myVMSecured \
    --image UbuntuLTS \
    --admin-username azureuser \
    --generate-ssh-keys \
    --custom-data cloud-init-secured.txt \
    --secrets "$vm_secret"
```

要创建 VM 需要几分钟，要安装软件包和启动应用程序。在 Azure CLI 将您的任务返回到提示之后，还有后台任务继续运行。可能再过几分钟就可以访问应用程

——— 部署 Linux 服务器到 Azure ———

序了。当创建了 VM 时，注意 Azure CLI 显示的 publicIpAddress。这个地址用于通过 Web 浏览器访问 Node.js 应用程序。

为了允许安全的 Web 流量到达您的虚拟机，使用 az vm open-port：从 Internet 打开端口 443：

```
az vm open-port \
    --resource-group myResourceGroupAutomate \
    --name myVMSecured \
    --port 443
```

12. 测试安全 Web 应用程序

现在，您可以打开 Web 浏览器并在地址栏中输入 https://，提供您自己的公共 IP 地址，如前面的 VM 创建过程的输出所示。如果使用自签名证书，请接受安全警告，如图 12 – 10 所示。

图 12 – 10 Safari 的警告信息

然后，您的安全 nginx 站点和 Node.js 应用程序将显示在下面的示例中，如图 12 – 11 所示。

图 12 – 11 注意到有一个锁图标，即为 https

12.5 用 Azure CLI 2.0 创建 Azure VM 的自定义虚拟机镜像

自定义镜像就像市场中的镜像，只是自己可以创建它们。自定义镜像可用于引导配置，例如预加载应用程序、应用程序配置和其他 OS 配置。在本教程中，将创建一个 Azure 虚拟机的自定义镜像。你将学会：
- 卸载与 VM 泛化；
- 创建自定义镜像；
- 从自定义镜像创建虚拟机；
- 列出订阅中的所有图片；
- 删除镜像。

1. 在你开始之前

下面的步骤详细说明了如何利用现有 VM 并将其转换成可用于创建新 VM 实例的可重用自定义镜像。

要完成本教程中的示例，必须有一个现有的虚拟机。如果需要，这个脚本示例可以为你创建一个脚本示例。当通过教程工作时，在需要时替换资源组和 VM 名称。

2. 创建自定义镜像

要创建虚拟机的镜像，需要通过去掉虚拟机、释放虚拟机，然后将源虚拟机标记为通用虚拟机来准备虚拟机。一旦虚拟机已经准备好，就可以创建一个镜像。

3. 撤销 VM

Debug 通过移除特定于机器的信息来概括 VM。这种泛化使得可以从单个图像部署多个 VM。在解除配置时，主机名被重置为 LoalHoal.LoalAdvices。SSH 主机密钥、名称服务器配置、根密码和缓存的 DHCP 租约也被删除。

若要撤销 VM，请使用 Azure VM 代理（WAAgent）。Azure VM 代理安装在 VM 上，并管理与 Azure 织物控制器的供应和交互。有关更多信息，请参阅 Azure Linux 代理用户指南。

使用 SSH 连接到 VM 并运行命令以撤销 VM。使用＋用户参数，最后提供的用户账户和任何关联的数据也被删除。用 VM 的公共 IP 地址替换示例 IP 地址。

4. SSH 到 VM

ssh azureuser@52.174.34.95

撤销(deprovision)VM

sudo waagent －deprovision＋user －force

关闭 SSH 会话

Exit

5. 回收并标记此 VM 已经通用化了

创建一个镜像，此 VM 需要回收。用 az vm deallocate 命令来回收此 VM：

az vm deallocate --resource-group myResourceGroup --name myVM

最后，设置此 VM 的状态为通用化的。使用 az vm generalize 命令来做通用化之后，Azure 平台就知道此 VM 已经通用化了。这样做的原因是智能地从一个通用化的 VM 创建一个 VM 镜像。

```
az vm generalize --resource-group myResourceGroup --name myVM
```

6. 创建镜像

现在你可以创建一个通用化 VM 的镜像,使用 az vm create 命令。下面的命令是从一个名为 myVM 的通用化 VM 创建一个镜像,名为 myImage:

```
az image create \
    -- resource-group myResourceGroup \
    -- name myImage \
    -- source myVM
```

7. 从镜像创建 VM

现在你有一个镜像,你可以从此镜像创建一个或多个新的 VM,使用 az vm create 命令。下面的例子使用名为 myImage 的 VM 创建一个 VM,命名为 myVMfromImage:

```
az vm create \
    -- resource-group myResourceGroup \
    -- name myVMfromImage \
    -- image myImage \
    -- admin-username azureuser \
    -- generate-ssh-keys
```

8. 镜像管理

这里是一些普通镜像管理的任务的样例。这些样例演示了如何用 Azure CLI 来完成常见的镜像管理。

列出所有的镜像名称列表:

```
az image list \
    -- resource-group myResourceGroup
```

Delete an image. This example deletes the image named myOldImage from the myResource-Group.

删除一个镜像,下面的例子是从 myResourcGroup 资源组删除名为 myOldImage 的镜像:

```
az image delete \
    -- name myOldImage \
    -- resource-group myResourceGroup
```

12.6 用 Azure CLI 2.0 创建和部署高可用虚拟机

在此你将学会如何使用可用性集合（Availability Sets，简称可用集）在 Azure 平台上增加你的 VM 可用性和可靠性。可用集保证你在 Azure 上部署的 VM 是多个物理分离分布式的硬件集群。这样做确保如果有一些硬件或者软件失败的情况下只有一部分的 VM 受到影响并且你的整体解决方案还是可用的和可操作的。

在这一节中，你将学会：
- 创建一个可用集；
- 在一个可用集里创建一个 VM；
- 检查可用的 VM 大小。

1. 可用集概览

一个可用集是指逻辑地将一些 VM 组织在一起，当这些 VM 在 Azure 数据中心中部署的时候，这些资源是物理隔离的。Azure 确保这些放入可用集中的 VM 在多个物理隔离的服务器上运行，计算块、存储单元和网络交换机等，都是物理隔离的。如果一个硬件或者 Azure 软件失败发生，只有一部分 VM 会受到影响，并且你的应用总体上还是可用并继续为你的客户提供服务。可用集是一个基本能够帮助你进行构建的可靠的云解决方案。

让我们考虑一个典型的基于 VM 的解决方案，其中您可能具有四个前端 Web 服务器，并使用两个后端 VM 来托管数据库。使用 Azure，您需要在部署 VM 之前定义两个可用集：一个用于 Web 层的可用集，一个用于数据库层的可用集。创建新 VM 时，可以将可用集指定为 az vm create 命令的参数，并且 Azure 自动确保在可用集内创建的 VM 跨多个物理硬件资源隔离。如果您的 Web 服务器或数据库服务器 VM 之一正在运行的物理硬件有问题，则 Web 服务器和数据库 VM 的其他实例仍然在运行，因为它们位于不同的硬件上。

当您想在 Azure 中部署可靠的基于 VM 的解决方案时，就使用可用集。

2. 创建可用集

可以使用 az vm availability-set create 创建可用集。在这个示例中，对于 myResourceGroupAvailability 资源组中的名为 myAvailabilitySet 的可用集，更新和故障域的数量设置为 2。

首先，使用 az group create 命令创建资源组，然后创建可用集：

```
az group create -- name myResourceGroupAvailability -- location eastus
```

```
az vm availability-set create \
    -- resource-group myResourceGroupAvailability \
```

```
--name myAvailabilitySet \
--platform-fault-domain-count 2 \
--platform-update-domain-count 2
```

可用集允许您隔离跨故障域和更新域的资源。故障域表示服务器＋网络＋存储资源的独立集合。在前面的示例中，当部署 VM 时，可用性集合分布在至少两个故障域上。可用集也分布在两个更新域上。两个更新域确保当 Azure 执行软件更新时，VM 资源是隔离的，从而防止在 VM 上运行的所有软件同时更新。

3. 在可用集内创建 VM

VM 必须在可用集内创建，以确保它们在硬件上正确分布。在创建了可用性集合之后，无法将现有的 VM 添加到可用性集合中。

当使用 az vm create 创建 VM 时，使用--availability-set 参数指定可用性集合的名称。

```
for i in `seq 1 2`; do
    az vm create \
        --resource-group myResourceGroupAvailability \
        --name myVM$i \
        --availability-set myAvailabilitySet \
        --size Standard_DS1_v2 \
        --vnet-name myVnet \
        --subnet mySubnet \
        --image UbuntuLTS \
        --admin-username azureuser \
        --generate-ssh-keys
done
```

现在有两个虚拟机在可用性集合内。因为它们在相同的可用性集中，所以 Azure 确保 VM 及其所有资源（包括数据磁盘）分布在隔离的物理硬件上。这种分布有助于确保整个 VM 解决方案的更高可用性。

可以通过转到 Resource Groups→myResourceGroupAvailab.→myAvailability-Set 在门户中查看可用性集分布。VM 分布在两个故障和更新域上，如图 12-12 所示。

4. 检查可用的 VM 大小

附加 VM 可以稍后添加到可用性集合中，其中 VM 大小在硬件上可用。使用 az vm availability-set list-sizes 列出所有可用的硬件集群上的可用性大小：

```
az vm availability-set list-sizes \
    --resource-group myResourceGroupAvailability \
    --name myAvailabilitySet \
    --output table
```

图 12-12　查看可用性集分布

12.7　用 Azure CLI 2.0 创建虚拟机自动伸缩集并在 Linux 上部署高度可用的应用程序

虚拟机规模集允许部署和管理一组相同的、自动伸缩的虚拟机。可以手动伸缩集合中的 VM 数量,或者根据资源的使用情况(如 CPU、内存需求或网络流量)定义自动伸缩的规则。在本节中,将在 Azure 中部署虚拟机自动伸缩集。你将学会:
- 如何使用云初始化创建一个应用程序;
- 创建虚拟机自动伸缩集;
- 增加或减少自动伸缩集中实例的数目;
- 创建自动伸缩集的规则;
- 查看自动伸缩集合实例的连接信息;
- 在自动伸缩集中使用数据磁盘。

1. 打开 Azure Cloud Shell

Azure Cloud Shell 是一个免费的交互式 Shell,可以使用它来运行本书中的步骤。常见的 Azure 工具预先安装和配置在 Cloud Shell 中,供您使用您的账户。只需选择复制按钮复制代码,粘贴到 Cloud Shell,然后按 Enter 键运行它即可。有几种方法可以打开 Cloud Shell,参见表 12-1。

如果选择安装和使用 CLI 本地,则本教程要求运行 Azure CLI 版本 2.0.30 或之后版本。运行 az --version 找到 Azure CLI 的版本。

2. 自动伸缩集概述

虚拟机自动伸缩集允许部署和管理一组相同的、自动伸缩的虚拟机。自动伸缩集中的 VM 分布在一个或多个放置组中的逻辑故障和更新域中。这些是类似配置的 VM 组,类似于可用集。

VM 是按比例设置的。您定义自动伸缩规则来控制如何以及何时从自动伸缩中添加或移除虚拟机。这些规则可以基于诸如CPU负载、内存使用或网络流量等度量来触发。

当使用Azure平台镜像时,自动伸缩集支持高达1 000个VM。对于具有显著安装或VM定制要求的工作负载,您可能希望创建自定义VM镜像。在使用自定义镜像时,可以创建一个比例为300的VM。

3. 创建可伸缩的应用程序

对于生产使用,您可能希望创建包括安装和配置应用程序的自定义VM镜像。在本教程中,让用户在第一次启动时定制VM,快速查看在动作中设置的比例。

在前一个教程中,学习了如何在第一次使用cloud-init引导下定制Linux虚拟机。您可以使用同一个cloud-init配置文件安装nginx并运行一个简单的"Hello World"Node.js应用程序。

在当前的Shell中,创建一个名为cloud-init.txt的文件,并粘贴以下配置。例如,在Cloud Shell中创建文件,而不是在本地机器上创建文件。输入sensible-editor cloud-init.txt以创建文件并查看可用编辑器的列表,确保整个cloud-init文件被正确复制,特别是第一行:

```
#cloud-config
package_upgrade: true
packages:
  - nginx
  - nodejs
  - npm
write_files:
  - owner: www-data:www-data
  - path: /etc/nginx/sites-available/default
    content: |
      server {
        listen 80;
        location / {
          proxy_pass http://localhost:3000;
          proxy_http_version 1.1;
          proxy_set_header Upgrade $http_upgrade;
          proxy_set_header Connection keep-alive;
          proxy_set_header Host $host;
          proxy_cache_bypass $http_upgrade;
        }
      }
  - owner: azureuser:azureuser
  - path: /home/azureuser/myapp/index.js
```

```
content: |
    var express = require('express')
    var app = express()
    var os = require('os');
    app.get('/', function (req, res) {
      res.send('Hello World from host ' + os.hostname() + '!')
    })
    app.listen(3000, function () {
      console.log('Hello world app listening on port 3000!')
    })
runcmd:
  - service nginx restart
  - cd "/home/azureuser/myapp"
  - npm init
  - npm install express -y
  - nodejs index.js
```

4. 创建自动伸缩集

在创建一个自动伸缩集之前,创建一个带有 az group create 的资源组。下面的示例是在 East US 位置创建一个名为 myResourceGroupScaleSet 的资源组:

```
az vmss create \
    --resource-group myResourceGroupScaleSet \
    --name myScaleSet \
    --image UbuntuLTS \
    --upgrade-policy-mode automatic \
    --custom-data cloud-init.txt \
    --admin-username azureuser \
    --generate-ssh-keys
```

创建和配置所有的比例集资源和虚拟机需要几分钟的时间。在 Azure CLI 将您的任务返回到提示之后,还有后台任务继续运行。可能再过几分钟就可以访问应用程序了。

5. 允许网络流量

作为虚拟机规模集的一部分,自动创建负载均衡器。负载均衡器使用负载平衡器规则在一组定义的 VMS 上分配业务。

为了允许流量到达 Web 应用程序,用 az network lb rule create 命令创建一个的规则。下面的示例是创建一个名为 myLoadBalancerRuleWeb 的规则:

```
az network lb rule create \
    --resource-group myResourceGroupScaleSet \
```

```
--name myLoadBalancerRuleWeb \
--lb-name myScaleSetLB \
--backend-pool-name myScaleSetLBBEPool \
--backend-port 80 \
--frontend-ip-name loadBalancerFrontEnd \
--frontend-port 80 \
--protocol tcp
```

6. 测试你的应用程序

在网络上看到你的Node.js应用程序,用az network public-ip show获得负载均衡器的公共IP地址。下面的示例是获得myScaleSetLBPublicIP创建的IP地址作为比例集的一部分:

```
az network public-ip show \
    --resource-group myResourceGroupScaleSet \
    --name myScaleSetLBPublicIP \
    --query [ipAddress] \
    --output tsv
```

在Web浏览器中输入公共IP地址,显示该应用程序,包括负载平衡器分发的VM的主机名,如图12-13所示。

图12-13 包含机器名的页面显示

要查看自动伸缩集如何伸缩,可以强制刷新Web浏览器,以查看负载平衡器在运行应用程序的所有VM之间分配的通信量。

7. 管理任务

在整个自动伸缩集的生命周期中,可能需要运行一个或多个管理任务。此外,可能需要创建自动化运行各种生命周期任务的脚本。Azure CLI 2提供了快速完成这些任务的方法。下面是一些常见的任务。

(1) 查看自动伸缩集中的VM

若要查看在自动伸缩集中运行的VM列表,请使用az vmss list-instances,如下:

```
az vmss list-instances \
    --resource-group myResourceGroupScaleSet \
    --name myScaleSet \
    --output table
```

输出类似于以下示例：

InstanceId	LatestModelApplied	Location	Name	ProvisioningState	ResourceGroup	VmId
1	True	eastus	myScaleSet_1	Succeeded	MYRESOURCEGROUPSCALESET	c72ddc34-6c41-4a53-b89e-dd24f27b30ab
3	True	eastus	myScaleSet_3	Succeeded	MYRESOURCEGROUPSCALESET	44266022-65c3-49c5-92dd-88ffa64f95da

(2) 手动增加或减少 VM 实例

若要查看当前在一个自动伸缩集中设置的实例数，请使用 az vmss show 对 sku.capacity 进行显示和查询：

```
az vmss show \
    --resource-group myResourceGroupScaleSet \
    --name myScaleSet \
    --query [sku.capacity] \
    --output table
```

然后，可以手动增加或减少自动伸缩集中的虚拟机数量，以及用 az vmss scale 命令设置的虚拟机数量。下面的示例是将伸缩集中的 VM 数量设置为 3：

```
az vmss scale \
    --resource-group myResourceGroupScaleSet \
    --name myScaleSet \
    --new-capacity 3
```

(3) 获取连接信息

若要在刻度集中获取有关 VM 的连接信息，请使用 az vmss list-instance-connection-info。这个命令为每个 VM 输出公共 IP 地址和端口，允许与 SSH 连接：

```
az vmss list-instance-connection-info \
    --resource-group myResourceGroupScaleSet \
    --name myScaleSet
```

(4) 使用带有刻度集的数据盘

可以创建和使用具有比例集的数据磁盘。在前面的教程中，学习了如何管理 Azure 磁盘，该磁盘概述了在数据磁盘而不是 OS 磁盘上构建应用程序的最佳实践和性能改进。

(5) 用数据盘创建自动伸缩集

若要创建自动伸缩集并附加数据磁盘，请将--data-disk-sizes-gb 参数添加到 az

vmss create 命令中。下面的示例创建了一个带有 50 GB 数据盘的比例集,每个数据盘连接到每个实例:

```
az vmss create \
    -- resource-group myResourceGroupScaleSet \
    -- name myScaleSetDisks \
    -- image UbuntuLTS \
    -- upgrade-policy-mode automatic \
    -- custom-data cloud-init.txt \
    -- admin-username azureuser \
    -- generate-ssh-keys \
    -- data-disk-sizes-gb 50
```

当从一个刻度集中删除实例时,任何附加的数据磁盘也被删除。

(6) 添加数据磁盘

若要将数据磁盘添加到刻度集中的实例中,请使用 az vmss disk attach。下面的示例是向每个实例添加一个 50 GB 的磁盘:

```
az vmss disk attach \
    -- resource-group myResourceGroupScaleSet \
    -- name myScaleSet \
    -- size-gb 50 \
    -- lun 2
```

(7) 分离数据磁盘

若要将数据磁盘移除到比例集的实例中,请使用 az vmss disk detach。下面的示例是从每个实例中删除 LUN 2 中的数据磁盘:

```
az vmss disk detach \
    -- resource-group myResourceGroupScaleSet \
    -- name myScaleSet \
    -- lun 2
```

12.8 用 Azure CLI 2.0 在 Azure 中负载平衡虚拟机,创建一个高度可用的应用程序

负载均衡通过在多个虚拟机上扩展传入请求提供更高的可用性级别。在本教程中,将了解 Azure 负载平衡器的不同组件,它们均衡分布通信量并提供高可用性。你将学会:

- 创建一个 Azure 负载均衡器;
- 创建负载均衡器健康探测器;

- 创建负载均衡器业务规则；
- 使用 cloud-init 创建基本的 Node.js 应用程序；
- 创建虚拟机并附加到负载均衡器；
- 查看负载均衡器的作用；
- 从负载均衡器中添加和移除 VM。

1. Azure 负载均衡器概述

Azure 负载均衡器是 4 层（TCP、UDP）负载均衡器，通过在健康的 VM 之间分配传入流量来提供高可用性。负载均衡器健康探针监视每个 VM 上的给定端口，并且仅将流量分配给操作 VM。

您定义了一个包含一个或多个公共 IP 地址的前端 IP 配置。这个前端 IP 配置允许您的负载均衡器和应用程序可以通过互联网访问。

虚拟机使用它们的虚拟网络接口卡（NIC）连接到负载均衡器。为了向 VM 分配流量，后端地址池包含连接到负载均衡器的虚拟（NIC）的 IP 地址。

为了控制网络流量，您可以为映射到 VM 的特定端口和协议定义负载均衡器规则。

如果您遵循前面的教程创建虚拟机自动伸缩集，则可以为您创建负载均衡器。所有这些组件都为您配置为比例集的一部分。

2. 创建 Azure 负载均衡器

本节详细说明如何创建和配置负载均衡器的每个组件。在创建负载均衡器之前，创建具有 az group create 的资源组。下面的示例是在 East US 位置创建一个名为 myResourceGroupLoadBalancer 的资源组：

```
az group create --name myResourceGroupLoadBalancer --location eastus
```

3. 创建一个公共 IP 地址

要访问 Internet 上的应用程序，需要负载均衡器的公共 IP 地址。使用 az network public-ip create 命令创建一个公共 IP 地址。下面的示例是在 myResourceGroupLoadBalancer 资源组中创建名为 myPublicIP 的公共 IP 地址：

```
az network public-ip create \
    --resource-group myResourceGroupLoadBalancer \
    --name myPublicIP
```

4. 创建负载均衡器

使用 az network lb create 命令创建负载均衡器。下面的示例是创建名为 myLoadBalancer 的负载平衡器，并将 myPublicIP 地址分配给前端 IP 配置：

```
az network lb create \
```

```
--resource-group myResourceGroupLoadBalancer \
--name myLoadBalancer \
--frontend-ip-name myFrontEndPool \
--backend-pool-name myBackEndPool \
--public-ip-address myPublicIP
```

5. 创建健康探测器

为了让负载均衡器监视应用程序的状态,使用健康探测器,健康探针根据其对健康检查的响应动态地从负载均衡器旋转中添加或移除 VMs。默认情况下,在 15 秒间隔的两次连续故障之后,从负载平衡器分布中移除 VM。可以根据应用程序的协议或特定的健康检查页创建健康探针。

下面的示例创建一个 TCP 探针。还可以创建自定义的 HTTP 探针以进行更细粒度的健康检查。使用自定义 HTTP 探针时,必须创建健康检查页,如 health-check.js。该探测器必须为负载均衡器返回 HTTP 200 OK 响应,以保持主机旋转。

要创建 TCP 健康探测,您需要使用 az network lb probe create 命令。下面的示例创建一个名为 myHealthProbe 的健康探针:

```
az network lb probe create \
--resource-group myResourceGroupLoadBalancer \
--lb-name myLoadBalancer \
--name myHealthProbe \
--protocol tcp \
--port 80
```

6. 创建负载均衡器规则

负载均衡器规则用于定义如何将流量分配给 VMs。可以为传入通信量定义前端 IP 配置,并为接收通信量定义后端 IP 池以及所需的源和目标端口。为了确保只有健康的虚拟机接收流量,还可以定义要使用的健康探针。

使用 az network lb rule create 命令创建负载均衡器规则。下面的示例是创建名为 myLoadBalancerRule 的规则,使用 myHealthProbe 健康探测,并平衡端口 80 上的通信量:

```
az network lb rule create \
--resource-group myResourceGroupLoadBalancer \
--lb-name myLoadBalancer \
--name myLoadBalancerRule \
--protocol tcp \
--frontend-port 80 \
--backend-port 80 \
--frontend-ip-name myFrontEndPool \
```

```
    --backend-pool-name myBackEndPool \
    --probe-name myHealthProbe
```

7. 配置虚拟网络

在部署一些 VM 并可以测试你的均衡器之前,应创建支持的虚拟网络资源。有关虚拟网络的更多信息,请参见管理 Azure 虚拟网络教程。

8. 创建网络资源

使用 az network vnet create 命令创建一个虚拟网络。下面的示例是使用名为 MySubNET 的子网创建一个名为 MyVNET 的虚拟网络:

```
az network vnet create \
    --resource-group myResourceGroupLoadBalancer \
    --name myVnet \
    --subnet-name mySubnet
```

要添加网络安全组,请使用 az network nsg create。下面的示例是创建名为 myNetworkSecurityGroup 的网络安全组:

```
az network nsg create \
    --resource-group myResourceGroupLoadBalancer \
    --name myNetworkSecurityGroup
```

用 az network nsg rule create 创建网络安全组规则。下面的示例是创建一个名为 myNetworkSecurityGroupRule 的网络安全组规则:

```
az network nsg rule create \
    --resource-group myResourceGroupLoadBalancer \
    --nsg-name myNetworkSecurityGroup \
    --name myNetworkSecurityGroupRule \
    --priority 1001 \
    --protocol tcp \
    --destination-port-range 80
```

虚拟 NIC 是用 az network nic create 命令来创建。下面的示例创建了三个虚拟 NIC(在下面的步骤中为每个应用程序创建一个虚拟 NIC)。可以在任何时候创建额外的虚拟 NIC 和 VMS,并将它们添加到负载均衡器中:

```
for i in `seq 1 3`; do
    az network nic create \
        --resource-group myResourceGroupLoadBalancer \
        --name myNic$i \
        --vnet-name myVnet \
        --subnet mySubnet \
```

```
    --network-security-group myNetworkSecurityGroup \
    --lb-name myLoadBalancer \
    --lb-address-pools myBackEndPool
done
```

当创建所有三个虚拟 NIC 时,继续进行下一步骤。

9. 创建 cloud-init 配置和虚拟机

(1) 创建 cloud-init 配置

在 12.4 节关于如何在第一次引导时定制 Linux 虚拟机中,学习了如何使用 cloud-init 自动进行 VM 定制。可以使用同一个 cloud-init 配置文件来安装 nginx, 并在下一个步骤中运行一个简单的"Hello World"Node.js 应用程序。要查看负载均衡器的作用,在教程的末尾,可以在 Web 浏览器中访问这个简单的应用程序。

在当前的 Shell 中,创建一个名为 cloud-init.txt 的文件,并粘贴以下配置。例如,在 Cloud Shell 中创建文件,而不是在本地机器上创建文件。输入 sensible-editor cloud-init.txt 以创建文件并查看可用编辑器的列表,确保整个 cloud-init 文件被正确复制,特别是第一行:

```
#cloud-config
package_upgrade: true
packages:
  - nginx
  - nodejs
  - npm
write_files:
  - owner: www-data:www-data
  - path: /etc/nginx/sites-available/default
    content: |
      server {
        listen 80;
        location / {
          proxy_pass http://localhost:3000;
          proxy_http_version 1.1;
          proxy_set_header Upgrade $http_upgrade;
          proxy_set_header Connection keep-alive;
          proxy_set_header Host $host;
          proxy_cache_bypass $http_upgrade;
        }
      }
  - owner: azureuser:azureuser
  - path: /home/azureuser/myapp/index.js
    content: |
```

```
            var express = require('express')
            var app = express()
            var os = require('os');
            app.get('/', function (req, res) {
                res.send('Hello World from host ' + os.hostname() + '!')
            })
            app.listen(3000, function () {
                console.log('Hello world app listening on port 3000!')
            })
runcmd:
    - service nginx restart
    - cd "/home/azureuser/myapp"
    - npm init
    - npm install express -y
    - nodejs index.js
```

(2) 创建虚拟机

为了提高应用程序的高可用性,将 VMs 放置在可用集中。有关可用集的更多信息,请参见前面如何创建高度可用的虚拟机教程。

使用 az vm availability-set create 创建可用性集。下面的示例是创建一个名为 myAvailabilitySet 的可用性集:

```
az vm availability-set create \
    --resource-group myResourceGroupLoadBalancer \
    --name myAvailabilitySet
```

现在可以使用 az vm create 创建 VM。下面的示例创建了三个 VMs,如果它们不存在,则生成 SSH 密钥:

```
for i in `seq 1 3`; do
    az vm create \
        --resource-group myResourceGroupLoadBalancer \
        --name myVM$i \
        --availability-set myAvailabilitySet \
        --nics myNic$i \
        --image UbuntuLTS \
        --admin-username azureuser \
        --generate-ssh-keys \
        --custom-data cloud-init.txt \
        --no-wait
done
```

在 Azure CLI 将您的任务返回到提示之后,还有后台任务继续运行。不等待参数、不等待所有任务完成,可能再过几分钟就可以访问应用程序了。负载均衡器健康探针自动检测应用程序在每个 VM 上运行的时间。一旦应用程序运行,负载均衡器规则就开始分配流量。

10. 测试负载均衡器

用 az network public-ip show 获得负载均衡器的公共 IP 地址。下面的示例获得早期创建的 myPublicIP 的 IP 地址:

```
az network public-ip show \
    --resource-group myResourceGroupLoadBalancer \
    --name myPublicIP \
    --query [ipAddress] \
    --output tsv
```

然后,可以将公共 IP 地址输入到 Web 浏览器中。记住,在负载均衡器开始分配流量之前,VMs 需要几分钟才能准备好。应用程序被显示,包括 VM 的主机名,负载均衡器将其传输到如下示例中,如图 12-14 所示。

图 12-14 负载均衡器

要查看负载均衡器在所有运行应用程序的三个虚拟机上分配的流量,可以强制刷新 Web 浏览器。

11. 删除和添加 VM

您可能需要对运行的应用程序的 VM 进行维护,例如安装 OS 更新。为了处理应用程序增加的流量,可能需要添加额外的 VM。下面展示如何从负载均衡器中移除或添加 VM。

(1) 从负载均衡器中删除 VM

您可以使用 az network nic ip-config address-pool remove 命令从后端地址池中移除一个 VM。下面的示例从 myLoadBalancer 删除 MyVM2 的虚拟 NIC:

```
az network nic ip-config address-pool remove \
    --resource-group myResourceGroupLoadBalancer \
    --nic-name myNic2 \
    --ip-config-name ipConfig1 \
    --lb-name myLoadBalancer \
    --address-pool myBackEndPool
```

为了看到负载均衡器在应用程序中运行的其余两个 VM 上分配的流量,可以强制刷新 Web 浏览器。现在可以对 VM 执行维护,例如安装 OS 更新或执行 VM 重启。

要查看虚拟 NIC 连接到负载均衡器的 VMS 列表,请使用 az network lb address-pool show。查询和过滤虚拟网卡的 ID 如下:

```
az network lb address-pool show \
    --resource-group myResourceGroupLoadBalancer \
    --lb-name myLoadBalancer \
    --name myBackEndPool \
    --query backendIpConfigurations \
    --output tsv | cut -f4
```

输出类似于下面的示例,它表明 VM 2 的虚拟 NIC 不再是后端地址池的一部分:

/subscriptions/<guid>/resourceGroups/myResourceGroupLoadBalancer/providers/Microsoft.Network/networkInterfaces/myNic1/ipConfigurations/ipconfig1
/subscriptions/<guid>/resourceGroups/myResourceGroupLoadBalancer/providers/Microsoft.Network/networkInterfaces/myNic3/ipConfigurations/ipconfig1

(2) 将 VM 添加到负载均衡器

在执行 VM 维护之后,或者如果需要扩展容量,可以使用 az network nic ip-config address-pool add 将一个 VM 添加到后端地址池。下面的示例将 myVM 2 的虚拟 NIC 添加到 myLoadBalancer:

```
az network nic ip-config address-pool add \
    --resource-group myResourceGroupLoadBalancer \
    --nic-name myNic2 \
    --ip-config-name ipConfig1 \
    --lb-name myLoadBalancer \
    --address-pool myBackEndPool
```

为了验证虚拟 NIC 连接到后端地址池,再次使用 az network lb address-pool show 命令查看。

第 13 章

部署 MongoDB 到 Azure

13.1 MongoDB 简介

MongoDB 是一个基于分布式文件存储的数据库,由 C++语言编写,旨在为 Web 应用提供可扩展的高性能数据存储解决方案。

MongoDB 是一个介于关系数据库和非关系数据库之间的产品,是非关系数据库当中功能最丰富、最像关系数据库的。MongoDB 是 NoSQL 的一种,其整个系统适合于比较大量的数据。

1. NoSQL 简介

NoSQL(Not Only SQL),意即"不仅仅是 SQL"。NoSQL 通常是为大数据量准备的。其存储方式、查询方式都与关系型数据库有所不同。NoSQL 是一项全新的数据库革命性运动,早期就有人提出,发展至 2009 年趋势越发高涨。NoSQL 的拥护者们提倡运用非关系型的数据存储,相对于铺天盖地的关系型数据库运用,这一概念无疑是一种全新思维的注入。

关系型数据库具有以下特点:
- 数据严格地进行结构化组织;
- 结构化查询语言 SQL;
- 数据和关系都存储在表中;
- 数据定义语言,数据操作语言;
- 数据的一致性、完整性保护机制;
- 基于事务的管理(ACID)。

NoSQL 的特点如下:
- 代表不仅仅是 SQL;
- 没有声明性查询语言;
- 没有预定义的模式;
- 灵活的数据类型,比如:键/值对存储、列存储、文档存储、图形存储等;
- 最终的数据一致性,而不是 ACID 特性;
- 非结构化,数据不可预知性;

- 符合分布式领域 CAP 理论；
- 高性能、高可用性、高可伸缩性。

2．NoSQL 的优点和缺点

（1）优　点

- 高可扩展性；
- 分布式计算；
- 低成本；
- 架构的灵活性，半结构化数据；
- 没有复杂的关系。

（2）缺　点

- 没有标准化；
- 有限的查询功能（到目前为止）；
- 最终一致，是不直观的程序。

3．分布式领域 CAP 理论

- Consistency（一致性），数据一致更新，所有数据变动都是同步的；
- Availability（可用性），好的响应性能；
- Partition tolerance（分区容忍性），好的可靠性。

定理：任何分布式系统只可同时满足两点，没法三者兼顾。

忠告：架构师不要将精力浪费在如何设计能满足三者的完美分布式系统上，而是应该进行取舍。选其中两个方面做到比较完美，而第三方面就可以适当进行妥协。

CA——单点集群，满足一致性、可用性的系统，通常在可扩展性上不太强大。这些通常是由关系型数据库管理系统 RDBMS 做的。

CP——满足一致性、分区容忍性的系统，通常性能不是特别强。MongoDB、Redis、Hbase 等就属于这一类。

AP——满足可用性、分区容忍性的系统，通常可能对一致性要求低一些。CouchDB、Cassandra、DynamoDB、Riak 等都属于这一类。

BASE：Basically available、Soft State、Eventually consistent 由 Eric Brewer 定义。

BASE 模型的反 ACID 模型，完全不同于 ACID 模型，牺牲高一致性，获得可用性或可靠性。

- Basically available（基本可用）：支持分区失败（如碎片划分数据库）。
- Soft state（软状态）：可以有一段时间不同步，异步。Soft state 可以理解为"无连接"的；而 Hard state（硬状态）是"面向连接"的。
- Eventually consistent（最终一致）：最终数据是一致的就可以了，而不是时时高一致。

BASE 思想实现的主要功能有：
① 按功能划分数据库；
② 按碎片划分数据库。

BASE 思想主要强调基本的可用性，如果需要 High 可用性，也就是纯粹的高性能，那么就要以一致性或容忍性为牺牲，BASE 思想的方案在性能上还是有潜力可挖的。NoSQL 数据库分类如表 13-1 所列。

表 13-1 NoSQL 数据库分类

类型	部分代表	特点
列存储	Hbase；Cassandra；Hypertable	顾名思义，是按列存储数据的。最大的特点是方便存储结构化和半结构化数据，方便做数据压缩，对针对某一列或者某几列的查询有非常大的 IO 优势
文档存储	MongoDB；CouchDB	文档存储一般用类似 json 的格式存储，存储的内容是文档型的。这样也就有机会对某些字段建立索引，实现关系数据库的某些功能
key-value 存储	MemcacheDB；Redis	可以通过 key 快速查询到其 value。一般来说，存储不管 value 的格式，照单全收。（Redis 包含了其他功能）
图存储	Neo4J；FlockDB	图形关系的最佳存储。使用传统关系数据库来解决的话性能低下，而且设计使用不方便
XML 数据库	Berkeley DB XML BaseX	高效地存储 XML 数据，并支持 XML 的内部查询语法，比如 XQuery、Xpath

MongoDB 提供了一种与 SQL 类似的语法来查询数据。下面将要学习如何在 Azure 云平台中使用 MongoDB 处理大数据。本章将要介绍在 Linux VM 上部署 MongoDB 的软件，以及利用 Azure CosmosDB 提供的 MongoDB API 来作为类似 MongoDB 的访问方式。

13.2 单个 MongoDB 部署

MongoDB 是一种流行的开源、高性能的 NoSQL 数据库。本节将介绍如何使用 Azure CLI 在 Linux VM 上安装和配置 MongoDB。示例演示了：
- 如何手动安装和配置基本的 MongoDB 实例；
- 使用资源管理器模板创建基本的 MongoDB 实例。

在 VM 上手动安装和配置 MongoDB

MongoDB 为 Linux 发行版本提供安装指令，包括 Red Hat/CENT OS、SUS-EU、Ubuntu 和 Debian。下面的示例是创建一个 CENTOS VM。要创建这个环境，需要安装最新的 Azure CLI 并使用 az login 登录到 Azure 账户。

创建具有 az group create 的资源组。下面的示例是在 East US 位置创建一个名为 myResourceGroup 的资源组：

az group create -- name myResourceGroup -- location eastus

使用 az vm create 命令创建一个虚拟机。下面的示例是使用 SSH 公钥认证创建一个名为 myVM 的 VM，该用户名为 azureuser。

```
az vm create \
    -- resource-group myResourceGroup \
    -- name myVM \
    -- image CentOS \
    -- admin-username azureuser \
    -- generate-ssh-keys
```

SSH 使用您自己的用户名和前面步骤输出中列出的 publicIpAddress 到 VM：

ssh azureuser@ <publicIpAddress>

若要添加 MongoDB 的安装源，请创建如下的 YUM 存储库文件：

sudo touch /etc/yum.repos.d/mongodb-org-3.6.repo

用 vi 或者 nano 打开 mongodb-org-3.6.repo 文件，编辑如下内容：

```
[mongodb-org-3.6]
name = MongoDB Repository
baseurl = https://repo.mongodb.org/yum/redhat/$releasever/mongodb-org/3.6/x86_64/
gpgcheck = 1
enabled = 1
gpgkey = https://www.mongodb.org/static/pgp/server-3.6.asc
```

使用 YUM 安装 MongoDB 如下：

sudo yum install -y mongodb-org

默认情况下，SELinux 在 CENT OS 镜像上强制执行，阻止您访问 MongoDB。安装策略管理工具并配置 SELinux，以允许 MongoDB 在其默认 TCP 端口 27017 上。操作如下：

```
sudo yum install -y policycoreutils-python
sudo semanage port -a -t mongod_port_t -p tcp 27017
```

启动 MongoDB 服务如下：

sudo service mongod start

通过使用本地 Mongo 客户端验证 MongoDB 安装：

mongo

现在通过添加一些数据,然后搜索。测试 MongoDB 实例如下:

```
> db
test
> db.foo.insert( { a:1 } )
> db.foo.find()
{ "_id":ObjectId("57ec477cd639891710b90727"), "a":1 }
> exit
```

如果需要,配置 MongoDB 在系统重启期间自动启动:

```
sudo chkconfig mongod on
```

至此,单机的 MongoDB 就安装好了。已经可以用本机的 Mongo 客户端访问本机的 MongoDB 服务了。

13.3 用 gz 发行包的方式安装 MongoDB

前面是采用增加 yum 库的方式来安装一个 MongoDB。这样在发出安装命令之后,yum 需要从 yum 库中指定的位置来下载安装包。这里再介绍一个基于 gz 发行包的方式安装 MongoDB。MongoDB 提供了 Linux 各发行版本 32 位/64 位的安装包,可以在官网下载安装包。MongoDB 分企业版和社区版。企业版是要收费的。如果不需要那些收费功能,则一般都选择社区版。下载网址如下:

https://www.mongodb.com/download-center#community。

下载 MongoDB 如图 13-1 所示。

① 下载 gz 文件。命令如下:

```
curl -O https://fastdl.mongodb.org/linux/mongodb-linux-x86_64-3.0.6.tgz
```

② 对 gz 文件进行解压。命令如下:

```
tar -zxvf mongodb-linux-x86_64-3.0.6.tgz
```

③ 此操作会解压出来一个目录,名为 mongodb-linux-x86_64-3.0.6,将其复制到 /usr/local/mongodb。命令如下:

```
mv mongodb-linux-x86_64-3.0.6/ /usr/local/mongodb
```

④ 将 MongoDB 安装目录中的 bin 放到 PATH 变量中,这样就会让系统能找到 MongoDB 的执行文件。命令如下:

```
export PATH = <mongodb-install-directory>/bin:$PATH
```

在这里就是:

```
export PATH = /usr/local/mongodb/bin:$PATH
```

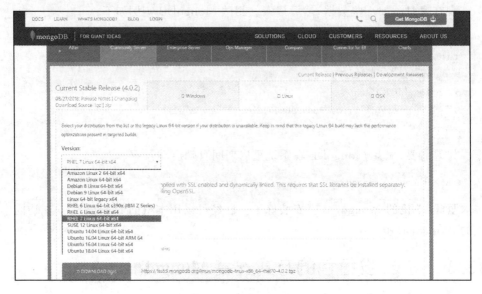

图 13-1　下载 MongoDB

1. 创建数据库数据文件所在的目录

MongoDB 默认使用/data/db 作为数据文件存放的目录。因为采用的是 gz 文件解压形式的安装,故这个数据文件的存储目录需要手工进行创建。以下实例中将 data 目录创建于根目录(/)下。采用如下命令：

```
mkdir -p /data/db
```

-p 是如果/data 不存在,则先创建/data 目录,创建好/data 目录后,再在/data 目录中创建 db 子目录。

MongoDB 采用--dbpath 启动参数来指定数据文件存放的目录,其默认值是/data/db。

2. 命令行中运行 MongoDB 服务

可以在命令行中执行 Mongo 安装目录中的 bin 目录,执行 Mongod 命令来启动 MongDB 服务。

```
$ ./mongod
2015 - 09 - 25T16:39:50.549 + 0800 I JOURNAL [initandlisten] journal dir = /data/db/journal
2015 - 09 - 25T16:39:50.550 + 0800 I JOURNAL [initandlisten] recover:no journal files present, no recovery needed
2015 - 09 - 25T16:39:50.869 + 0800 I JOURNAL [initandlisten] preallocateIsFaster = true 3.16
2015 - 09 - 25T16:39:51.206 + 0800 I JOURNAL [initandlisten] preallocateIsFaster = true 3.52
2015 - 09 - 25T16:39:52.775 + 0800 I JOURNAL [initandlisten] preallocateIsFaster = true 7.7
```

看到类似的信息则说明 MongoDB 已经启动起来了。

13.4 MongoDB 的管理

表 13-2 是关系型数据库概念与 MongoDB 基本概念对照表。

表 13-2 关系型数据库概念与 MongoDB 基本概念对照表

关系型数据库概念	MongoDB 基本概念	说　明
database	database	数据库
table	collection	数据库表/集合
row	document	数据记录行/文档
column	field	字段/域
index	index	索引
table joins		表连接，MongoDB 不支持
primary key	primary key	关系型数据库可以指定多个字段为主键，MongoDB 是自带的_id 字段指定为主键

MongoDB 的记录用 json 格式表示，如：

{

"_id":"ObjectId("5604ff74a274a611b0c990aa")",

"student_name":"Scott Tiblier",

"birth_date":"1990-05-23",

"address":"256, 1st Ave, Beaverton, Oragon",

"age":"28"

},

{

"_id":"ObjectId("5604ff74a274a611b0c990ab")",

"student_name":"Mike Ribbon",

"birth_date":"2008-01-09",

"address":"12, 2nd Ave, Beaverton, Oragon",

"age":"18"

}

…

…

它实际是一系列如此的 document 构成的集合。这每一个 document，都类似于一个表的记录。在这样的基础上可以进行搜索、数据的管理等。

关系型数据库的表格是一个二维的表格，与 MongoDB 的文档是有差别的。MongoDB 的存储引擎存储这些数据的方式与关系型数据库的存储方式差别较大。

1. 数据库

一个 MongoDB 中可以建立多个数据库。MongoDB 的默认数据库为 db，该数据库存储在 data 目录中。MongoDB 的单个实例可以容纳多个独立的数据库，每一个都有自己的集合和权限，不同的数据库也放置在不同的文件中。

show dbs 命令可以显示所有数据的列表。

```
$ ./mongo
MongoDB shell version:3.0.6
connecting to:test
> show dbs
local   0.078GB
test    0.078GB
>
```

执行 db 命令可以显示当前数据库对象或集合。

```
> db
test
>
```

运行 use 命令，可以转到一个指定的数据库，作为当前数据库。如果此数据库不存在，则会新建此数据库。

```
> use local
switched to db local
> db
local
>
```

use 命令后面的数据库名的要求如下：
- 不能是空字符串("")。
- 不得含有" '"(空格)、"."、"$"、"/"、"\"和"\0"(空字符)。
- 应全部小写。
- 最多 64 字节。
- 不能使用保留的数据库名：admin、local、config。这几个都是保留的数据库名。

① admin 数据库是最高权限的数据库，如果将用户加入此数据库，则此用户自动继承所有数据库的权限。有一些命令只能针对 admin 数据库来运行，比如列出所有数据库或者关闭 MongoDB 服务。

② local 数据库是一个本地数据库，它不会被 MongoDB 管理系统复制，它只在运行 MongoDB 的服务器上驻留，适合用来存储只在本地单台服务器中的文档。

如前所述，文档相当于关系型数据库的记录。文档以 json 格式记录一些 Field 和对应的值。多个文档有相同的 Field，对应于关系型数据库就是多个记录有相同的字段。

与关系型数据库不同的是，MongoDB 下，同一个 Field 在不同文档中可以有不同数据类型。而关系型数据库则多个记录的相同字段必须具有同样的数据类型，如：

```
{
"_id":"ObjectId("5604ff74a274a611b0c990aa")",
"student_name":"Scott Tiblier",
"birth_date":"1990-05-23",
"address":"256, 1st Ave, Beaverton, Oragon",
"age":28
},
{
"_id":"ObjectId("5604ff74a274a611b0c990ab")",
"student_name":"Mike Ribbon",
"birth_date":"2008-01-09",
"address":"12, 2nd Ave, Beaverton, Oragon",
"age":"18"
}
```

注意到上面的 age field 可以有不同类型的值。用过关系型数据库系统的，都会知道上述的特性在关系型数据库中是不可能具有的，即在关系型数据库之下这是不允许的。如果真那么做了，就不是关系型数据库了。

需要注意的是：
- 文档中的键/值对是有序的。
- 文档中的值不仅可以是在双引号里面的字符串，还可以是其他几种数据类型（甚至可以是整个嵌入的文档）。
- MongoDB 区分类型和大小写。
- MongoDB 的文档不能有重复的键。
- 文档的键是字符串。除了少数例外情况，键可以使用任意 UTF-8 字符。

文档键命名规范：
- 键不能含有"\0"(空字符)。这个字符用来表示键的结尾。
- "."和"＄"有特别的意义，只有在特定环境下才能使用。
- 以下划线"_"开头的键是保留的(不是严格要求的)。

2. 集　合

集合就是 MongoDB 文档的集合，类似于 RDBMS(关系数据库管理系统：Relational Database Management System)中的表格 Table。

集合存在于数据库中，集合没有固定的结构，这意味着在集合中可以插入不同

Field 和类型的数据,但通常情况下我们插入集合的数据都会有一定的关联性。可以有的文档 Field 多,有的少。同样的 Field 数值类型也不一样。

比如,我们可以将以下不同数据结构的文档插入到集合中:

{"site":"www.google.com","name":"Google"}
{"site":"www.baidu.com", visits:56000}
{"site":"www.sohu.com","name":"sohu","num";5}

3. 合法的集合名

- 集合名不能是空字符串""""。
- 集合名不能含有"\0"字符(空字符),这个字符表示集合名的结尾。
- 集合名不能以"system."开头,这是为系统集合保留的前缀。

用户创建的集合名字不能含有保留字符。有些驱动程序的确支持在集合名里面包含,这是因为某些系统生成的集合中包含该字符。除非要访问这种系统创建的集合,否则千万不要在名字里出现"$"。

一般地,集合不限定文档的个数,只受限于存储空间的大小。

4. Capped collections

Capped collections 就是固定大小的 collection。

Capped collections 有很强的性能,并且可以维护文档的时效。若过期,则按照插入的顺序来丢弃过期的文档。

Capped collections 可高性能地自动维护对象的插入顺序。创建 collection 时必须显示地支持 Capped,并用 size 指定其大小。这个 size 参数可以根据实际需要来提供。大部分情况下的尺寸能满足需求才可以,单位是字节。collection 的数据存储空间值是提前分配的。

要注意的是,指定的存储大小包含了数据库的头信息:

db.createCollection("mycoll", {capped:true, size:100000})。

在 Capped collections 中,能添加新的对象,能进行更新,然而对象不会增加存储空间。如果增加,更新就会失败。

数据库不允许进行删除。使用 drop() 方法删除 collection 所有的行。

注意:删除之后,必须显式地重新创建这个 collection。

在 32 bit 机器中,Capped collections 最大存储为 $1e9(1 \times 10^9)$ 个字节。

5. 元数据

数据库的信息是存储在集合中的。它们使用了系统的命名空间:

dbname.system.*。

表 13-3 是在 MongoDB 数据库中的集合命名空间,dbname.system.* 是包含多种系统信息的特殊集合(collection)。

表 13-3 集合命名空间及其描述

集合命名空间	描述
dbname.system.namespaces	列出所有名字空间
dbname.system.indexes	列出所有索引
dbname.system.profile	包含数据库概要(profile)信息
dbname.system.users	列出所有可访问数据库的用户
dbname.local.sources	包含复制对端(slave)的服务器信息和状态

对于修改系统集合中的对象有如下限制：

- 在{{system.indexes}}中只能插入数据(创建索引)，除此之外该表信息是不可变的(特殊的 drop index 命令将自动更新相关信息)。
- {{system.users}}是可修改的，可增加、删除、修改用户等。
- {{system.profile}}是可删除的。

6. MongoDB 数据类型

表 13-4 为 MongoDB 中常用的几种数据类型。

表 13-4 数据类型及其描述

数据类型	描述
String	字符串。在 MongoDB 中，默认使用 UTF-8 编码。它是存储数据中最常见的数据类型
Integer	用于存储整型数值。根据所采用的服务器，可分为 32 位或 64 位
Boolean	用于存储布尔值(true/false)
Double	用于存储双精度浮点值
Min/Max keys	将一个值与 BSON(二进制的 JSON)元素的最低值和最高值相对比
Array	数组，用于将多个元素存储为一个整体。应用程序使用的时候可以遍历它
Timestamp	时间戳。记录文档修改或添加的具体时间。 BSON 有一个特殊的时间戳类型用于 MongoDB 内部使用，与普通的日期类型不相关。时间戳值是一个 64 位的值。其中： 前 32 位是一个 time_t 值(与 Unix 新纪元相差的秒数)； 后 32 位是在某秒中操作的一个递增的序数。 在单个 Mongod 实例中，时间戳值通常是唯一的。 在复制集中，oplog 有一个 ts 字段。这个字段中的值使用 BSON 时间戳表示了操作时间。 BSON 时间戳类型主要在 MongoDB 内部使用。在大多数情况下的应用开发中，可以使用 BSON 日期类型

续表 13-4

数据类型	描 述
Object	用于内嵌文档。前文中提到，一个文档可以作为另一个文档的某个 Field 的内容
Null	空值。与关系型数据库中的空值类似
Symbol	符号。该数据类型基本上等同于字符串类型，但不同的是，它一般用于采用特殊符号类型的语言
Date	日期时间。可以指定自己的日期时间，创建 Date 对象，传入年月日信息，按照服务器的 timezone 来存储。 表示当前距离 Unix 新纪元(1970 年 1 月 1 日)的毫秒数。日期类型是有符号的，负数表示 1970 年之前的日期。 ＞ var mydate1＝new Date()　　//格林尼治 GMT 时间 ＞ mydate1 ISODate("2018-03-04T14:58:51.233Z") ＞ typeof mydate1 object ＞ var mydate2＝ISODate()　　//格林尼治 GMT 时间 ＞ mydate2 ISODate("2018-03-04T15:00:45.479Z") ＞ typeof mydate2 object 这样创建的时间是日期类型，可以使用 JS 中的 Date 类型的方法。 返回一个时间类型的字符串： ＞ var mydate1str＝mydate1.toString() ＞ mydate1str Sun Mar 04 2018 14:58:51 GMT＋0000（UTC） ＞ typeof mydate1str string 或者 ＞ Date() Sun Mar 04 2018 15:02:59 GMT＋0000（UTC）
Object ID	对象 ID。用于创建文档的 ID。这是一个集合的主键，一般用于_id field。 ObjectId 类似唯一主键，可以很快地去生成和排序，包含 12 个字节，含义是： 前 4 个字节表示创建 unix 时间戳，格林尼治时间、UTC 时间，比北京时间晚了 8 个小时。 接下来的 3 个字节是机器标识码。 紧接的 2 个字节由进程 ID 组成 PID。 最后 3 个字节是随机数

续表 13-4

数据类型	描　述
| Object ID | 如图 13-2 所示。<br><br>```
 0 | 1 | 2 | 3 | 4 | 5 | 6 | 7 | 8 | 9 | 10 | 11
   时间戳        机器        PID      计数器
```<br><br>图 13-2　ObjectID<br><br>MongoDB 中存储的文档必须有一个 _id 键。这个键的值可以是任何类型的，默认是个 ObjectId 对象。<br>由于 ObjectId 中保存了创建的时间戳，所以不需要为文档保存时间戳字段，可以通过 getTimestamp 函数来获取文档的创建时间：<br>&gt; var newObject=ObjectId()<br>&gt; newObject.getTimestamp()<br>ISODate("2017-11-25T07:21:10Z")<br>ObjectId 转为字符串<br><br>&gt; newObject.str<br>5a1919e63df83ce79df8b38f |
Binary Data	二进制数据类型。用于存储二进制数据
Code	代码类型。用于在文档中存储 JavaScript 代码
Regular expression	正则表达式类型。用于存储正则表达式

7. 后台管理

如果需要进入 MongoDB 后台进行后台管理，则需要使用 mongo 命令来连接 MongoDB 服务器端。mongo 命令是 MongoDB 中自带的客户端程序，便于使用者使用命令行的方式发送命令给 MongoDB 服务器端来执行。

```
$ cd /usr/local/mongodb/bin
$ ./mongo
MongoDB shell version: 3.0.6
connecting to: test
Welcome to the MongoDB shell.
……
```

此命令行 Shell 自带一个 Javascript 引擎，可以执行一些简单运算。如：

> 3+6
9

8. MongoDB 连接

./mongo 是用来在服务器上连接本机的 MongoDB 服务的。如果需要从别的
```

linux 机器上连接 MongoDB 服务器,则需要如下的命令:

mongodb://[username:password@]host1[:port1][,host2[:port2],...[,hostN[:portN]]][/[database][?options]]

- "mongodb://"是固定的格式,必须要指定,表明是一个 MongoDB 连接。
- "username:password@"是可选项,如果设置,则在连接数据库服务器之后,驱动都会尝试登录这个数据库,在需要登录时,从用户名和密码就是需要的。
- host1 必须指定至少一个 host,host1 是这个 URI 唯一要填写的。它指定了要连接服务器的地址。如果要连接复制集,请指定多个主机地址。
- portX 是可选的指定端口,如果不填,则默认为是 27017。
- /database 是指,如果指定 username:password@,则连接并验证登录指定数据库;若不指定,则默认打开 test 数据库。
- ?options 是连接选项。如果不使用/database,则前面需要加上"/"。所有连接选项都是键/值对 name=value,键/值对之间通过"&"或";"(分号)隔开。

replicaSet=name 是验证 replica set 的名称。命令如下:

Impliesconnect = replicaSet.

(1) slaveOk=true|false

true:在 connect=direct 模式下,驱动会连接第一台机器,即使这台服务器不是主服务器。在 connect=replicaSet 模式下,驱动会发送所有的写请求到主服务器并且把读取操作分布在其他从服务器上。

false:在 connect=direct 模式下,驱动会自动寻找主服务器。在 connect=replicaSet 模式下,驱动仅仅连接主服务器,并且所有的读/写命令都连接到主服务器。

(2) safe=true|false

true:在执行更新操作之后,驱动都会发送 getLastError 命令来确保更新成功。(还要参考 wtimeoutMS。)

false:在每次更新之后,驱动都不会发送 getLastError 来确保更新成功。

w=n 是驱动添加 {w:n} 到 getLastError 的命令,应用于 safe=true。

wtimeoutMS=ms 是驱动添加{ wtimeout:ms }到 getlasterror 的命令,应用于 safe=true。

(3) fsync=true|false

true:驱动添加 { fsync:true } 到 getlasterror 命令,应用于 safe=true。

false:驱动不会添加到 getLastError 命令中。

journal=true|false 是如果设置为 true,则同步到 journal(在提交到数据库前写入到实体中),应用于 safe=true。

connectTimeoutMS=ms 是可以打开连接的时间。

socketTimeoutMS=ms 是发送和接收 sockets 的时间。

按照上面的 MongoDB 连接的格式说明,举一个样例:

mongodb://fred:tiger@example1.com:27017/? connect = direct;slaveOk = true

用用户名和密码直接连接名为 example1.com 的第一个服务器,无论是 replica set 的一部分还是主服务器或者从服务器。

9. 创建数据库/打开数据库

use DATABASE_NAME

如果此名字的数据库不存在的话,这样的命令会自动创建一个此名字的数据库;如果存在,就会将此数据库打开。

10. 创建集合

db.createCollection(name, options)

name:要创建的集合名称。

options:可选参数,指定有关内存大小及索引的选项。

options 可以是如下参数,如表 13 - 5 所列。

表 13 - 5 db.createCollection 的参数

字段	类型	说明
capped	布尔	(可选)如果为 true,则创建固定集合。当该值为 true 时,必须指定 size 参数。固定集合是指有着固定大小的集合,当达到最大值时,它会自动覆盖最早的文档。当该值为 false 时,为非固定的集合,则 size 参数不必提供
autoIndexId	布尔	(可选)默认为 false。如果为 true,则自动在 _id 字段创建索引
size	数值	(可选)配合 capped 字段使用。如果 capped 为 true,则需要指定该字段。该数值用来指定固定集合的大小
max	数值	(可选)指定固定集合中包含文档的最大数量

用下面的命令创建一个集合(相当于 table):

db.createCollection("testdb")

现在让我们插入一些简单的数据,并对插入的数据进行检索。

(1) MongoDB 插入文档

```
> db.testdb.insert({count:10})
WriteResult({ "nInserted":1 })
> db.testdb.find()
{ "_id":ObjectId("5678ea74a265a611b0c880bb"), "count":10 }
>
```

第一个命令将数字 10 插入到 testdb 集合的 count 字段中。后面可以继续插入数据(文档)。后面插入的数据可以跟第一个插入的数据有不同的 Field。比如比第一个数据多一些 Field。

```
> db.testdb.insert({title: 'MongoDB 教程',
    description: 'MongoDB 是一个 Nosql 数据库',
    by: '菜鸟教程',
    url: 'http://selflearn',
    tags: ['mongodb', 'database', 'NoSQL'],
    likes: 100,
    count:100
})
```

(2) MongoDB 更新文档

MongoDB 使用 update()和 save()方法来更新集合中的文档。接下来详细看一下两个函数的应用及其区别。

1) update()方法

update()方法用于更新已存在的文档。语法格式如下：

```
db.collection.update(
    <query>,
    <update>,
    {
        upsert: <boolean>,
        multi: <boolean>,
        writeConcern: <document>
    }
)
```

参数说明：

query:update 的查询条件,类似 sql update 查询内 where 后面的部分。

update:update 的对象和一些更新的操作符(如 $,$ inc...)等,也可以理解为 sql update 查询内 set 后面的部分。

upsert:可选,这个参数的意思是,如果不存在 update 的记录,是否插入 objNew。true 为插入；默认是 false,不插入。

multi:可选,mongodb 默认是 false,只更新找到的第一条记录。如果这个参数为 true,就把按条件查出来的多条记录全部更新。

writeConcern:可选,抛出异常的级别。

实例：在集合 col 中插入如下数据：

```
>db.col.insert({
    title: 'MongoDB 教程',
    description: 'MongoDB 是一个 Nosql 数据库',
```

```
        by:'菜鸟教程',
        url:'http://selflearn',
        tags:['mongodb','database','NoSQL'],
        likes:100
})
```

接着通过 update() 方法来更新标题(title)：

```
> db.col.update({'title':'MongoDB 教程'},{ $ set:{'title':'MongoDB'}})
WriteResult({ "nMatched":1, "nUpserted":0, "nModified":1 })   ＃输出信息
> db.col.find().pretty()
{
    "_id": ObjectId("56064f89ade2f21f36b03136"),
    "title": "MongoDB",
    "description": "MongoDB 是一个 Nosql 数据库",
    "by": "菜鸟教程",
    "url": "selflearn",
    "tags": [
  "mongodb",
  "database",
  "NoSQL"
    ],
    "likes": 100
}
>
```

可以看到标题(title)由原来的"MongoDB 教程"更新为了"MongoDB"。

以上语句只会修改第一条发现的文档，如果你要修改多条相同的文档，则需要设置 multi 参数为 true。

```
> db.col.update({'title':'MongoDB 教程'},{ $ set:{'title':'MongoDB'}},{multi:true})
```

2) save()方法

save()方法通过传入的文档来替换已有的文档。语法格式如下：

```
db.collection.save(
    <document>,
    {
        writeConcern: <document>
    }
)
```

参数说明：

document:文档数据。

writeConcern：可选，抛出异常的级别。

实例。以下实例是替换了_id 为 76084f89adf3f21f36b13136 的文档数据：

```
>db.col.save({
    "_id":ObjectId("76084f89adf3f21f36b13136"),
    "title":"MongoDB",
    "description":"MongoDB 是一个 Nosql 数据库",
    "by":"selflearn",
    "url":"http://selflearn",
    "tags":[
        "mongodb",
        "NoSQL"
    ],
    "likes":110
})
```

替换成功后，可以通过 find() 命令来查看替换后的数据。

```
>db.col.find().pretty()
{
    "_id":ObjectId("76084f89adf3f21f36b13136"),
    "title":"MongoDB",
    "description":"MongoDB 是一个 Nosql 数据库",
    "by":"selflearn",
    "url":"http://selflearn",
    "tags":[
        "mongodb",
        "NoSQL"
    ],
    "likes":110
}
>
```

11. MongoDB 删除文档

MongoDB remove() 函数用来移除集合中的数据。在执行 remove() 函数前先执行 find() 命令来判断执行的条件是否正确，这是一个比较好的习惯。

remove() 方法的基本语法格式如下所示。

如果你的 MongoDB 是 2.6 版本以后的，语法格式如下：

```
db.collection.remove(
    <query>,
    {
        justOne: <boolean>,
```

```
        writeConcern: <document>
    }
)
```

参数说明：

query:(可选)删除的文档的条件。

justOne:(可选)如果设为 true 或 1,则只删除一个文档。

writeConcern :(可选)抛出异常的级别。

实例。以下文档执行两次插入操作：

```
>db.col.insert({title: 'MongoDB 教程 ',
    description: 'MongoDB 是一个 Nosql 数据库 ',
    by: ' 菜鸟教程 ',
    url: 'http://selflearn',
    tags: ['mongodb', 'database', 'NoSQL'],
    likes: 100
})
```

使用 find()函数查询数据：

```
> db.col.find()
{ "_id":ObjectId("67086169ade2f21f36b03137"), "title":"MongoDB 教程", "description":"MongoDB 是一个 Nosql 数据库", "by":"菜鸟教程", "url":"http://selflearn", "tags":["mongodb", "database", "NoSQL" ], "likes":100 }
{ "_id":ObjectId("6708616dade2f21f36b03138"), "title":"MongoDB 教程", "description":"MongoDB 是一个 Nosql 数据库", "by":"菜鸟教程", "url":"http://selflearn", "tags":["mongodb", "database", "NoSQL" ], "likes":100 }
```

接下来是移除 title 为 'MongoDB 教程 ' 的文档：

```
>db.col.remove({'title':'MongoDB 教程 '})
WriteResult({ "nRemoved":2 })      #删除了两条数据
>db.col.find()
……                                 #没有数据
```

一般地,可以用_id Field 的值类进行查找,即 query 部分可以用_id Field。命令如下：

```
>db.col.remove({ "_id":ObjectId("67086169ade2f21f36b03137")})
```

因为_id Field 作为 MongoDB 的主键,可以用来唯一定位一个文档。这样找到的文档总是唯一的。

如果你只想删除第一条找到的记录,可以设置 justOne 为 1,如下所示：

```
>db.COLLECTION_NAME.remove(DELETION_CRITERIA,1)
```

如果想删除所有数据,可以使用以下方式(类似常规 SQL 的 truncate 命令):

```
>db.col.remove({})
>db.col.find()
>
```

12. MongoDB 查询文档

MongoDB 查询文档使用 find()方法。无参数的 find()方法则表示查询出所有文档,就会以非结构化的方式来显示所有文档。find()方法也允许带参数。

MongoDB 查询数据的语法格式如下:

db.collection.find(query, projection)

query:可选,使用查询操作符指定查询条件,这和其他操作类似。可以指定一些条件来过滤文档,符合条件的文档才会出现在结果集合里。

projection:可选,用来做投影操作,即输出文档的部分 Field。因为是可选参数,所有默认情况下都没有投影操作,即输出文档的所有 Field。如果有需要做投影操作,则需要带上 projection 部分的参数。其分为包含模式或者排除模式(要么全部 0,要么全部 1)。全部 1 的是包含模式,让一些 Field=1,则表示这些字段是需要输出的。如果是让一些 Field=0,则表示这些字段是需要从结果中排除的。例如:

{url=0,tags=0}是一个排除模式的投影操作,即从结果中排除 url 和 tags 两个 Field。

{url=1,tags=1}是一个包含模式的投影操作,即在结果中只包含 url、tags 两个 Field。

如果需要以易读的方式来读取数据,可以使用 pretty()方法,语法格式如下:

```
>db.col.find().pretty()
```

pretty()方法以格式化的方式来显示所有文档。

实例。以下实例查询了集合 col 中的数据:

```
>db.col.find().pretty()
{
    "_id":ObjectId("36903717ade2f21f36b03133"),
    "title":"MongoDB 教程",
    "description":"MongoDB 是一个 Nosql 数据库",
    "by":"菜鸟教程",
    "url":"http://selflearn",
    "tags":[
        "mongodb",
        "database",
        "NoSQL"
```

],
"likes":100
}

除了 find() 方法之外,还有一个 findOne() 方法,它只返回一个文档。

13. MongoDB 与 RDBMS Where 语句比较

如果你熟悉常规的 SQL 数据,通过表 13-6 可以更好地理解 MongoDB 的条件语句查询。

表 13-6 条件语句对比

操 作	范 例	关系型数据库中的类似语句
等于	db.col.find({"by":"菜鸟教程"})	where by='菜鸟教程'
小于	db.col.find({"test":{$lt:50}})	where test <50
小于或等于	db.col.find({"test":{$lte:50}})	where test <=50
大于	db.col.find({"test":{$gt:50}})	where test> 50
大于或等于	db.col.find({"test":{$gte:50}})	where test>=50
不等于	db.col.find({"test":{$ne:50}})	where test!=50

14. MongoDB AND 条件

AND 条件的 query,用逗号分隔开,即为 AND 条件。

语法格式如下:

>db.col.find({key1:value1, key2:value2}).pretty()

实例。以下实例通过 by 和 title 键来查询菜鸟教程中 MongoDB 教程的数据:

>db.col.find({"by":"菜鸟教程","title":"MongoDB 教程"}).pretty()
{
    "_id":ObjectId("56063f17ade2f21f36b03133"),
    "title":"MongoDB 教程",
    "description":"MongoDB 是一个 Nosql 数据库",
    "by":"菜鸟教程",
    "url":"http://selflearn",
    "tags":[
        "mongodb",
        "database",
        "NoSQL"
    ],
    "likes":100
}

以上实例中类似于 WHERE 语句：WHERE by=' 菜鸟教程 'AND title='MongoDB 教程'。

15. MongoDB OR 条件

MongoDB OR 条件语句使用了关键字 $or,语法格式如下：

```
>db.col.find(
   {
      $or: [
         {key1: value1}, {key2:value2}
      ]
   }
).pretty()
```

实例。以下实例中,演示了查询键 by 值为菜鸟教程或键 title 值为 MongoDB 教程的文档：

```
>db.col.find({ $or:[{"by":"菜鸟教程"},{"title": "MongoDB 教程"}]}).pretty()
{
   "_id":ObjectId("56063f17ade2f21f36b03133"),
   "title":"MongoDB 教程",
   "description":"MongoDB 是一个 Nosql 数据库",
   "by":"菜鸟教程",
   "url":"http://selflearn",
   "tags":[
      "mongodb",
      "database",
      "NoSQL"
   ],
   "likes":100
}
>
```

16. AND 和 OR 联合使用

以下实例演示了 AND 和 OR 联合使用,类似常规 SQL 语句：'where likes > 50 AND (by=' 菜鸟教程 'OR title='MongoDB 教程')'。

```
>db.col.find({"likes": {$gt:50}, $or: [{"by":"菜鸟教程"},{"title": "MongoDB 教程"}]}).pretty()
{
   "_id":ObjectId("56063f17ade2f21f36b03133"),
   "title":"MongoDB 教程",
   "description":"MongoDB 是一个 Nosql 数据库",
```

```
    "by":"菜鸟教程",
    "url":"http://selflearn",
    "tags":[
        "mongodb",
        "database",
        "NoSQL"
    ],
    "likes":100
}
```

17. MongoDB $ type 操作符

$type 操作符用来得到 Field 的类型。这个类型可以在某些 query 中作为条件查询。例如：

```
db.col.find({"title":{$type:'string'}})
```

此语句查询 title Field，如果为 string 就输出出来。前文说过，多个文档的同一个 Field 的数据类型可以不同，因为它是非结构化的数据。当某一部分的文档该 Field 的类型为某一种，比如数组或者整型时，可以用 $type 操作符来查询出数据的类型。

18. MongoDB limit 与 skip 方法

limit 方法可以帮助我们读取指定数量的数据记录。limit() 方法接受一个数字参数，该参数指定从集合中读取的记录条数。而使用 skip() 方法来跳过指定数量的数据，skip 方法同样接受一个数字参数作为跳过的记录条数。

limit 语法：

```
> db.COLLECTION_NAME.find().limit(NUMBER)
```

skip 语法：

```
> db.COLLECTION_NAME.find().skip(NUMBER)
```

limit 方法和 skip 方法，可以结合起来用，实现分页查询。以下是分页查询的做法。

```
> db.myCollection.find().sort({"ID":1}).skip(10).limit(10)
```

通过 db.myCollection.find().sort({"ID":1}).skip(10).limit(10) 命令，将其根据 ID 排序后，跳过 10，查询 10 条，结果为 10～19 条的数据。

- .sort({"ID":1}) 是按 ID 排序。这个排序不是必需的，可以有，也可以无。
- skip(10) 是跳过 10 条记录。
- limit(10) 是取 10 条记录。

这个是类似于 EntityFramework 框架下的 skip() 和 take() 方法的操作。

假设每页是 10 条记录,要查询第 15 页。命令如下:

db.myCollection.find().skip((15-1) * 10).limit(10)

即跳过前面 14×10 个记录。然后取 10 个记录,即第 15 页的 10 条记录。

19. MongoDB 排序

MongoDB sort()方法

在 MongoDB 中使用 sort()方法对数据进行排序,sort()方法可以通过参数指定排序的字段,并使用 1 和 -1 来指定排序的方式,其中 1 为升序排列,而 -1 是用于降序排列。

sort()方法的基本语法如下所示:

>db.COLLECTION_NAME.find().sort({KEY:1})

>db.col.find().sort({"test":-1})

说明:-1 表示降序,1 表示升序。多个排序字段之间用逗号分隔,如:

>db.col.find().sort({"test":-1,"name":1})

20. MongoDB 索引

索引是 MongoDB 在文档之外存储的信息,它可帮助提高查询的效率。如果没有索引,MongoDB 在读取数据时必须扫描集合中的每个文件并选取那些符合查询条件的记录。这种扫描全集合的查询效率是非常低的,这跟关系型数据库的全表扫描一样,特别慢。特别在处理大量的数据时,查询可以要花费几十秒甚至几分钟,这对网站的性能是非常致命的。没有用户愿意等那么长时间。索引是特殊的数据结构,索引存储在一个易于遍历读取的数据集合中,索引是对数据库表中一列或多列的值进行排序的一种结构。

MongoDB 使用 createIndex()方法来创建索引。

注意,在 3.0.0 版本前创建索引的方法为 db.collection.ensureIndex(),之后的版本使用了 db.collection.createIndex()方法,ensureIndex()还能用,但只是 createIndex()的别名。

语句格式如下:

>db.collection.createIndex(keys, options)

- keys 是用来创建索引的 Field 列表。它的格式类似于 sort。多个 Field 之间用逗号分隔。
- options 是可有可无的。没有提供 options 时用的是默认值,如:

>db.col.createIndex({"title":1,"description":-1})

options 的值如表 13-7 所列。

表 13-7 db.col.createIndex 的参数

参 数	类 型	描 述
background	Boolean	background 可指定以后台方式创建索引,即增加 background 可选参数。background 默认值为 false
unique	Boolean	是否 true 唯一索引。默认值为 false
sparse	Boolean	对文档中不存在的字段数据不启用索引;默认值为 false 前文说过,部分文档可能没有某些 Field,如果这个参数设置为 true,则在索引 Field 中不会查询出不包含对应 Field 的文档
dropDups	Boolean	3.0+版本已废弃。在建立唯一索引时是否删除重复记录,指定 true 创建唯一索引。默认值为 false
name	string	索引的名称
weights	document	索引权重值,数值在 1~99 999 之间,表示该索引相对于其他索引字段的得分权重
v	index version	索引的版本号。默认的索引版本取决于 mongoDB 创建索引时运行的版本
expireAfterSeconds	integer	指定一个以秒为单位的数值,完成 TTL 设定,设定集合的生存时间
default_language	string	对于文本索引,该参数决定了停用词及词干和词基的规则的列表。默认为英语
language_override	string	对于文本索引,该参数指定了包含在文档中的字段名,语言覆盖默认的 language,默认值为 language

例如,在后台创建索引:

db.col.createIndex({open: 1, close: 1}, {background: true})

13.5 MongoDB 集群的部署

在本节将学会使用资源管理模板创建具有副本集的复杂 MongoDB 共享集群。您可以使用 GITHUB 的以下 Azure 快速启动模板创建复杂的 MongoDB 共享集群。该模板遵循 MangoDB 共享集群最佳实践以提供冗余和高可用性。模板创建两个碎片,每个副本集中有三个节点;还创建了一个配置服务器副本集,该副本集具有三个节点;另外,还创建了两个 Mongos 路由器服务器,以便为来自不同碎片的应用程序提供一致性。

https://github.com/Azure/azure-quickstart-templates/tree/master/mongodb-sharding-centos

https://raw.githubusercontent.com/Azure/azure-quickstart-templates/master/mongodb-sharding-centos/azuredeploy.json

要创建这个环境,您需要安装最新的 Azure CLI 并使用 az login 登录到 Azure 账户。首先,创建一个带有 az group create 的资源组。下面的示例是在 East US 位置创建一个名为 myResourceGroup 的资源组:

```
az group create -- name myResourceGroup -- location eastus
```

接下来,部署带有 az group deployment create 的 MongoDB 模板。在方定义自己的资源名称和大小,例如 mongoAdminUsername,sizeOfDataDiskInGB,and configNodeVmSize:

```
az group deployment create -- resource - group myResourceGroup \
    -- parameters '{"adminUsername": {"value": "azureuser"},
"adminPassword": {"value": "P@ssw0rd!"},
"mongoAdminUsername": {"value": "mongoadmin"},
"mongoAdminPassword": {"value": "P@ssw0rd!"},
"dnsNamePrefix": {"value": "mypublicdns"},
"environment": {"value": "AzureCloud"},
"numDataDisks": {"value": "4"},
"sizeOfDataDiskInGB": {"value": 20},
"centOsVersion": {"value": "7.0"},
"routerNodeVmSize": {"value": "Standard_DS3_v2"},
"configNodeVmSize": {"value": "Standard_DS3_v2"},
"replicaNodeVmSize": {"value": "Standard_DS3_v2"},
"zabbixServerIPAddress": {"value": "Null"}}' \
    -- template-uri https://raw.githubusercontent.com/Azure/azure-quickstart-templates/master/mongodb-sharding-centos/azuredeploy.json \
    -- name myMongoDBCluster \
    -- no-wait
```

部署和配置所有 VM 实例可能需要一个多小时。在 Azure 平台接受模板部署之后,在前面的命令末尾使用 --no-wait 标志将控制返回到命令提示符。然后可以用 az group deployment show 查看部署状态。下面的示例查看 myResourceGroup 资源组中 myMongoDBCluster 集群部署的状态:

```
az group deployment show \
    -- resource-group myResourceGroup \
    -- name myMongoDBCluster \
    -- query [properties.provisioningState] \
    -- output tsv
```

这样就根据模板将 MongoDB VM 建立好了。MongoDB 集群像一个逻辑 MongoDB 一样运行。

13.6　Azure Cosmos DB 提供的 MongoDB API

　　Azure Cosmos DB 是微软开发的一个适合大数据的非关系型数据库。它提供了 MongoDB API 来供客户端访问。Cosmos DB 的存储方式与 MongoDB 有所不同。但是其提供的接口与 MongoDB 的对外接口是兼容的。客户端感知不到服务器端是否为真正的 MongoDB，只要按照 MongoDB 的方式访问就可以了。

　　Azure Cosmos DB 是微软面向任务关键应用的全球分布、多模型数据库服务。Azure Cosmos DB 提供交钥匙全球分布、全球吞吐量和存储的弹性扩展、第 99％ 的单位数字毫秒延迟，以及保证高可用性，所有这些都由行业领先的 SLA 支持。Azure Cosmos DB 自动索引数据，而不需要处理模式和索引管理。它是多模型的，支持文档、键/值对、图形和柱状数据模型。

　　Azure Cosmos DB 数据库可以用作 MongoDB 编写的应用程序的数据存储。此功能意味着通过使用现有驱动程序，为 MongoDB 编写的应用程序现在可以与 Azure Cosmos DB 进行通信，并使用 Azure Cosmos DB 数据库而不是 MongoDB 数据库。在许多情况下，可以通过简单地更改连接字符串来使用 MongoDB 切换到 Azure Cosmos DB。使用此功能，可以在 Azure 云中使用 Azure Cosmos DB 及其全面的行业领先 SLA 轻松构建和运行 MongoDB 全球分布式数据库应用程序，同时继续使用 MongoDB 的熟悉技能和工具。

　　MongoDB 兼容性：可以使用现有的 MongoDB 专长、应用程序代码和工具，因为 Azure Cosmos DB 实现了 MongoDB 有线协议。可以使用 MongoDB 开发应用程序，并使用完全管理和全球分布的 Azure Cosmos DB 服务将它们部署到生产中。有关支持的版本的更多信息，请参见 MongoDB 协议支持。

　　使用 MongoDB 应用程序的 Azure Cosmos DB 有什么好处？

　　① 弹性可伸缩的吞吐量和存储：通过方便地放大或缩小 MangoDB 数据库满足应用程序需求。您的数据存储在固态磁盘（SSD）上，用于低可预测延迟。Azure Cosmos DB 支持 MongoDB 集合，可以扩展到几乎无限的存储大小和提供的吞吐量。随着应用程序的增长，可以无缝地缩放 Azure Cosmos DB，具有可预测的性能。

　　② 多区域复制：Azure Cosmos DB 透明地将数据复制到与 MongoDB 账户关联的所有区域，使您可以开发需要对数据进行全局访问的应用程序，同时提供一致性、可用性和性能之间的折衷，所有这些都使用相应的保证。Azure Cosmos DB 提供了具有多归属 API 的透明区域故障转移，以及跨全球弹性地扩展吞吐量和存储的能力，可以了解分布式数据在全球范围内的更多信息。

　　③ 没有服务器管理：你不必管理和伸缩你的 MongoDB 数据库。Azure Cosmos DB 是完全管理的服务，这意味着不必自己管理任何基础设施或虚拟机。Azure Cosmos DB 可在 30 个以上 Azure Region 中使用。

④ 可调一致性级别:因为 Azure Cosmos DB 支持多模型 API,所以一致性设置适用于账户级别,并且一致性的强制由每个 API 控制。在 MongoDB 3.6 之前,还没有会话一致性的概念,所以如果设置 MongoDB API 账户以使用会话一致性,那么在使用 MongoDB API 时,一致性会降级到最终。如果 MongoDB API 账户需要自读自写保证,则账户的默认一致性级别应该设置为强过期或有限过期。一致性级别对照表如表 13-8 所列。

表 13-8 一致性级别对照表

Azure Cosmos DB 默认一致性级别	MongoDB API 3.4
最终	最终
一致的前缀	最终一致
会话	最终一致
有限过期	强
强	强

想学习更多信息,可参考:https://docs.microsoft.com/en-us/azure/cosmos-db/consistency-levels。

自动索引:默认情况下,Azure Cosmos DB 自动对 MongoDB 数据库中的文档内的所有属性进行索引,并且不期望也不需要任何模式或创建辅助索引。此外,唯一的索引功能支持对 Azure Cosmos DB 中已经自动索引的任何文档字段的唯一性约束。

企业级:Azure Cosmos DB 支持多个本地副本,以便在本地和区域故障时提供 99.99% 的可用性和数据保护。Azure Cosmos DB 具有企业级遵从性认证和安全特性。

13.6.1 Azure Cosmos DB:移植一个已有的 Node.js MongoDB Web 应用

Azure Cosmos DB 是微软的全球分布式多模型数据库服务。可以快速创建和查询文档、键/值对和图形数据库,所有这些数据库都受益于 Azure Cosmos DB 核心的全局分布和水平伸缩能力。

这个快速启动演示如何使用 Node.js 编写的现有 MongoDB 应用程序并将其连接到 Azure Cosmos DB 数据库,该数据库通过使用 MongoDB API 支持 MongoDB 客户端连接。换句话说,Node.js 应用程序只知道使用 MongoDB API 连接到数据库。数据存储在 Azure Cosmos DB 中,对应用程序是透明的。

完成后,将在 Azure Cosmos DB 上运行一个 MEAN 应用程序(MongoDB、Express、Angular 和 Node.js),如图 13-3 所示。

图 13 - 3　一个 MEAN 应用

打开 Azure Cloud Shell，操作方式前文已述。

首先得有一个 Azure Cosmos DB，在 Azure Portal，选择 All services，如图 13 - 4 所示。

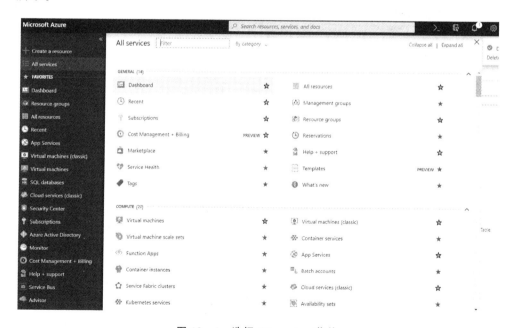

图 13 - 4　选择 All services 菜单

输入 Cosmos，如图 13 - 5 所示。

Azure Cosmos DB 出来了。点击它，如图 13 - 6 所示。

图 13-5 查找 Azure Cosmos DB

图 13-6 查看 Azure Cosmos DB 页

选择 Create Azure Cosmos DB Account，如图 13-7 所示。

Azure Cosmos DB 支持好几种 API，SQL 是传统的 SQL92 查询方式，MongoDB 就是兼容 MongoDB 的客户端，Cassandra 就是兼容 Cassandra 的客户端。Azure Table 就是兼容 Azure storage table 的访问方式。Gremlin 就是兼容 Gremlin 图形数据库的方式。这里需要讲解 MongoDB 的兼容，所以用的是 MongoDB。

选择 Location，就近选择一个数据中心，如图 13-8 所示。

如果选择 Geo-Redundancy，Enable，则会在物理地址不同的地方选择多个数据中心来创建此 Cosmos DB 的备份。这样会有一些额外的存储和网络带宽，所以费用上要更贵。对于生产环境，这样的安排是必需的，可以保证数据不会丢失，数据丢失的话，是无法承受的。

部署 MongoDB 到 Azure

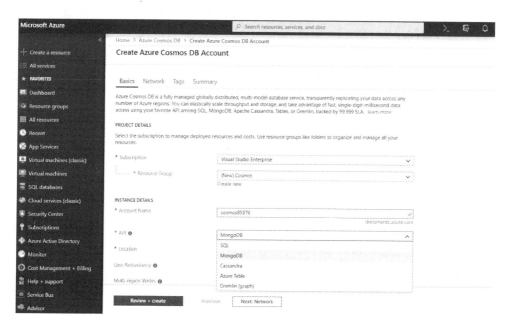

图 13-7 选择 Create Azure Cosmos DB

图 13-8 选择就近的数据中心

如果选择 Multi-region Writes，Enable，则会在多个 region 同时写入数据，这样可以保证数据同步性比较好。

如果没有其他的设定，则可以点击 Review+Create 按钮。

如果有其他的一些特殊要求，比如虚拟 VPN 之类，就需要进入 Network 页取设置 VPN 等设置。

进入 Summary 页，审视这些配置是否正常，确认正常，就单击 Create 按钮创建此 Cosmos DB。

Cosmos DB 的可伸缩性和其他性能是相当不错的。物理上的多地备份保证了数据不会丢失，大大保证了数据的安全性。

1. 克隆样本应用程序

假定已经安装 node.js。如果没有安装 node.js，请安装 node.js。运行以下命令克隆示例存储库。此示例存储库包含默认的 Mean.js 应用程序。

① 打开命令提示符，创建一个名为 Git 示例的新文件夹，然后关闭命令提示符。

```
md "C:\git-samples"
```

② 打开 Git 终端窗口，如 Git BASH，并使用 CD 命令更改到新文件夹，以安装示例应用程序。

```
cd "C:\git-samples"
```

运行以下命令克隆示例存储库。此命令在计算机上创建示例应用程序的副本。

```
git clone https://github.com/prashanthmadi/mean
```

③ 运行应用程序，安装所需的软件包并启动应用程序。

```
cd mean
npm install
npm start
```

应用程序将尝试连接到 MongoDB 源并失败，在输出返回"[MongoError: connect ECONNREFUSED 127.0.0.1:27017]"时继续退出应用程序。

2. 登录 Azure

如果您使用的是已安装的 Azure CLI，请使用 az login 命令登录到您的 Azure 订阅，并遵循屏幕上的说明。如果您使用的是 Azure 的 Cloud Shell，则可以跳过这一步。

```
az login
```

3. 添加 Azure Cosmos DB 模块

如果您使用的是已安装的 Azure CLI，请检查是否已经运行了 AZ 命令安装了 Cosmo DB 组件。如果 Cosmo DB 在基本命令列表中，则继续执行下一个命令。如果您使用的是 Azure 的 Cloud Shell,则可以跳过这一步。

如果 Cosmo DB 不在基本命令列表中,请重新安装 Azure CLI。

4. 创建资源组

创建具有 az group create 的资源组。Azure 资源组是一个逻辑容器,其中可部署和管理 Web 应用程序、数据库和存储账户等 Azure 资源。

下面的示例在西欧地区创建了一个资源组,为资源组选择唯一的名称。

如果您正在使用 Azure Cloud Shell,请单击"尝试它",按照屏幕上的提示进行登录,然后将命令复制到命令提示符中。

```
az group create -- name myResourceGroup -- location "West Europe"
```

5. 用 Azure CLI 创建 Azure Cosmos DB 账户

用 AZ Cosmos DB 命令创建一个 Azure Cosmos DB 账户。这和之前在 Portal 中创建 Cosmos DB 是一样的,只是 Azure CLI 是命令行。

在下面的命令中,请替换您自己的唯一 Azure Cosmos DB 账户名,其中可以看到<.osdb-name>占位符。此唯一名称将用作 Azure Cosmos DB 端点的一部分(https://<.osdb-name>...azure.com/),因此该名称需要在 Azure 的所有 Azure Cosmos DB 账户中是唯一的。

```
az cosmosdb create -- name <cosmosdb-name> -- resource-group myResourceGroup -- kind MongoDB
```

注:--kind MongoDB 选择的是 MongoDB。

当创建 Azure Cosmos DB 账户时,Azure CLI 显示与下面的示例类似的信息。

```
{
"databaseAccountOfferType": "Standard",
"documentEndpoint": "https://<cosmosdb-name>.documents.azure.com:443/",
"id": "/subscriptions/00000000-0000-0000-0000-000000000000/resourceGroups/myResourceGroup/providers/Microsoft.DocumentDB/databaseAccounts/<cosmosdb-name>",
"kind": "MongoDB",
"location": "West Europe",
"name": "<cosmosdb-name>",
"readLocations": [
    {
"documentEndpoint": "https://<cosmosdb-name>-westeurope.documents.azure.com:443/",
"failoverPriority": 0,
"id": "<cosmosdb-name>-westeurope",
"locationName": "West Europe",
```

```
           "provisioningState": "Succeeded"
        }
    ],
    "resourceGroup": "myResourceGroup",
    "type": "Microsoft.DocumentDB/databaseAccounts",
    "writeLocations": [
        {
            "documentEndpoint": "https://<cosmosdb-name>-westeurope.documents.azure.com:443/",
            "failoverPriority": 0,
            "id": "<cosmosdb-name>-westeurope",
            "locationName": "West Europe",
            "provisioningState": "Succeeded"
        }
    ]
}
```

6. 将 Node.js 应用程序连接到数据库

在此步骤中,使用 MongoDB 连接字符串将 MEAN.js 示例应用程序连接到刚刚创建的 Azure Cosmos DB 数据库。

7. 在 Node.js 应用程序中配置连接字符串

在您的 MEAN.js 库中,打开 config/env/local-development.js。用以下代码替换此文件的内容。请务必用 Azure Cosmos DB 账户名替换两个 <cosmosdb-name> 占位符。

```
'use strict';

module.exports = {
  db: {
    uri: 'mongodb://<cosmosdb-name>:<primary_master_key>@<cosmosdb-name>.documents.azure.com:10255/mean-dev?ssl=true&sslverifycertificate=false'
  }
};
```

8. 检索密钥

为了连接到 Azure Cosmos DB 数据库,需要数据库密钥。使用 az cosmosdb list-keys 命令检索主键。

```
az cosmosdb list-keys --name <cosmosdb-name> --resource-group myResourceGroup --query "primaryMasterKey"
```

这个方法是在 CLI 中查询 key,也可以在 Portal 中查询 Cosmos DB 的 key,如

图 13-9 所示。

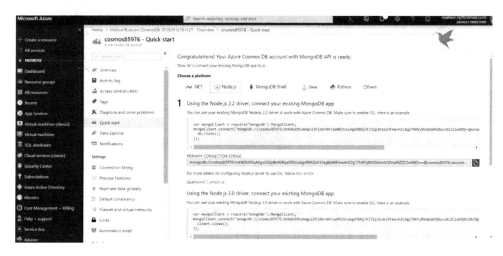

图 13-9 查询 Azure Cosmos DB 的 key

Node.js 的程序可以查询到其链接串。Portal 中的方法和 CLI 的方式是一致的。

Azure CLI 输出的信息与下面的示例相似。

"RUayjYjixJDWG5xTqIiXjC..."

复制 primaryMasterKey 的值。将它粘贴到 local-development.js 中的 <primary_master_key>。保存更改。再次运行应用程序。再次运行 npm start。

npm start

控制台消息现在应该告诉您开发环境正在启动和运行。

在浏览器中导航到 http://localhost:3000。单击顶部菜单中的注册并尝试创建两个虚拟用户。

JS 示例应用程序将用户数据存储在数据库中。如果您成功则意味着 JS 自动登录到创建的用户，那么您的 Azure Cosmos DB 连接正在工作，如图 13-10 所示。

9. 在数据资源管理器中查看数据

Azure Cosmos DB 存储的数据可用于在 Azure 门户中查看、查询和运行业务逻辑。要查看、查询和使用上一步创建的用户数据，请登录到 Web 浏览器中的 Azure 门户。

在顶部搜索框中，键入 Azure Cosmos DB。当您的宇宙 DB 账户打开时，选择您的 Cosmos DB 账户。在左侧导航中，单击数据资源管理器。在集合窗格中展开集合，然后可以查看集合中的文档、查询数据，甚至创建和运行存储过程、触发器和 UDF，如图 13-11 所示。

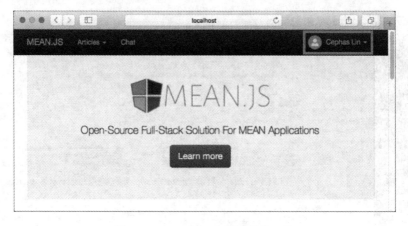

图 13-10　MEAN 应用能正常工作

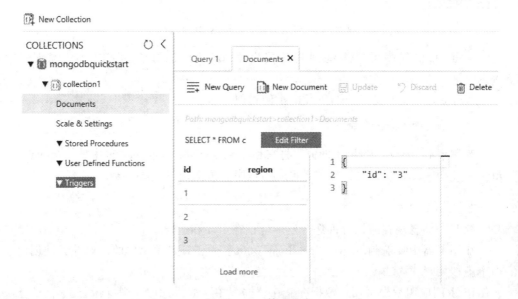

图 13-11　管理集合

10. 将 Node.js 应用程序部署到 Azure

在这个步骤中,您将 MangoDB 连接的 Node.js 应用程序部署到 Azure Cosmos DB。您可能已经注意到,先前更改的配置文件是针对开发环境(/config/env/local-development.js)的。当您将应用程序部署到应用程序服务时,默认情况下它将在生产环境中运行,所以现在需要对相应的配置文件进行相同的更改。

在 MEAN.js 库中,打开 config/env/production.js。

在 DB 对象中,替换 URI 的值,如以下示例中所示。务必像以前那样替换占位符。

```
'mongodb://<cosmosdb-name>:<primary_master_key>@<cosmosdb-name>.documents.
azure.com:10255/mean? ssl=true&sslverifycertificate=false',
```

在终端中,将所有更改提交给 Git。可以复制两个命令一起运行。

```
git add .
git commit -m "configured MongoDB connection string"
```

13.6.2 将数据迁移到 Azure Cosmos DB MangoDB API 账户

如前文所述获得连接串。

1. 使用 MongoDB 导入迁移数据

若要将数据导入到 Azure Cosmos DB 账户,请使用以下模板。用特定于您账户的值填写主机、用户名和密码。

模板:

```
mongoimport.exe --host <your_hostname>:10255 -u <your_username> -p <your_password> --db <your_database> --collection <your_collection> --ssl --sslAllowInvalidCertificates --type json --file "C:\sample.json"
```

例如:

```
mongoimport.exe --host cosmosdb-mongodb-account.documents.azure.com:10255 -u cosmosdb-mongodb-account -p tkvaVkp4Nnaoirnouenrgisuner2435qwefBH0z256Na24frio34-LN-Qasfaefarfernoimczciqis AXw== --ssl --sslAllowInvalidCertificates --db sampleDB --collection sampleColl --type json --file "C:\Users\admin\Desktop\*.json"
```

2. 使用 Mavigestor 迁移数据

若要将数据恢复到 MyGoDB 账户的 API,请使用以下模板执行导入。用特定于您账户的值填写主机、用户名和密码。

模板:

```
mongorestore.exe --host <your_hostname>:10255 -u <your_username> -p <your_password> --db <your_database> --collection <your_collection> --ssl --sslAllowInvalidCertificates <path_to_backup>
```

例如:

```
mongorestore.exe --host cosmosdb-mongodb-account.documents.azure.com:10255 -u cosmosdb-mongodb-account -p tkvaVkp4Nnaoirnouenrgisuner2435qwefBH0z256Na24frio34-LN-Qasfaefarfernoimczciqis AXw== --ssl --sslAllowInvalidCertificates ./dumps/dump-2016-12-07
```

第 14 章

Azure 人工智能

14.1 人工智能基础

　　人工智能(Artificial Intelligence)，英文缩写为 AI，是人类利用计算机资源模拟人类的智力进行的活动，是一门集理论、方法、技术及应用系统于一体的技术科学。人工智能一词最初是在 1956 年 Dartmouth 学会上提出的。总的来说，人工智能研究的一个主要目标是使机器能够胜任一些通常需要人类智能才能完成的复杂工作。人工智能包括十分广泛的学科，它由不同的领域组成，如机器学习、计算机视觉等。

　　早期的模拟人类智能的活动包括图像识别、语音识别、机器人和专家系统。这些研究活动获得了一些有益的成果，包括一些算法、一些较低级的机器人和一些比较简单的专家系统。比如数字的识别算法，如何二值化图像的算法，在有干扰的情况下如何分开每个数字的算法；还有如语音识别，产生了隐马尔科夫模型算法，可以用来识别非特定人的语音，从语音中得到文字信息；还有就是早期的机器人，能完成的动作有限，所以最先应用的是工业机器人，在生产线中一些固定岗位，需要完成的动作非常有限，比如只是拧螺丝，在某个角度安装零件，这些都可以由初级工业机器人来完成，其智能化程度较低，只是简单过程的不断重复，可以避免人类的疲劳、精力不集中、走神等错误，人类工人需要吃饭，需要沟通，需要休息，才能保持好的工作态度，专注地去完成精神高度紧张、强度大、单调的工作，这方面工业机器人的优势就比较大。工业机器人可以 24 小时不间断地工作，只需要提供电力，少量的维护，比如加机油等，就可长时间地工作，而且效率也比较高。另外，早期的专家系统，多半是一些比较简单的存储知识的系统，比如疾病诊断，某个方面的诊断，可以通过问一系列的相关问题来给病人进行诊断，病人只需要在可能的回答中选择一个就可以了。通过回答一系列相关问题，可以导航到最后的定位诊断，判断出病人得的是什么病。但是这样的专家系统更多的是知识的存储，相当于是让病人做选择题，先确定病的大致所属范围，然后再通过更细致的问题来区别病更小的所属范围，最后确定具体病症名称。这样的系统有不少缺陷，只能适用于特定的领域，比如某一类疾病，而且要在病人的理解能力没有问题，且对自己的病情了解得比较清楚的情况下才能实施此类方式的问答。

人工智能的科学也在不断发展,从最开始试图解决的一些问题,到后来一些更难、更有挑战性的难题,比如自然语言理解、现场口译、自动驾驶汽车等。2016年1月,Google旗下的深度学习团队Deepmind开发的人工智能围棋软件AlphaGo,以5∶0战胜了围棋欧洲冠军樊麾。这是人工智能第一次战胜职业围棋手。2017年5月,AlphaGo又跟中国围棋九段选手柯洁进行三番棋大战,以3∶0战胜了柯洁。这都说明近年来人工智能领域又有了一些新的进展。

1. 实际应用

实际应用包括指纹识别、人脸识别、视网膜识别、虹膜识别、掌纹识别、专家系统、智能搜索、定理证明、博弈、自动程序设计、智能控制、机器人学、语言和图像理解、遗传编程等。

2. 涉及学科

涉及学科包括哲学和认知科学、数学、神经生理学、心理学、计算机科学、信息论、控制论、不定性论。

3. 研究范畴

研究范畴包括语言的学习与处理、知识表现、智能搜索、推理、规划、机器学习、知识获取、组合调度问题、感知问题、模式识别、逻辑程序设计、软计算、不精确和不确定的管理、人工生命、神经网络、复杂系统、遗传算法、人类思维方式等,最关键的难题还是机器的自主创造性思维能力的塑造与提升。

4. 意识和人工智能

人工智能就其本质而言,是对人的思维的信息过程的模拟。

对于人的思维模拟可以从两个方向进行:一是结构模拟,仿照人脑的结构机制,制造出"类人脑"的机器;二是智能模拟,暂时撇开人脑的内部结构,而从其功能过程进行模拟。现代电子计算机的产生便是对人脑思维功能的模拟,是对人脑思维的信息过程的模拟。人工智能不是人的智能,更不会超过人的智能。

14.2 Azure人工智能简介

Azure平台提供了一整套服务、基础架构和工具来供开发者使用。Azure平台提供的服务包括一些预先构建好的服务,如视觉、语音、语言文字理解、知识、搜索等。用户可以在自己的App中调用这些已经构建好的服务,而且只需要一些比较少的代码就可以完成调用。如果用户有一些特殊需要,则可以用一些自己的数据来对这些服务进行培训,培训出一些特有的行为和识别。默认来说这些预先构建好的服务都是经过数据培训的,可以识别相当比例的数据模型。比如语音识别已经经过不少数据的培训,可以识别一些人的语音。如果用户有需求,需要对一些方言的腔调进行适应,则可以用一些方言的数据来进行调校,使得语音识别的概率更高一些,能够识别一些方言的语音。

除了预先构建好的服务,如视觉、语音、语言文字理解、知识、搜索等外,还有 Azure 机器学习(Machine Learning),有了 Azure Machine Learning 就可以帮助用户更好地做决定,任何规模的数据,Azure Machine Learning 对于数据分析师和开发者来说都是受管理的云服务,能够帮助用户准备数据,构建和训练你的模型,以获得最好的结果,而且可以在你需要的时候运行它。因为用户的数据是在企业级的安全保护下,故用户不必担心。

Azure 平台可以快速地构建原型,然后很容易地扩展到多个虚拟机或者是扩展到 Spark 集群;能够非常积极地管理模型的性能,找出最好性能的模型,用数据驱动的视野来驱动它;可以在任何地方部署和管理你的模型。用 Docker 容器去部署模型到生产环境中,要么是云环境,要么是 On-Premise,要么是更新的技术。它可以提交最好表现的训练模型到生产环境并在任何需要的时候进行再次训练。

Azure 另外还有一种会话机器人的服务。可以将这样的会话机器人嵌入到系统中,比如一个网站,提供人工智能服务的机器人,可以聆听客户的话语,试图懂得客户的要求或者投诉,然后再相应地进行一些处理;可以是文字的形式,也可以是语音、视频的形式。这里面就牵涉到自然语言理解。客户输入文字信息,一行行的文字,会话机器人得理解客户所说的是什么,分词、断句、了解客户在说什么。理解了客户问题以后,再去找相应系统或者人员来处理。如果不是很清楚,还会再提出问题,进一步进行澄清,对客户的需求进一步了解,这样可以更好地服务客户。同时,机器人可以 24 小时值班,不管客户在全球哪里,都可以进行服务。这样的服务比那些固定的菜单就高级多了。如果是以语音或者视频的方式进行服务,则还有语音识别的问题,从语音中识别出文字,然后对文字进行内容理解,理解客户需求之后再进行相应的服务。这样的难度更高。

Azure 可以构建、连接、部署和管理智能机器人,以便在网站、应用程序、Cortana、Microsoft Teams、Skype、Slack、Facebook Messenger 等上自然地与用户交互。用一个完整的 BOT 构建环境快速启动,使用 C♯、JavaScript、Python 和 Java 语言。

14.3 在 Azure 门户中创建认知服务账户

1. 创建和订阅 Azure 认知服务资源

登录到 Azure 门户,单击"+Create a resource",如图 14-1 所示。

在 Azure 市场下,选择 AI+Machine Learning。如果您没有看到感兴趣的服务,请单击 See all 查看认知服务 API 的整个目录,如图 14-2 所示。

在创建页面上,提供如下信息,如表 14-1 所列。

Azure 人工智能

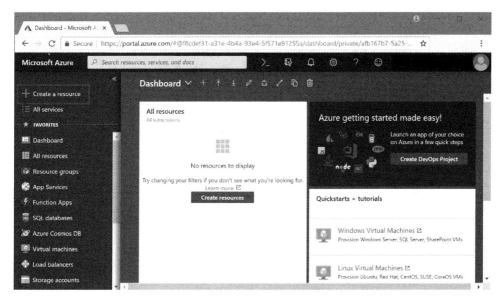

图 14-1　单击"＋Create a resource"

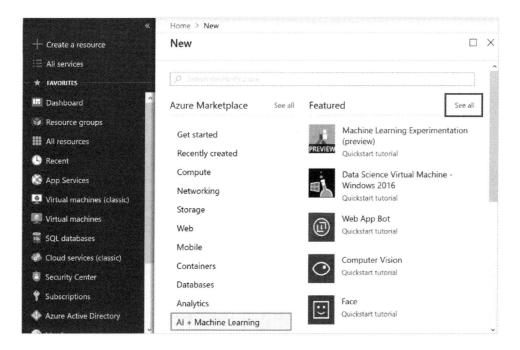

图 14-2　AI 和机器学习的资源

表 14-1 创建认知服务资源所需输入信息

输入信息	描述
Name	您的认知服务资源的描述性名称。我们建议使用描述性名称,例如 MyNameFaceAPIAccount
Subscription	选择一个可用的 Azure 订阅
Location	您的认知服务实例的位置。不同的位置可能会引入延迟,但对资源运行时的可用性没有影响
Pricing tier	您的认知服务账户的成本取决于您选择的选项和使用情况
Resource group	将包含认知服务资源的 Azure 资源组。可以创建一个新组或将其添加到预先存在的组中

将表 14-1 的信息输入到如图 14-3 所示的界面。

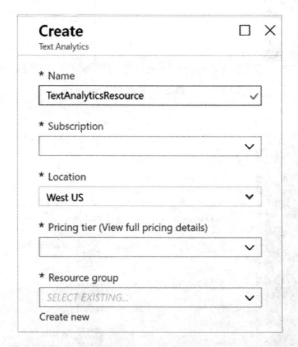

图 14-3 输入表 14-1 的信息到此界面

2. 访问资源

注释:订阅所有者可以通过应用 Azure 策略、分配"不允许资源类型"策略定义以及指定 Microsoft.CognitiveServices/accounts 作为目标资源类型来禁用为资源组和订阅创建认知服务账户。

在创建资源之后,如果您将它固定在 Azure 仪表板上,就可以访问它;否则,可以在资源组中找到它。

在 Cognitive Services 资源中,可以使用 Overview 部分中的 Endpoint URL 在

应用程序中开始进行 API 调用,如图 14-4 所示。

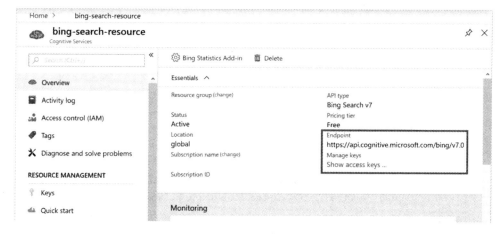

图 14-4 bing-search-resource 界面

14.4 第一个 Azure 人工智能程序——在计算机视觉中使用 REST API 和 Python 分析远程图像

在这个快速启动中,可以通过使用计算机视觉的 REST API 来分析远程存储的图像并提取视觉特征。通过分析图像,可以提取基于图像内容的视觉特征。

可以使用 MyBinder 上的 Jupyter 笔记本来一步一步地运行这个快速启动。要启动绑定器 Binder,请选择以下按钮,如图 14-5 所示。

图 14-5 两个按钮

要求已安装一个 Python,这样才能在本地运行。

1. 创建并运行示例

若要创建和运行示例,请执行以下步骤:

① 将以下代码复制到文本编辑器中。在需要的地方进行代码的下列更改:用订阅密钥替换订阅值的值。

如果需要,则将 Azure 区域中的计算机视觉资源的端点 URL 替换为 vision_base_url 的值,在那里得到订阅密钥。

可选地,用要分析的不同图像的 URL 替换 image_url 的值。将代码保存为具有 .py 扩展名的文件,例如 analyze-image.py。

② 打开命令提示符窗口。在提示符下,使用 Python 命令运行示例,例如:pythonanalyze-image.py。

```
import requests
# If you are using a Jupyter notebook, uncomment the following line.
```

```python
# %matplotlib inline
import matplotlib.pyplot as plt
import json
from PIL import Image
from io import BytesIO

# Replace <Subscription Key> with your valid subscription key.
subscription_key = "<Subscription Key>"
assert subscription_key

# You must use the same region in your REST call as you used to get your
# subscription keys. For example, if you got your subscription keys from
# westus, replace "westcentralus" in the URI below with "westus".
#
# Free trial subscription keys are generated in the westcentralus region.
# If you use a free trial subscription key, you shouldn't need to change
# this region.
vision_base_url = "https://westcentralus.api.cognitive.microsoft.com/vision/v2.0/"

analyze_url = vision_base_url + "analyze"

# Set image_url to the URL of an image that you want to analyze.
image_url = "https://upload.wikimedia.org/wikipedia/commons/thumb/1/12/" + \
    "Broadway_and_Times_Square_by_night.jpg/450px-Broadway_and_Times_Square_by_night.jpg"

headers = {'Ocp-Apim-Subscription-Key': subscription_key}
params = {'visualFeatures': 'Categories,Description,Color'}
data = {'url': image_url}
response = requests.post(analyze_url, headers=headers, params=params, json=data)
response.raise_for_status()

# The 'analysis' object contains various fields that describe the image. The most
# relevant caption for the image is obtained from the 'description' property.
analysis = response.json()
print(json.dumps(response.json()))
image_caption = analysis["description"]["captions"][0]["text"].capitalize()

# Display the image and overlay it with the caption.
image = Image.open(BytesIO(requests.get(image_url).content))
plt.imshow(image)
plt.axis("off")
_ = plt.title(image_caption, size="x-large", y=-0.1)
plt.show()
```

2. 检查响应

JSON 中返回一个成功的响应。示例网页在命令提示符窗口中解析并显示成功的响应，类似于以下示例：

```
{
"categories": [
    {
        "name": "outdoor_",
        "score": 0.00390625,
        "detail": {
        "landmarks": []
        }
    },
    {
        "name": "outdoor_street",
        "score": 0.33984375,
        "detail": {
        "landmarks": []
        }
        }
    ],
    "description": {
    "tags": [
    "building",
    "outdoor",
    "street",
    "city",
    "people",
    "busy",
    "table",
    "walking",
    "traffic",
    "filled",
    "large",
    "many",
    "group",
    "night",
    "light",
    "crowded",
    "bunch",
    "standing",
```

```
            "man",
            "sign",
            "crowd",
            "umbrella",
            "riding",
            "tall",
            "woman",
            "bus"
        ],
        "captions": [
            {
                "text": "a group of people on a city street at night",
                "confidence": 0.9122243847383961
            }
        ]
    },
    "color": {
        "dominantColorForeground": "Brown",
        "dominantColorBackground": "Brown",
        "dominantColors": [
            "Brown"
        ],
        "accentColor": "B54316",
        "isBwImg": false
    },
    "requestId": "c11894eb-de3e-451b-9257-7c8b168073d1",
    "metadata": {
        "height": 600,
        "width": 450,
        "format": "Jpeg"
    }
}
```

我们可以看到这是一个比较复杂的场景，有很多元素，而且此照片还是在运动中拍的，有的地方比较模糊。此视觉识别出很多东西，如图14-6所示。

如：这些

```
"building",
"outdoor",
"street",
"city",
"people",
```

图 14-6 样例图,用来识别

```
"busy",
"table",
"walking",
"traffic",
"filled",
"large",
"many",
"group",
"night",
"light",
"crowded",
"bunch",
"standing",
"man",
```

```
"sign",
"crowd",
"umbrella",
"riding",
"tall",
"woman",
"bus"
```

人类来看都不一定能发现某些细节。

14.5 利用认知服务语音 SDK 在浏览器中识别 JavaScript 中的语音

在本节中,将了解如何使用认知服务语音 SDK 的 JavaScript 绑定来创建网站,以将语音转录为文本。该应用基于微软认知服务语音 SDK。

1. 先决条件

- 语音服务的订阅密钥。免费试用语音服务。
- PC 或麦克风,带有工作麦克风。
- 文本编辑器。
- 当前版本的 Chrome 或微软边缘。
- 可选地,支持宿主 PHP 脚本的 Web 服务器。

2. 创建一个新的网站文件夹

创建一个新的空文件夹。如果想在 Web 服务器上托管该示例,请确保 Web 服务器可以访问该文件夹。

3. 将 JavaScript 的语音 SDK 解压缩到该文件夹中

将语音 SDK 下载为 .zip 包 https://aka.ms/csspeech/jsbrowserpackage,并将其解压缩到新创建的文件夹中。这将导致两个文件被解包出来,即 microsoft.cognitiveservices.speech.sdk.bundle.js 和 microsoft.cognitiveservices.speech.sdk.bundle.js.map。后一个文件是可选的,并用于帮助调试到 SDK 代码,如果必要,则查看下载的压缩包文件,将其解压,如图 14-7 所示。

图 14-7 待解压的样例程序

4. 创建一个索引页 index.html

在文件夹中创建一个新的文件,命名为 index.html,并用文本编辑器打开该文件。创建以下 HTML 骨架:

```html
<html>
<head>
<title> Microsoft Cognitive Service Speech SDK JavaScript Quickstart </title>
</head>
<body>
<!-- UI code goes here -->

<!-- SDK reference goes here -->

<!-- Optional authorization token request goes here -->

<!-- Sample code goes here -->
</body>
</html>
```

将下面的 UI 代码添加到文件中，在第一个注释下面：

```html
<div id = "warning">
<h1 style = "font-weight:500;"> Speech Recognition Speech SDK not found (microsoft.cognitiveservices.speech.sdk.bundle.js missing). </h1>
</div>

<div id = "content" style = "display:none">
<table width = "100%">
<tr>
<td> </td>
<td> <h1 style = "font-weight:500;"> Microsoft Cognitive Services Speech SDK JavaScript Quickstart </h1> </td>
</tr>
<tr>
<td align = "right"> <a href = "https://docs.microsoft.com/azure/cognitive-services/speech-service/get-started" target = "_blank"> Subscription </a> : </td>
<td> <input id = "subscriptionKey" type = "text" size = "40" value = "subscription"> </td>
</tr>
<tr>
<td align = "right"> Region </td>
<td> <input id = "regionKey" type = "text" size = "40" value = "YourSubscriptionServiceRegion"> </td>
</tr>
<tr>
<td> </td>
<td> <button id = "startRecognizeOnceAsyncButton"> Start recognition </button> </
```

```
td>
    </tr>
    <tr>
        <td align = "right" valign = "top"> Results </td>
        <td> <textarea id = "phraseDiv" style = "display: inline - block;width:500px;height:200px"> </textarea> </td>
    </tr>
</table>
</div>
```

5. 添加对语音 SDK 的引用

```
<!-- Speech SDK reference sdk. -->
<script src = "microsoft.cognitiveservices.speech.sdk.bundle.js"> </script>
```

时间处理机制将用于把 UI 按钮定义的识别按钮、识别结果和订阅相关字段联系起来：

```
<!-- Speech SDK USAGE -->
<script>
  // status fields and start button in UI
  var phraseDiv;
  var startRecognizeOnceAsyncButton;

  // subscription key and region key for speech services.
  var subscriptionKey, regionKey;
  var authorizationToken;
  var SpeechSDK;
  var recognizer;

  document.addEventListener("DOMContentLoaded", function () {
      startRecognizeOnceAsyncButton = document.getElementById("startRecognizeOnceAsyncButton");
      subscriptionKey = document.getElementById("subscriptionKey");
      regionKey = document.getElementById("regionKey");
      phraseDiv = document.getElementById("phraseDiv");

      startRecognizeOnceAsyncButton.addEventListener("click", function () {
          startRecognizeOnceAsyncButton.disabled = true;
          phraseDiv.innerHTML = "";

          // if we got an authorization token, use the token. Otherwise use the provided subscription key
```

```javascript
            var speechConfig;
            if (authorizationToken) {
                speechConfig = SpeechSDK.SpeechConfig.fromAuthorizationToken(authorizationToken, regionKey.value);
            } else {
                if (subscriptionKey.value === "" || subscriptionKey.value === "subscription") {
                    alert("Please enter your Microsoft Cognitive Services Speech subscription key!");
                    return;
                }
                speechConfig = SpeechSDK.SpeechConfig.fromSubscription(subscriptionKey.value, regionKey.value);
            }

            speechConfig.speechRecognitionLanguage = "en-US";
            var audioConfig = SpeechSDK.AudioConfig.fromDefaultMicrophoneInput();
            recognizer = new SpeechSDK.SpeechRecognizer(speechConfig, audioConfig);

            recognizer.recognizeOnceAsync(
                function (result) {
                    startRecognizeOnceAsyncButton.disabled = false;
                    phraseDiv.innerHTML += result.text;
                    window.console.log(result);

                    recognizer.close();
                    recognizer = undefined;
                },
                function (err) {
                    startRecognizeOnceAsyncButton.disabled = false;
                    phraseDiv.innerHTML += err;
                    window.console.log(err);

                    recognizer.close();
                    recognizer = undefined;
                });
        });

        if (!!window.SpeechSDK) {
            SpeechSDK = window.SpeechSDK;
            startRecognizeOnceAsyncButton.disabled = false;
```

```
      document.getElementById('content').style.display = 'block';
      document.getElementById('warning').style.display = 'none';

      // in case we have a function for getting an authorization token, call it.
      if (typeof RequestAuthorizationToken === "function") {
        RequestAuthorizationToken();
      }
    }
  });
</script>
```

6. 创建令牌源(可选)

万一想在 Web 服务器上承载网页,那么可以为演示应用程序提供一个令牌源。这样,您的订阅密钥永远不会离开服务器,同时允许用户在不输入任何授权代码的情况下使用语音功能。

创建一个名为 token.php 的新文件。在这个例子中,假设您的 Web 服务器支持 PHP 脚本语言。输入以下代码:

```
<?php
header('Access-Control-Allow-Origin: ' . $_SERVER['SERVER_NAME']);

// Replace with your own subscription key and service region (e.g., "westus").
$subscriptionKey = 'YourSubscriptionKey';
$region = 'YourSubscriptionServiceRegion';

$ch = curl_init();
curl_setopt($ch, CURLOPT_URL, 'https://' . $region . '.api.cognitive.microsoft.com/sts/v1.0/issueToken');
curl_setopt($ch, CURLOPT_POST, 1);
curl_setopt($ch, CURLOPT_POSTFIELDS, '{}');
curl_setopt($ch, CURLOPT_HTTPHEADER, array('Content-Type: application/json', 'Ocp-Apim-Subscription-Key: ' . $subscriptionKey));
curl_setopt($ch, CURLOPT_RETURNTRANSFER, 1);
echo curl_exec($ch);
?>
```

编辑 index.html 文件,并将下面的代码添加到文件中:

```
<!-- Speech SDK Authorization token -->
<script>
// Note: Replace the URL with a valid endpoint to retrieve
//       authorization tokens for your subscription.
var authorizationEndpoint = "token.php";
```

```
function RequestAuthorizationToken() {
    if (authorizationEndpoint) {
        var a = new XMLHttpRequest();
        a.open("GET", authorizationEndpoint);
        a.setRequestHeader("Content-Type", "application/x-www-form-urlencoded");
        a.send("");
        a.onload = function() {
            var token = JSON.parse(atob(this.responseText.split(".")[1]));
            regionKey.value = token.region;
            authorizationToken = this.responseText;
            subscriptionKey.disabled = true;
            subscriptionKey.value = "using authorization token (hit F5 to refresh)";
            console.log("Got an authorization token: " + token);
        }
    }
}
</script>
```

注释：授权令牌只有有限的生存期。这个简化的示例没有显示如何自动刷新授权令牌。作为用户，可以手动重新加载页面或按 F5 键刷新。

7. 本地生成和运行样本

要启动应用程序，双击索引文件.html 或者打开你喜欢的 Web 浏览器的索引文件。它将呈现一个简单的 GUI，允许输入订阅密钥和区域，并使用麦克风触发识别。

8. 通过 Web 服务器构建和运行示例

要启动应用程序，请打开您最喜欢的网络浏览器，并将其指向托管文件夹的公共 URL，输入您的区域，并使用麦克风触发识别。如果配置，它将从令牌来源获取令牌。

9. 获取示例

对于最新的示例，请参见 GITHUB 上的认知服务语音 SDK 示例代码存储库 https://aka.ms/csspeech/samples。在 quickstart/js-browser 文件夹中查找此示例。

14.6 利用 LUIS 根据文字内容理解用户所需

LUIS 是语言理解交互系统 Language Understand Interactive System 的缩写。LUIS 是微软提供的认知服务的一部分，可以帮助用户来使用此服务，去理解文字信息中的具体含义，然后相应地做出反应。比如一个常见的应用就是 24 小时无人值守的智能聊天客服。

在这个快速启动过程中,将话语传递给 LUIS 端点并获得意图和实体。

在这个快速入门中,使用一个可用的公共 LUIS 应用程序从会话文本中确定用户的意图。将用户的意图作为文本发送到公共应用程序的 HTTP 预测端点。在端点处,LUIS 应用公共应用程序的模型来分析自然语言文本的含义,确定总体意图和提取与应用程序的主题域相关的数据。

这个快速启动使用端点 REST API。有关更多信息,请参见端点 API 文档。

对于这篇文章,您需要一个免费的 LUIS 账户。

1. 先决条件

- Python 3.6 或更高版本。
- VisualStudio Code。

注释:完整的解决方案可从 LUIS 样例 GITHUB 存储库获得:https://github.com/Microsoft/LUIS-Samples/blob/master/documentation-samples/quickstarts/analyze-text/。

2. 获取 LUIS 密钥

使用端点密钥提供对预测端点的访问。为了这个快速入门的目的,使用与 LUIS 账户相关联的自由启动键。

登录使用您的 LUIS 账户,如图 14-8 所示。

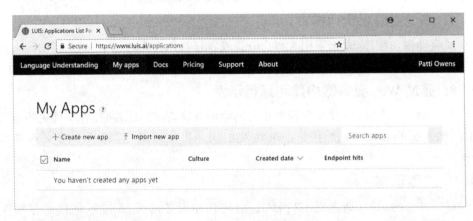

图 14-8　LUIS 应用中注册 App

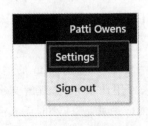

图 14-9　Settings 菜单

在右上菜单中选择你的名字,然后选择设置,如图 14-9 所示。

复制创作键的值。稍后你会在快速入门中使用它,出现如图 14-10 所示界面。

授权密钥允许对创作 API 的免费无限请求,以及针对所有 LUIS 应用程序的每个月对预测端点 API 的至多 1 000 个查询。

图 14 - 10　获取 Key

3. 用浏览器获取意图

若要了解 LUIS 预测端点返回的内容，请在 Web 浏览器中查看预测结果。为了查询公共应用程序，您需要自己的密钥和应用程序 ID。公共 IoT 应用程序 ID df67dcdb-c37d-46af-88e1-8b97951ca1c2 在步骤 1 中作为 URL 的一部分提供。

GET 端点请求的 URL 格式是：

https:// <region>. api. cognitive. microsoft. com/luis/v2. 0/apps/ <appID > ? subscription-key = <YOUR-KEY > &q = <user-utterance>

公共 IOT 应用程序的端点是这种格式：

https://westus. api. cognitive. microsoft. com/luis/v2. 0/apps/df67dcdb-c37d-46af-88e1-8b97951ca1c2? subscription-key = <YOUR_KEY > &q = turn on the bedroom light

4. 复制 URL 并替换您的键值 <YOUR_KEY>

将 URL 粘贴到浏览器窗口中并按 Enter 键。浏览器显示一个 JSON 结果，表明 LUIS 检测到 HomeAutomation. TurnOn 意图作为最高意图，HomeAutomation. room 实体具有值：bedroom。

```
{
"query": "turn on the bedroom light",
"topScoringIntent": {
"intent": "HomeAutomation.TurnOn",
"score": 0.809439957
},
"entities": [
```

```
        {
            "entity": "bedroom",
            "type": "HomeAutomation.Room",
            "startIndex": 12,
            "endIndex": 18,
            "score": 0.8065475
        }
    ]
}
```

更改 URL 中 q=参数的值,turn off the living room light,然后按 Enter 键。结果已经表明,LUIS 发现 HomeAutomation.TurnOff 意图为最高意图和 HomeAutomation.Room 实体有一个 living room 的值。

```
{
    "query": "turn off the living room light",
    "topScoringIntent": {
        "intent": "HomeAutomation.TurnOff",
        "score": 0.984057844
    },
    "entities": [
        {
            "entity": "living room",
            "type": "HomeAutomation.Room",
            "startIndex": 13,
            "endIndex": 23,
            "score": 0.9619945
        }
    ]
}
```

5. 以编程方式获得意图

您可以使用 Python 来访问前面一步中浏览器窗口中看到的相同结果。
将下面的代码段复制到一个名为 quickstart-call-endpoint.py 的文件中:

(1) Python 2.7 代码

```
########### Python 2.7 #############
import httplib, urllib, base64

headers = {
    # Request headers includes endpoint key
    # You can use the authoring key instead of the endpoint key.
    # The authoring key allows 1000 endpoint queries a month.
```

```python
        'Ocp-Apim-Subscription-Key': 'YOUR-KEY',
    }

    params = urllib.urlencode({
        # Text to analyze
        'q': 'turn on the left light',
        # Optional request parameters, set to default values
        'verbose': 'false',
    })

    # HTTP Request
    try:
        # LUIS endpoint HOST for westus region
        conn = httplib.HTTPSConnection('westus.api.cognitive.microsoft.com')

        # LUIS endpoint path
        # includes public app ID
        conn.request("GET", "/luis/v2.0/apps/df67dcdb-c37d-46af-88e1-8b97951ca1c2?%s" % params, "{body}", headers)

        response = conn.getresponse()
        data = response.read()

        # print HTTP response to screen
        print(data)
        conn.close()
    except Exception as e:
        print("[Errno {0}] {1}".format(e.errno, e.strerror))

###################################
```

(2) Python 3.6 代码

```python
########## Python 3.6 ##############
import requests

headers = {
    # Request headers
    'Ocp-Apim-Subscription-Key': 'YOUR-KEY',
}

params = {
    # Query parameter
```

```python
        'q': 'turn on the left light',
        # Optional request parameters, set to default values
        'timezoneOffset': '0',
        'verbose': 'false',
        'spellCheck': 'false',
        'staging': 'false',
    }

    try:
        r = requests.get('https://westus.api.cognitive.microsoft.com/luis/v2.0/apps/df67dcdb-c37d-46af-88e1-8b97951ca1c2',headers = headers, params = params)
        print(r.json())

    except Exception as e:
        print("[Errno {0}] {1}".format(e.errno, e.strerror))

###############################
```

用LUIS端点键替换Ocp-Apim-Subscription-Key字段的值。用pip install安装依赖项。用python ./quickstart-call-endpoint.py运行脚本。它显示在浏览器窗口前面看到的同一个JSON中。

6. LUIS钥匙

为了方便起见,这个快速启动使用创作密钥。密钥主要用于编写模型,但允许少量(1 000个)端点请求。当您准备好在测试阶段或生产环境中接受更多端点请求时,请在Azure门户中创建Language Understanding资源,并将其分配给LUIS门户中的LUIS应用程序。

14.7 用Python调用文本分析认知服务

本演练向您展示如何用Python和文本分析API来检测语言、分析情绪和提取关键短语。

可以通过点击启动绑定标记来运行MyBinder上的Jupyter笔记本。https://mybinder.org/,输入一个github的源码地址,即可将其变成运行Python的笔记本。可以单步运行,如图14-11所示。

图14-11 两个按钮

1. 先决条件

必须有一个带有文本分析API的认知服务API账户。您可以使用5 000个事务/月的自由层完成此演练。还必须具有在注册过程中为您生成的端点和访问密钥。

若要继续此演练,请用先前获得的有效订阅密钥替换订阅 subscription_key。

subscription_key=None
assert subscription_key

接下来,验证 text_analytics_base_url 中的区域是否与设置服务时使用的区域相对应。如果您使用的是免费试用键,则不需要更改任何内容。

text_analytics_base_url = "https://westcentralus.api.cognitive.microsoft.com/text/analytics/v2.0/"

2. 检测语言

语言检测 API 使用检测语言的方法检测文本文档的语言。您的区域的语言检测 API 的服务端点可通过以下 URL 获得:

language_api_url=text_analytics_base_url+"languages"
print(language_api_url)

https://westcentralus.api.cognitive.microsoft.com/text/analytics/v2.0/languages

API 的有效载荷包括一个文档 documents 列表,每个文档又包含一个 ID 和一个文本 text 属性。文本属性存储要分析的文本。

将文档字典替换为任何其他用于语言检测的文本。

```
documents = { 'documents': [
    { 'id': '1', 'text': 'This is a document written in English.' },
    { 'id': '2', 'text': 'Este es un document escrito en Espa?ol.' },
    { 'id': '3', 'text': '这是一个用中文写的文件' }
]}
```

接下来的几行代码使用 Python 中的请求库调用语言检测 API 来确定文档中的语言。

```
import requests
from pprint import pprint
headers = {"Ocp-Apim-Subscription-Key": subscription_key}
response = requests.post(language_api_url, headers=headers, json=documents)
languages = response.json()
pprint(languages)
```

```
{'documents': [{'detectedLanguages': [{'iso6391Name': 'en',
                                        'name': 'English',
                                        'score': 1.0}],
                'id': '1'},
               {'detectedLanguages': [{'iso6391Name': 'es',
                                        'name': 'Spanish',
```

```
                                        'score': 1.0}],
                      'id': '2'},
                     {'detectedLanguages': [{'iso6391Name': 'zh_chs',
                                             'name': 'Chinese_Simplified',
                                             'score': 1.0}],
                      'id': '3'}],
 'errors': []}
```

下面的代码行将 JSON 数据呈现为 HTML 表。

```
from IPython.display import HTML
table = []
for document in languages["documents"]:
    text = next(filter(lambda d: d["id"] == document["id"], documents["documents"]))["text"]
    langs = ", ".join(["{0}({1})".format(lang["name"], lang["score"]) for lang in document["detectedLanguages"]])
    table.append("<tr><td>{0}</td><td>{1}</td>".format(text, langs))
HTML("<table><tr><th>Text</th><th>Detected languages(scores)</th></tr>{0}</table>".format("\n".join(table)))
```

3. 分析情绪

情感分析 API 利用情感方法来检测一组文本记录的情感。下面的例子是两个文件,一个是英语的,另一个是西班牙语的。

用于情感分析的服务端点可通过以下 URL 提供给您的区域:

```
sentiment_api_url = text_analytics_base_url + "sentiment"
print(sentiment_api_url)
```

https://westcentralus.api.cognitive.microsoft.com/text/analytics/v2.0/sentiment

与语言检测示例一样,向服务提供带有文档列表的文档键的字典。每个文档是由 ID、要分析的文本和文本的语言组成的元组。可以使用前一节中的语言检测 API 来填充该字段。

```
documents = {'documents' : [
    {'id': '1', 'language': 'en', 'text': 'I had a wonderful experience! The rooms were wonderful and the staff was helpful.'},
    {'id': '2', 'language': 'en', 'text': 'I had a terrible time at the hotel. The staff was rude and the food was awful.'},
    {'id': '3', 'language': 'es', 'text': 'Los caminos que llevan hasta Monte Rainier son espectaculares y hermosos.'},
    {'id': '4', 'language': 'es', 'text': 'La carretera estaba atascada. Hab a mucho tr fico el d a de ayer.'}
```

]}

情感 API 可以用来分析文档的情感。

```
headers = {"Ocp-Apim-Subscription-Key": subscription_key}
response = requests.post(sentiment_api_url, headers=headers, json=documents)
sentiments = response.json()
pprint(sentiments)


{'documents': [{'id': '1', 'score': 0.7673527002334595},
               {'id': '2', 'score': 0.18574094772338867},
               {'id': '3', 'score': 0.5}],
 'errors': []}
```

一份文件的情感得分在 0 美元到 1 美元之间；得分更高，表示更积极的情绪。

4. 提取关键短语

关键短语提取 API 使用关键字短语方法从文本文档中提取关键短语。本节的演练提取了英语和西班牙语文档的关键短语。

通过以下 URL 访问关键短语提取服务的服务端点：

```
key_phrase_api_url = text_analytics_base_url + "keyPhrases"
print(key_phrase_api_url)


https://westcentralus.api.cognitive.microsoft.com/text/analytics/v2.0/keyPhrases
```

文档的收集与情感分析所用的文档相同。

```
documents = {'documents': [
    {'id': '1', 'language': 'en', 'text': 'I had a wonderful experience! The rooms were wonderful and the staff was helpful.'},
    {'id': '2', 'language': 'en', 'text': 'I had a terrible time at the hotel. The staff was rude and the food was awful.'},
    {'id': '3', 'language': 'es', 'text': 'Los caminos que llevan hasta Monte Rainier son espectaculares y hermosos.'},
    {'id': '4', 'language': 'es', 'text': 'La carretera estaba atascada. Hab a mucho tr fico el d a de ayer.'}
]}
headers = {'Ocp-Apim-Subscription-Key': subscription_key}
response = requests.post(key_phrase_api_url, headers=headers, json=documents)
key_phrases = response.json()
pprint(key_phrases)


{'documents': [
    {'keyPhrases': ['wonderful experience', 'staff', 'rooms'], 'id': '1'},
```

```
    {'keyPhrases': ['food', 'terrible time', 'hotel', 'staff'], 'id': '2'},
    {'keyPhrases': ['Monte Rainier', 'caminos'], 'id': '3'},
    {'keyPhrases': ['carretera', 'tr fico', 'd a'], 'id': '4'}],
'errors': []
}
```

JSON 对象可以使用以下代码行再次呈现为 HTML 表：

```
from IPython.display import HTML
table = []
for document in key_phrases["documents"]:
    text = next(filter(lambda d: d["id"] == document["id"], documents["documents"]))["text"]
    phrases = ",".join(document["keyPhrases"])
    table.append("<tr> <td> {0} </td> <td> {1} </td> ".format(text, phrases))
HTML(" <table> <tr> <th> Text </th> <th> Key phrases </th> </tr> {0} </table> ".format("\n".join(table)))
```

5. 识别实体

实体 API 使用实体方法标识文本文档中的已知实体。下面的示例标识了英文文档的实体。

通过以下 URL 访问实体链接服务的服务端点：

```
entity_linking_api_url = text_analytics_base_url + "entities"
print(entity_linking_api_url)
```

https://westcentralus.api.cognitive.microsoft.com/text/analytics/v2.1-preview/entities

文件的收集如下：

```
documents = {'documents': [
    {'id': '1', 'text': 'Jeff bought three dozen eggs because there was a 50% discount.'},
    {'id': '2', 'text': 'The Great Depression began in 1929. By 1933, the GDP in America fell by 25%.'}
]}
```

现在，文档可以发送到文本分析 API 以接收响应。

```
headers = {"Ocp-Apim-Subscription-Key": subscription_key}
response = requests.post(entity_linking_api_url, headers = headers, json = documents)
entities = response.json()
```

```
{
    "Documents": [
```

```json
{
    "Id": "1",
    "Entities": [
        {
            "Name": "Jeff",
            "Matches": [
                {
                    "Text": "Jeff",
                    "Offset": 0,
                    "Length": 4
                }
            ],
            "Type": "Person"
        },
        {
            "Name": "three dozen",
            "Matches": [
                {
                    "Text": "three dozen",
                    "Offset": 12,
                    "Length": 11
                }
            ],
            "Type": "Quantity",
            "SubType": "Number"
        },
        {
            "Name": "50",
            "Matches": [
                {
                    "Text": "50",
                    "Offset": 49,
                    "Length": 2
                }
            ],
            "Type": "Quantity",
            "SubType": "Number"
        },
        {
            "Name": "50%",
            "Matches": [
                {
```

```
                    "Text": "50%",
                    "Offset": 49,
                    "Length": 3
                }
            ],
            "Type": "Quantity",
            "SubType": "Percentage"
        }
    ]
},
{
    "Id": "2",
    "Entities": [
        {
            "Name": "Great Depression",
            "Matches": [
                {
                    "Text": "The Great Depression",
                    "Offset": 0,
                    "Length": 20
                }
            ],
            "WikipediaLanguage": "en",
            "WikipediaId": "Great Depression",
            "WikipediaUrl": "https://en.wikipedia.org/wiki/Great_Depression",
            "BingId": "d9364681-98ad-1a66-f869-a3f1c8ae8ef8"
        },
        {
            "Name": "1929",
            "Matches": [
                {
                    "Text": "1929",
                    "Offset": 30,
                    "Length": 4
                }
            ],
            "Type": "DateTime",
            "SubType": "DateRange"
        },
        {
            "Name": "By 1933",
            "Matches": [
```

```json
            {
                "Text": "By 1933",
                "Offset": 36,
                "Length": 7
            }
        ],
        "Type": "DateTime",
        "SubType": "DateRange"
    },
    {
        "Name": "Gross domestic product",
        "Matches": [
            {
                "Text": "GDP",
                "Offset": 49,
                "Length": 3
            }
        ],
        "WikipediaLanguage": "en",
        "WikipediaId": "Gross domestic product",
        "WikipediaUrl": "https://en.wikipedia.org/wiki/Gross_domestic_product",
        "BingId": "c859ed84-c0dd-e18f-394a-530cae5468a2"
    },
    {
        "Name": "United States",
        "Matches": [
            {
                "Text": "America",
                "Offset": 56,
                "Length": 7
            }
        ],
        "WikipediaLanguage": "en",
        "WikipediaId": "United States",
        "WikipediaUrl": "https://en.wikipedia.org/wiki/United_States",
        "BingId": "5232ed96-85b1-2edb-12c6-63e6c597a1de",
        "Type": "Location"
    },
    {
        "Name": "25",
        "Matches": [
```

```
                "Text": "25",
                "Offset": 72,
                "Length": 2
            }
        ],
        "Type": "Quantity",
        "SubType": "Number"
    },
    {
        "Name": "25%",
        "Matches": [
            {
                "Text": "25%",
                "Offset": 72,
                "Length": 3
            }
        ],
        "Type": "Quantity",
        "SubType": "Percentage"
    }
  ]
 }
],
"Errors": []
}
```

14.8 用 Python 调用 Bing Web 搜索 API

用此快速启动程序,可以在不到 10 分钟的时间内首次调用 Bing Web 搜索 API,并接收 JSON 响应。

这个快速启动需要一个 Azure 认知服务账户与 Bing 搜索 API。如果您没有账户,可以使用免费试用获得订阅密钥。

这个例子是在 Mybinder 上运行的 Jupyter 笔记本。单击发射 binder 按钮,如图 14-11 所示。

1. 定义变量

用 Azure 账户中的有效订阅密钥替换 subscription_key 值。

```
subscription_key="YOUR_ACCESS_KEY"
assert subscription_key
```

声明 Bing Web 搜索 API 终结点。如果遇到任何授权错误,请在 Azure 仪表板中与 Bing 搜索端点重复检查此值。

```
search_url = "https://api.cognitive.microsoft.com/bing/v7.0/search"
```

可以通过替换 search_term 的值来自定义搜索查询。

```
search_term = "Azure Cognitive Services"
```

2. 提出请求

此块使用请求库来调用 Bing Web 搜索 API,并将结果作为 JSON 对象返回。在密钥字典中传递 API 密钥,并在 PARAMS 字典中传递搜索项和查询参数。请参阅 Bing Web 搜索 API V7 文档以获取选项和参数的完整列表。

```
import requests

headers = {"Ocp-Apim-Subscription-Key" : subscription_key}
params = {"q": search_term, "textDecorations":True, "textFormat":"HTML"}
response = requests.get(search_url, headers=headers, params=params)
response.raise_for_status()
search_results = response.json()
```

3. 格式化并显示响应

搜索结果对象包括搜索结果和元数据,如相关查询和页面。此代码使用 IPython.display 来格式化并在浏览器中显示响应。

```
from IPython.display import HTML

rows = "\n".join(["""<tr>
            <td> <a href =\"{0}\"> {1} </a> </td>
            <td> {2} </td>
            </tr>""".format(v["url"],v["name"],v["snippet"]) \
            for v in search_results["webPages"]["value"]])
HTML(" <table> {0} </table> ".format(rows))
```

完整的样例代码在此地址:https://github.com/Azure-Samples/cognitive-services-REST-api-samples/blob/master/nodejs/Search/BingWebSearchv7.js。

14.9 用 Bing Web 搜索 API 创建单页应用程序

这个单页应用程序演示了如何从 Bing Web 搜索 API 检索、解析和显示搜索结果。本教程使用样板 HTML 和 CSS,并专注于 JavaScript 代码。HTML、CSS 和 JS 文件可以在 GITHUB 上使用 QuaskScript 指令。

此示例应用程序可以:

- 调用具有搜索选项的 Bing Web 搜索 API;

- 显示网页、图像、新闻和视频、分页结果；
- 管理订阅密钥；
- 处理错误。

要使用这个应用程序，需要使用 Bing 搜索 API 的 Azure 认知服务账户。如果您没有账户，可以使用免费试用获得订阅密钥。

1. 先决条件

这里有一些需要运行的应用程序：Node.js 8 或更高版本。

2. 订阅密钥

获取源代码并安装依赖项。先用示例应用程序的源代码克隆存储库。

git clone https://github.com/Azure-Samples/cognitive-services-REST-api-samples.git

然后运行 npm install。对于本教程，Express.js 是唯一的依赖项。

cd <path-to-repo>/cognitive-services-REST-api-samples/Tutorials/Bing-Web-Search

npm install

3. 应用程序组件

我们正在构建的样本应用程序由四个部分组成：

① bing-web-search.js：我们的 Express.js 应用程序。它处理请求/响应逻辑和路由。

② public/index.html：我们的应用程序的骨架，它定义了如何向用户呈现数据。

③ public/css/styles.css：定义页面样式，如字体、颜色、文本大小。

④ public/js/scripts.js：包含向 Bing Web 搜索 API 发出请求、管理订阅密钥、处理和解析响应以及显示结果的逻辑。

本教程着重介绍 Script.js 和调用 Bing Web 搜索 API 所需的逻辑并处理响应。

4. HTML 表单

索引。HTML 包含一个表单，使用户可以搜索和选择搜索选项。OnSt 递交属性在提交表单时，调用 Script.js 中定义的 bingWebSearch() 方法。它需要三个参数：搜索查询、选择的选项、订阅密钥。

<form name = "bing" onsubmit = "return bingWebSearch(this.query.value, bingSearchOptions(this), getSubscriptionKey())">

5. 查询选项

HTML 表单包括映射到 Bing Web 搜索 API V7 中的查询参数的选项。该表提供了用户如何使用示例应用程序过滤搜索结果的故障。

参数描述如下：

query：文本字段以输入查询字符串。

where：在下拉菜单中选择市场（位置和语言）。

what：复选框来促进特定的结果类型。例如，提升图像在搜索结果中的排序。

when：下拉菜单允许用户将搜索结果限制到今天、本周或本月。

safe：一个复选框，使 Bing SuffeSeCH 能够过滤成人内容。

count：隐藏字段。在每个请求上返回的搜索结果的数量。更改此值以显示每页较少或更多的结果。

offset：隐藏字段。请求中的第一搜索结果的偏移，用于寻呼。使用新的请求将其重置为 0。

注释：Bing Web 搜索 API 提供了额外的查询参数以帮助细化搜索结果。此示例仅使用少量。bingSearchOptions()函数将这些选项转换为与 Bing 搜索 API 所需的格式相匹配。

```
// Build query options from selections in the HTML form.
function bingSearchOptions(form) {

    var options = [];
    // Where option.
    options.push("mkt = " + form.where.value);
    // SafeSearch option.
    options.push("SafeSearch = " + (form.safe.checked ? "strict" : "off"));
    // Freshness option.
    if (form.when.value.length) options.push("freshness = " + form.when.value);
    var what = [];
    for (var i = 0; i <form.what.length; i ++ )
        if (form.what[i].checked) what.push(form.what[i].value);
    // Promote option.
    if (what.length) {
        options.push("promote = " + what.join(","));
        options.push("answerCount = 9");
    }
    // Count option.
    options.push("count = " + form.count.value);
    // Offset option.
    options.push("offset = " + form.offset.value);
    // Hardcoded text decoration option.
    options.push("textDecorations = true");
    // Hardcoded text format option.
    options.push("textFormat = HTML");
```

```
        return options.join("&");
}
```

SafeSearch 可以设置为严格的 strict、适度的 moderate 或关闭的 off，moderate 是 Bing Web 搜索的默认设置。此表单使用复选框，它具有两种状态。在这个片段中，SafeSarch 设置为 strict 或 off，不使用 moderate。

如果选择了任何一个推广 Promote 复选框，则将 answerCount 参数添加到查询中。在使用 promote 参数时，answerCount 参数是必需的。在这个代码片段中，该值设置为 9，以返回所有可用的结果类型。

注释：推广 Promote 结果类型并不保证它会包含在搜索结果中；相反，Promote 增加了这些类型的成绩相对于它们平时的排名。若要将搜索限制为特定类型的结果，请使用 responseFilter 查询参数，或调用更具体的端点，如 Bing Image Search 或 Bing News Search。

textDecoration 和 textFormat 查询参数被硬编码到脚本中，并导致搜索结果中的搜索项被粗体显示。这些参数不是必需的。

6. 管理订阅密钥

为了避免硬编码 Bing Search API 订阅密钥，这个示例应用程序使用浏览器的持久存储来存储订阅密钥。如果没有存储订阅密钥，则提示用户输入一个。如果订阅密钥被 API 拒绝，则提示用户重新输入订阅密钥。

getSubscriptionKey() 函数使用 storeValue 和 retrieveValue 函数存储和检索用户的订阅密钥。这些函数使用 LocalStorage 对象，如果支持，或者也用 Cookie。

```
// Cookie names for stored data.
API_KEY_COOKIE = "bing-search-api-key";
CLIENT_ID_COOKIE = "bing-search-client-id";

BING_ENDPOINT = "https://api.cognitive.microsoft.com/bing/v7.0/search";

// See source code for storeValue and retrieveValue definitions.

// Get stored subscription key, or prompt if it isn't found.
function getSubscriptionKey() {
    var key = retrieveValue(API_KEY_COOKIE);
    while (key.length! == 32) {
        key = prompt("Enter Bing Search API subscription key:", "").trim();
    }
    // Always set the cookie in order to update the expiration date.
    storeValue(API_KEY_COOKIE, key);
    return key;
}
```

正如我们之前看到的,当提交表单时,onsubmit 启动,调用 bingWebSearch。此函数初始化并发送请求。在每个请求提交上调用 getSubscriptionKey 来验证请求。

7. 调用 Bing 网络搜索

给定查询、选项字符串和订阅键,BingWebSearch 函数创建一个 XMLHttpRequest 对象来调用 BingWebSearch 端点。

```
// Perform a search constructed from the query, options, and subscription key.
function bingWebSearch(query, options, key) {
    window.scrollTo(0, 0);
    if (!query.trim().length) return false;

    showDiv("noresults", "Working. Please wait.");
    hideDivs("pole", "mainline", "sidebar", "_json", "_http", "paging1", "paging2", "error");

    var request = new XMLHttpRequest();
    var queryurl = BING_ENDPOINT + "? q = " + encodeURIComponent(query) + "&" + options;

    // Initialize the request.
    try {
        request.open("GET", queryurl);
    }
    catch (e) {
        renderErrorMessage("Bad request (invalid URL)\n" + queryurl);
        return false;
    }

    // Add request headers.
    request.setRequestHeader("Ocp-Apim-Subscription-Key", key);
    request.setRequestHeader("Accept", "application/json");
    var clientid = retrieveValue(CLIENT_ID_COOKIE);
    if (clientid) request.setRequestHeader("X-MSEdge-ClientID", clientid);

    // Event handler for successful response.
    request.addEventListener("load", handleBingResponse);

    // Event handler for errors.
    request.addEventListener("error", function() {
        renderErrorMessage("Error completing request");
    });
```

```javascript
// Event handler for an aborted request.
request.addEventListener("abort", function() {
    renderErrorMessage("Request aborted");
});

// Send the request.
request.send();
return false;
}
```

在成功请求之后,加载事件处理程序启动并调用 handleBingResponse 函数。handleBingResponse 解析结果对象,显示结果,并包含失败请求的错误逻辑。

```javascript
function handleBingResponse() {
    hideDivs("noresults");

    var json = this.responseText.trim();
    var jsobj = {};

    // Try to parse results object.
    try {
        if (json.length) jsobj = JSON.parse(json);
    } catch(e) {
        renderErrorMessage("Invalid JSON response");
        return;
    }

    // Show raw JSON and the HTTP request.
    showDiv("json", preFormat(JSON.stringify(jsobj, null, 2)));
    showDiv("http", preFormat("GET " + this.responseURL + "\n\nStatus: " + this.status + " " +
        this.statusText + "\n" + this.getAllResponseHeaders()));

    // If the HTTP response is 200 OK, try to render the results.
    if (this.status === 200) {
        var clientid = this.getResponseHeader("X-MSEdge-ClientID");
        if (clientid) retrieveValue(CLIENT_ID_COOKIE, clientid);
        if (json.length) {
            if (jsobj._type === "SearchResponse"&&"rankingResponse" in jsobj) {
                renderSearchResults(jsobj);
            } else {
                renderErrorMessage("No search results in JSON response");
            }
        }
```

```
            } else {
                renderErrorMessage("Empty response (are you sending too many requests too
quickly?)");
            }
        }

        // Any other HTTP response is considered an error.
        else {
            // 401 is unauthorized; force a re-prompt for the user's subscription
            // key on the next request.
            if (this.status === 401) invalidateSubscriptionKey();

            // Some error responses don't have a top-level errors object, if absent
            // create one.
            var errors = jsobj.errors || [jsobj];
            var errmsg = [];

            // Display the HTTP status code.
            errmsg.push("HTTP Status " + this.status + " " + this.statusText + "\n");

            // Add all fields from all error responses.
            for (var i = 0; i <errors.length; i++) {
                if (i) errmsg.push("\n");
                for (var k in errors[i]) errmsg.push(k + ": " + errors[i][k]);
            }

            // Display Bing Trace ID if it isn't blocked by CORS.
            var traceid = this.getResponseHeader("BingAPIs-TraceId");
            if (traceid) errmsg.push("\nTrace ID " + traceid);

            // Display the error message.
            renderErrorMessage(errmsg.join("\n"));
        }
    }
}
```

注意：成功的 HTTP 请求并不意味着搜索本身就成功了。如果在搜索操作中发生错误，则 Bing Web Search API 返回非 200HTTP 状态代码，并在 JSON 响应中包括错误信息。如果请求是速率受限的，则 API 返回空响应。

上述两个函数中的大部分代码都专门用于错误处理。表 14-2 列出了错误可能发生的阶段。

错误是通过调用 renderErrorMessage() 来处理的。如果响应通过所有错误测试,则调用 renderSearchResults()来显示搜索结果。

表 14-2 错误发生的阶段、可能的原因及处理

阶 段	可能的原因	错误处理
Building the request object（建立请求对象）	Invalid URL（非法 URL）	try/catch（块）
Making the request（发出请求）	Network errors, aborted connections（网路错误,中断连接）	error and abort event handlers（错误或者中断时间处理者）
Performing the search（执行搜索）	Invalid request, invalid JSON, rate limits（非法请求,非法 JSON,比率限制）	Tests in load event handler（在load事件中处理者）

8. 显示搜索结果

Bing Web 搜索 API 返回的结果有使用和显示要求。由于响应可能包括各种结果类型,因此不足以遍历顶级网页集合。取而代之的是,样例应用程序使用 RankingResponse 来将结果排序为规格。

注释:如果你只想要一个结果的类型,则使用 responseFilter 参数;或考虑使用另一个搜索的端点,如 Bing 图片搜索。

每个响应都有一个 RankingResponse 对象,它可以包括多达三个集合:极点 pole、主线 mainline 和侧栏 sidebar。极,如果存在,是最相关的搜索结果,并且必须突出显示。主线包含大部分搜索结果,并在极点之后立即显示。侧栏包括辅助搜索结果。如果可能,这些结果应该显示在侧栏中。如果屏幕限制使侧栏不实用,则这些结果应该出现在主线结果之后。

每个 RankingResponse 应包括一个 RankingItem 数组,指定如何排序结果。我们的示例应用程序使用 answerType 和 resultIndex 参数来标识结果。

注释:还有其他方法来识别和排列结果。有关更多信息,请参见使用排序来显示结果。

让我们看一下代码:

```
// Render the search results from the JSON response.
function renderSearchResults(results) {

    // If spelling was corrected, update the search field.
    if (results.queryContext.alteredQuery)
        document.forms.bing.query.value = results.queryContext.alteredQuery;

    // Add Prev / Next links with result count.
    var pagingLinks = renderPagingLinks(results);
    showDiv("paging1", pagingLinks);
```

```
        showDiv("paging2", pagingLinks);

        // Render the results for each section.
        for (section in {pole: 0, mainline: 0, sidebar: 0}) {
            if (results.rankingResponse[section])
                showDiv(section, renderResultsItems(section, results));
        }
    }
```

renderResultsItems()函数迭代每个 RankingResponse 集合中的项,使用.Type 和 resultIndex 值将每个排名结果映射到搜索结果,并调用适当的呈现函数来生成 HTML。如果没有为项指定 resultIndex,则 renderResultsItems()迭代该类型的所有结果,并为每个项调用呈现函数。生成的 HTML 插入到 index.html 中的适当的 <div> 元素中。

```
// Render search results from the RankingResponse object per rank response and
// use and display requirements.
function renderResultsItems(section, results) {

    var items = results.rankingResponse[section].items;
    var html = [];
    for (var i = 0; i <items.length; i++) {
        var item = items[i];
        // Collection name has lowercase first letter while answerType has uppercase
        // e.g. 'WebPages' RankingResult type is in the 'webPages' top-level collection.
        var type = item.answerType[0].toLowerCase() + item.answerType.slice(1);
        if (type in results && type in searchItemRenderers) {
            var render = searchItemRenderers[type];
            // This ranking item refers to ONE result of the specified type.
            if ("resultIndex" in item) {
                html.push(render(results[type].value[item.resultIndex], section));
            // This ranking item refers to ALL results of the specified type.
            } else {
                var len = results[type].value.length;
                for (var j = 0; j <len; j++) {
                    html.push(render(results[type].value[j], section, j, len));
                }
            }
        }
    }
    return html.join("\n\n");
}
```

9. 审查渲染器功能

在我们的示例应用程序中,searchItemRenderers 对象包括为每种类型的搜索结果生成 HTML 的函数。

```
// Render functions for each result type.
searchItemRenderers = {
    webPages: function(item) { ... },
    news: function(item) { ... },
    images: function(item, section, index, count) { ... },
    videos: function(item, section, index, count) { ... },
    relatedSearches: function(item, section, index, count) { ... }
}
```

注意:样例应用程序具有网页、新闻、图像、视频和相关搜索的渲染器。您的应用程序将需要渲染器来渲染它可能接收的任何类型的结果,包括计算、拼写建议、实体、时区和定义。

一些渲染函数只接受项目参数。其他则接受附加参数,这些参数可以用于基于上下文来不同地呈现项目。不使用此信息的渲染器不需要接受这些参数。表 14-3 列出了上下文参数及描述。

表 14-3 上下文参数

参 数	描 述
section	项目出现的结果部分(极点、主线或侧栏)
Index count	当 RankingResponse 项指定要显示给定集合中的所有结果时,可用;否则未定义。您可以使用集合中的项的索引和该集合中的项的总数等信息对结果进行编号,以生成第一个或最后一个结果的不同 HTML 等

在样例应用程序中,images 和 relatedSearches 渲染器都使用上下文参数来定制生成的 HTML。让我们仔细看看 images 渲染器:

```
searchItemRenderers = {
    // Render image result with thumbnail.
    images: function(item, section, index, count) {
        var height = 60;
        var width = Math.round(height * item.thumbnail.width/item.thumbnail.height);
        var html = [];
        if (section === "sidebar") {
            if (index) html.push(" <br> ");
        } else {
            if (!index) html.push(" <p class = 'images'> ");
```

```
            html.push(" <a href = '" + item.hostPageUrl + "'> ");
            var title = escape(item.name) + "\n" + getHost(item.hostPageDisplayUrl);
            html.push(" <img src = '" + item.thumbnailUrl + "&h = " + height + "&w = " + width +
"' height = " + height + " width = " + width + " title = '" + title + "' alt = '" + title + "'> ");
            html.push(" </a> ");
            return html.join("");
        },
        // Other renderers are omitted from this sample...
}
```

10. 图像渲染器

- 计算图像缩略图的大小(宽度变化,而高度固定在60像素)。
- 插入基于上下文的图像结果之前的HTML。
- 建立链接到包含图像的页面的HTML <a>标签。
- 建立HTML <img>标签以显示图像缩略图。

图像渲染器使用区段和索引变量以不同的方式显示结果,这取决于它们出现的位置。在侧栏中的图像结果之间插入断线(
标签),以便侧栏显示图像列。在其他部分中,第一个图像结果(index==0)前面有一个<p>标签。

缩略图大小用于缩略图 URL 中的<img>tag 和 h、w 字段。title 和 alt 属性(图像的文本描述)由图像的名称和 URL 中的主机名构成。

下面是应用程序中如何显示图像的示例,如图 14-12 所示。

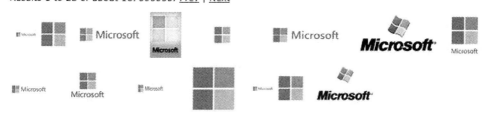

图 14-12　应用程序中如何显示图像的示例

11. 保存客户端 ID

Bing 搜索 API 的响应可以包括一个 X-MSEdge-ClientID 头部,该头部应该随每个连续请求一起发送回 API。如果您的应用程序使用了多个 BingSearchAPI,请确保每个请求跨服务发送相同的客户端 ID。

提供 X-MSEdge-ClientID 标头允许 Bing API 关联用户的搜索。首先,它允许 Bing 搜索引擎将过去的上下文应用到搜索中,以找到更好地满足请求的结果。例如,如果用户先前搜索过与航行相关的术语,那么稍后搜索"节"可以优先返回关于航行中使用的节的信息。其次，Bing 可以随机选择用户体验新的特征，然后才被广泛使用。在每个请求中提供相同的客户端 ID，确保被选中的用户可以看到它。如果没有客户端 ID，用户可能会看到一个在搜索结果中看起来是随机的特征。

浏览器安全策略，例如跨源资源共享（Cross Origin Resource Sharing，CORS），可能会阻止示例应用程序访问 X-MSEdge-ClientID 头部。当搜索响应与请求的页面有不同的原点时，会发生这种限制。在生产环境中，应该通过托管在与 Web 页面相同的域上执行 API 调用的服务器端脚本来解决此策略。由于脚本与 Web 页面具有相同的原点，因此 X-MSEdge-ClientID 头可用于 JavaScript。

注释：在生产 Web 应用程序中，无论如何都应该执行请求服务器端；否则，您的 Bing Search API 订阅密钥必须包括在网页中，任何查看源代码的人都可以使用该密钥。您在 API 订阅密钥下为所有使用情况计费，甚至包括未授权方发出的请求，因此不要公开密钥非常重要。

为了开发的目的，可以通过 CORS 代理进行请求。此类代理的响应具有 Access-Control-Expose-Headers 头部，该头部将响应头部白名单，并使它们可用于 JavaScript。

很容易安装一个 CORS 代理来允许我们的示例应用程序访问客户 ID 标头。运行此命令：

```
npm install -g cors-proxy-server
```

接下来，将 script.js 中的 Bing Web 搜索端点更改为

```
http://localhost:9090/https://api.cognitive.microsoft.com/bing/v7.0/search
```

用这个命令启动 CORS 代理：

```
cors-proxy-server
```

在使用示例应用程序时，让命令窗口打开；关闭窗口停止代理。在搜索结果下面的可扩展 HTTP 报头部分中，X-MSEdge-ClientID 报头应该是可见的。验证每个请求是否相同。

14.10　利用 Azure 门户开始 Azure 机器学习

在这个快速入门中，利用 Azure 门户创建 Azure 机器学习工作区。这个工作区是云中的基本块，利用它来实验、培训和部署机器学习模型。

在这个快速启动中，可以：

- 在 Azure 订阅中创建工作区。
- 将 Python 在 Azure 笔记本中进行试用，并在多次迭代中记录值。
- 查看工作区中的日志值。

当区域可用时,下列 Azure 资源将自动添加到您的工作空间中:
- Azure Container Registry;
- Azure Storage;
- Azure Application Insights;
- Azure Key Vault。

您创建的资源可以作为其他机器学习服务教程和如何使用的文章的先决条件。与其他 Azure 服务一样,与机器学习相关的某些资源也存在限制。一个例子是 Azure Batch AI 集群大小。

1. 创建工作区

使用 Azure 订阅的凭据登录到 Azure 门户。如果你没有 Azure 订阅,那么现在创建一个免费账户。

门户的工作区仪表板仅支持在边缘、Chrome 和 Firefox 浏览器上,如图 14-13 所示。

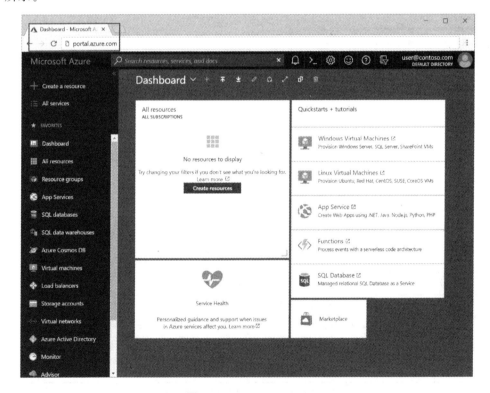

图 14-13　Azure Portal

在门户的左上角,选择创建一个资源,如图 14-14 所示。

在搜索栏中,输入机器学习。选择机器学习服务工作区搜索结果,如图 14-15 所示。

图 14-14　Create a resource 菜单

图 14-15　选择机器学习服务工作区

在 Machine Learning service workspace(preview)工作区窗格中,滚动到底部并选择 Create 开始。

在 ML 服务工作区窗格中,配置工作区。

表 14-4 为创建机器学习服务工作区所需输入的字段及其描述。

表 14-4　所需输入的字段及其描述

字　段	描　述
Workspace name	输入标识工作区的唯一名称。这里我们使用 DOCS WS。资源组中的名称必须是唯一的。使用一个易于回忆和区别于其他人创建的工作空间的名称
Subscription	即你用的订阅号。如果有多个订阅,请选择一个正确的订阅
Resource group	资源组帮我们很好地管理资源。新建一个或者选择一个已有的资源组
Location	选择一个较近的数据中心。网络会好一些

输入以上信息到如图 14-16 所示的界面。

若要启动创建过程,请选择创建。创建工作空间可能需要一些时间。

若要检查部署的状态,请选择工具栏上的通知图标(Bell)。

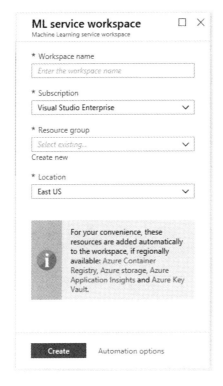

图 14-16　创建机器学习服务工作区

当进程完成时,出现部署成功消息,它也存在于通知部分。若要查看新工作区,请选择转到资源。

在工作区页面上,选择 Explore your Azure Machine Learning service workspace,如图 14-17 所示。

Getting Started

Explore your Azure Machine Learning service workspace

Explore your Machine Learning service workspace to run and track experiments, compare model performance, and deploy models.

图 14-17　Explore your Azure Machine Learning service workspace 链接

2. 使用工作区

现在看看工作空间如何帮助您管理机器学习脚本。在这一节,可以学会:

- 在 Azure 笔记本上打开笔记本。
- 运行创建一些日志值的代码。
- 查看工作区中的日志值。

这个示例展示了工作区如何帮助您跟踪脚本中生成的信息。

3. 打开笔记本

Azure Notebooks 为 Jupyter 笔记本电脑提供了一个免费的云平台,它预先配置了运行 Machine Learning 所需的所有东西。

选择打开 Azure 笔记本 Open Azure Notebooks,尝试你的第一个实验,如图 14-18 所示。

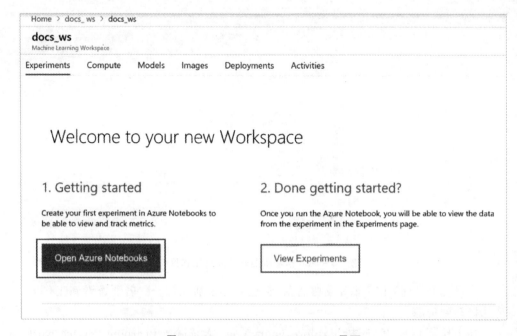

图 14-18　Open Azure Notebooks 界面

您的组织可能需要管理员同意才能签到。登录后,打开一个新标签,出现一个克隆库提示符。选择克隆。

4. 运行笔记本

与两个笔记本一起,您可以看到一个 config.json 文件。此配置文件包含有关您创建的工作区的信息。

选择 01.run-experiment.ipynb 打开笔记本。

要一次运行单元格,请使用 Shift+Enter 键。或选择单元格→运行全部以运行整个笔记本。你可以看到一个星号"*"旁边的一个单元格,它正在运行。在该单元的代码完成后,出现一个数字。

可能会提示您登录。复制消息中的代码。然后选择链接,并将代码粘贴到新窗口中。一定不要在代码之前或之后复制一个空间。在 Azure 门户中使用同一个账户登录,如图 14-19 所示。

```
Performing interactive authentication. Please follow the instructions on the terminal.
To sign in, use a web browser to open the page https://microsoft.com/devicelogin and enter the code FMNB4SCPP to authenticate.
```

图 14 - 19　Azure 中的提示

在笔记本中，第二个单元格从 config.json 读取，连接到您的工作区。

ws = Workspace.from_config()

代码的第三个单元格以"my-first-experiment."这个名称开始一个实验。

experiment = Experiment(workspace_object = ws, name = "my - first - experiment")

在笔记本的最后一个单元格中，注意写入日志文件的值。

```
# Log final results
run.log("Final estimate: ",pi_estimate)
run.log("Final error: ",math.pi - pi_estimate)
```

代码运行后，您可以在工作区中查看这些值。

5. 查看日志值

运行笔记本中的所有单元格后，返回门户页面。

选择 View Experiments，如图 14 - 20 所示。

图 14 - 20　View Experiments 界面

关闭 Reports 弹窗。选择 my-first-experiment。

请参阅刚刚执行的运行信息。向下滚动页面以查找运行表。选择运行编号链接，如图 14 - 21 所示。

您可以看到记录值自动创建的绘图，如图 14 - 22 所示。

由于近似 π 的代码使用随机值，所以您的绘图将显示不同的值。

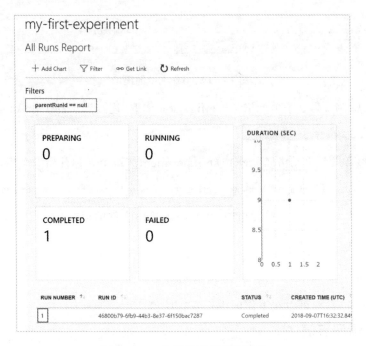

图 14-21 查看运行编号的详情

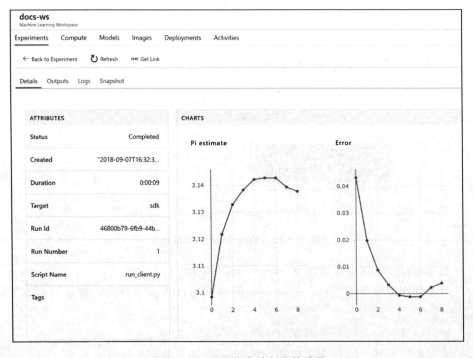

图 14-22 记录值自动创建的绘图

14.11 利用 Azure 机器学习服务训练图像分类模型

在本教程中,您将在本地和远程计算资源上训练机器学习模型。您将使用 Python Jupyter 笔记本上的 Azure 机器学习服务(预览)的培训和部署工作流。然后,可以用笔记本作为模板,用自己的数据来训练自己的机器学习模型。本教程是两部分教程系列中的第一部分。

本教程使用 MNIST 数据集和 scikit-learn 与 Azure 机器学习服务进行简单的逻辑回归。MNIST 是一个流行的数据集,由 70 000 幅灰度图像组成。每个图像是一个 28×28 像素的手写数字,代表从 0~9 的数字。目标是创建一个多类分类器来识别给定图像所代表的数字。

1. 获得笔记本

为了方便起见,本教程可作为 Jupyter 笔记本使用。运行 01.train-models.ipynb 笔记本,要么在 Azure 笔记本上,要么在自己的 Jupyter 笔记本服务器上运行。

(1) Azure Notebooks——在 Azure 中的免费 Jupyter 笔记本

SDK 已经安装在 Azure 笔记本上,并为您配置。完成快速入门,创建工作区并启动 Azure 笔记本。

(2) 去 Azure 笔记本

在快速启动过程中创建的起始样例库中,转到教程文件夹。打开笔记本。

(3) 你自己的 Jupyter 笔记本服务器

用 Python SDK 快速入门完成安装 SDK 并创建工作区。克隆 GITHUB 存储库。将快速启动期间创建的 aml_config 目录复制到克隆目录中。从克隆目录启动笔记本服务器。转到教程文件夹。打开笔记本。

注释:本教程使用 Azure 机器学习 SDK 版本 0.168 进行了测试。

2. 建立开发环境

您的开发工作的所有设置都可以在 Python 笔记本中完成。设置包括:

- 导入 Python 软件包;
- 连接到工作区以实现本地计算机和远程资源之间的通信;
- 创建一个跟踪所有运行的实验;
- 创建用于训练的远程计算目标;
- 导入包。

在这个会话中导入您需要的 Python 程序包,还显示了 Azure 机器学习 SDK 版本。

```
% matplotlib inline
import numpy as np
```

```
import matplotlib
import matplotlib.pyplot as plt

import azureml
from azureml.core import Workspace, Run

# check core SDK version number
print("Azure ML SDK Version: ", azureml.core.VERSION)
```

3. 连接到工作空间

从现有工作区创建工作区对象。Workspace.from_config()读取文件config.json 并将详细信息加载到一个名为 WS 的对象中。

```
# load workspace configuration from the config.json file in the current folder.
ws = Workspace.from_config()
print(ws.name, ws.location, ws.resource_group, ws.location, sep = '\t')
```

4. 创造实验

创建一个实验来跟踪工作区中的运行。工作区可以进行多个实验。

```
experiment_name = 'sklearn-mnist'

from azureml.core import Experiment
exp = Experiment(workspace = ws, name = experiment_name)
```

5. 创建远程计算目标

Azure Batch AI 是一个管理服务，它使数据科学家能够在 Azure 虚拟机集群上训练机器学习模型，包括支持 GPU 的 VM。在本教程中，您创建一个 Azure 批量 AI 集群作为您的培训环境。如果代码不存在于工作区中，则此代码为您创建集群。

集群的创建大约需要 5 分钟。如果集群已经在工作区中，则该代码使用它并跳过创建过程。

```
from azureml.core.compute import ComputeTarget, BatchAiCompute
from azureml.core.compute_target import ComputeTargetException

# choose a name for your cluster
batchai_cluster_name = "traincluster"

try:
    # look for the existing cluster by name
    compute_target = ComputeTarget(workspace = ws, name = batchai_cluster_name)
    if type(compute_target) is BatchAiCompute:
```

```
            print('found compute target {}, just use it.'.format(batchai_cluster_name))
        else:
            print('{} exists but it is not a Batch AI cluster. Please choose a different
name.'.format(batchai_cluster_name))
    except ComputeTargetException:
        print('creating a new compute target...')
        compute_config = BatchAiCompute.provisioning_configuration(vm_size = "STANDARD_
D2_V2", # small CPU-based VM
                                                                   # vm_priority = 'lowpriority', # optional
                                                                   autoscale_enabled = True,
                                                                   cluster_min_nodes = 0,
                                                                   cluster_max_nodes = 4)

        # create the cluster
        compute_target = ComputeTarget.create(ws, batchai_cluster_name, compute_config)

        # can poll for a minimum number of nodes and for a specific timeout.
        # if no min node count is provided it uses the scale settings for the cluster
        compute_target.wait_for_completion(show_output = True, min_node_count = None,
timeout_in_minutes = 20)

        # Use the 'status' property to get a detailed status for the current cluster.
        print(compute_target.status.serialize())
```

6. 探索数据

在训练模型之前,需要了解用来训练它的数据。还需要将数据复制到云中,以便云计算环境可以访问该数据。在本节中,您将学习:

- 下载 MNIST 数据集;
- 显示一些样本图像;
- 将数据上传到云。

7. 下载 MNIST 数据集

下载 MNIST DataSet 并将文件保存到本地的数据目录中。下载培训和测试的图像及标签。

```
import os
import urllib.request

os.makedirs('./data', exist_ok = True)

urllib.request.urlretrieve('http://yann.lecun.com/exdb/mnist/train-images-idx3-
ubyte.gz', filename = './data/train-images.gz')
```

```
urllib.request.urlretrieve('http://yann.lecun.com/exdb/mnist/train-labels-idx1-
ubyte.gz', filename = './data/train-labels.gz')
urllib.request.urlretrieve('http://yann.lecun.com/exdb/mnist/t10k-images-idx3-
ubyte.gz', filename = './data/test-images.gz')
urllib.request.urlretrieve('http://yann.lecun.com/exdb/mnist/t10k-labels-idx1-
ubyte.gz', filename = './data/test-labels.gz')
```

8. 显示一些样本图像

将压缩文件加载到 NUMPY 数组中,然后用 matplotlib 从数据集上绘制 30 个随机图像,加入上面的标签。注意,这个步骤需要一个包含在 UTILY 文件中的 load_data 函数。此文件包含在示例文件夹中。请确保它放在与本笔记本相同的文件夹中。load_data 函数将压缩文件解析为 NUMPY 数组。

```
# make sure utils.py is in the same directory as this code
from utils import load_data

# note we also shrink the intensity values (X) from 0-255 to 0-1. This helps the model converge faster.
X_train = load_data('./data/train-images.gz', False) / 255.0
y_train = load_data('./data/train-labels.gz', True).reshape(-1)

X_test = load_data('./data/test-images.gz', False) / 255.0
y_test = load_data('./data/test-labels.gz', True).reshape(-1)

# now let's show some randomly chosen images from the traininng set.
count = 0
sample_size = 30
plt.figure(figsize = (16, 6))
for i in np.random.permutation(X_train.shape[0])[:sample_size]:
    count = count + 1
    plt.subplot(1, sample_size, count)
    plt.axhline('')
    plt.axvline('')
    plt.text(x = 10, y = -10, s = y_train[i], fontsize = 18)
    plt.imshow(X_train[i].reshape(28, 28), cmap = plt.cm.Greys)
plt.show()
```

随机图像显示如图 14-23 所示。

现在你已经知道这些图像的样子和预测结果了。

9. 将数据上传到云

现在,通过将数据从本地计算机上传到 Azure,使得数据能够远程访问,以便进

```
8 9 4 7 9 2 9 4 7 8 4 8 8 2 3 2 4 6 1 1 3 8 8 1 3 1 8 4 1 8
```

图 14 - 23　随机图像

行远程培训。数据存储是与工作空间相关联的一种方便的结构,用于上传/下载数据,并从远程计算目标与它交互。它是由 Azure Blob 存储账户支持的。

MNIST 文件被上传到一个名为 MNIST 的目录中,在数据存储的根目录下。

```
ds = ws.get_default_datastore()
print(ds.datastore_type, ds.account_name, ds.container_name)

ds.upload(src_dir = './data', target_path = 'mnist', overwrite = True, show_progress = True)
```

你现在拥有了开始训练模型所需的一切。

10. 在本地训练模型

从 scikit-learn 学习一个简单的逻辑回归模型。根据您的计算机配置,本地培训可能需要一两分钟。

```
% %time
from sklearn.linear_model import LogisticRegression

clf = LogisticRegression()
clf.fit(X_train, y_train)
```

接下来,使用测试集进行预测并计算精度。

```
y_hat = clf.predict(X_test)
print(np.average(y_hat == y_test))
```

局部模型精度显示:

0.9202

只需几行代码,就可以达到 92% 的准确率。

11. 在远程集群上进行训练

现在可以通过建立一个具有不同正则化率的模型来扩展这个简单的模型。这次您将在远程资源上训练模型。

对于此任务,请将作业提交到您先前设置的远程培训集群。提交一份工作:
- 创建目录;
- 创建培训脚本;
- 创建估计器;
- 提交工作。

(1) 创建目录

创建一个目录,将必要的代码从计算机传递到远程资源。

```
import os
script_folder = './sklearn-mnist'
os.makedirs(script_folder, exist_ok=True)
```

(2) 创建训练脚本

要将作业提交到集群,首先要创建一个培训脚本。运行下面的代码,在刚刚创建的目录中创建名为 train.py 的训练脚本。这种训练增加了训练算法的正则化速率,因此产生了与局部版本稍有不同的模型。

```
%%writefile $script_folder/train.py

import argparse
import os
import numpy as np

from sklearn.linear_model import LogisticRegression
from sklearn.externals import joblib

from azureml.core import Run
from utils import load_data

# let user feed in 2 parameters, the location of the data files (from datastore), and the regularization rate of the logistic regression model
parser = argparse.ArgumentParser()
parser.add_argument('--data-folder', type=str, dest='data_folder', help='data folder mounting point')
parser.add_argument('--regularization', type=float, dest='reg', default=0.01, help='regularization rate')
args = parser.parse_args()

data_folder = os.path.join(args.data_folder, 'mnist')
print('Data folder:', data_folder)

# load train and test set into numpy arrays
# note we scale the pixel intensity values to 0-1 (by dividing it with 255.0) so the model can converge faster.
X_train = load_data(os.path.join(data_folder, 'train-images.gz'), False) / 255.0
X_test = load_data(os.path.join(data_folder, 'test-images.gz'), False) / 255.0
y_train = load_data(os.path.join(data_folder, 'train-labels.gz'), True).reshape(-1)
y_test = load_data(os.path.join(data_folder, 'test-labels.gz'), True).reshape(-1)
```

```python
print(X_train.shape, y_train.shape, X_test.shape, y_test.shape, sep = '\n')

# get hold of the current run
run = Run.get_context()

print('Train a logistic regression model with regularizaion rate of', args.reg)
clf = LogisticRegression(C = 1.0/args.reg, random_state = 42)
clf.fit(X_train, y_train)

print('Predict the test set')
y_hat = clf.predict(X_test)

# calculate accuracy on the prediction
acc = np.average(y_hat == y_test)
print('Accuracy is', acc)

run.log('regularization rate', np.float(args.reg))
run.log('accuracy', np.float(acc))

os.makedirs('outputs', exist_ok = True)
# note file saved in the outputs folder is automatically uploaded into experiment record
joblib.dump(value = clf, filename = 'outputs/sklearn_mnist_model.pkl')
```

注意学习脚本如何获取数据并保存模型。

训练脚本读取一个参数来查找包含数据的目录。当您稍后提交作业时,将指向此参数的数据存储:parser.add_argument('--data-folder', type = str, dest = 'data_folder', help = 'data directory mounting point')。

训练脚本将模型保存到一个名为"输出"的目录中。

```python
joblib.dump(value = clf, filename = 'outputs/sklearn_mnist_model.pkl')
```

在这个目录中写的任何东西都会自动上传到您的工作空间。稍后将在教程中从该目录访问您的模型。

从训练脚本引用文件 utils.py 来正确加载数据集。将此脚本复制到脚本文件夹中,以便可以与远程资源上的训练脚本一起访问该脚本。

```python
import shutil
shutil.copy('utils.py', script_folder)
```

(3) 创建估计器

估计对象用于提交运行。通过运行以下代码创建您的估计器来定义:
- 估计对象的名称。

- 包含脚本的目录。该目录中的所有文件都被上传到集群节点中执行。
- 计算目标。在这种情况下,您将使用您创建的批 AI 集群。
- 培训脚本名称。
- 训练脚本所需的参数。
- Python 软件包需要的培训。

在本教程中,这个目标是 Batch AI 集群。将项目目录中的所有文件上载到集群节点中执行。data_folder 文件夹被设置为使用 datastore (ds.as_mount())。

```python
from azureml.train.estimator import Estimator

script_params = {
    '--data-folder': ds.as_mount(),
    '--regularization': 0.8
}

est = Estimator(source_directory = script_folder,
         script_params = script_params,
         compute_target = compute_target,
         entry_script = 'train.py',
         conda_packages = ['scikit-learn'])
```

(4) 提交工作

通过提交估计器对象来运行实验。

```
run = exp.submit(config = est)
Run
```

由于调用是异步的,所以一旦作业启动,它就会返回准备状态或运行状态。

12. 监视远程运行

总的来说,第一次运行大约需要 10 分钟。但是对于后续的运行,只要脚本依赖关系没有改变,就会重用相同的映像,因此容器启动时间要快得多。

下面是您等待的时候发生的事情:

① 图像创建:创建一个匹配由估计器指定的 Python 环境的 DOCKER 图像。图像被上传到工作区。图像创建和上传大约需要 5 分钟。

这个阶段对于每个 Python 环境都会发生一次,因为容器是为后续运行缓存的。在图像创建期间,日志被流到运行历史。您可以使用这些日志监视图像创建进度。

② 伸缩:如果远程集群需要比当前可用的节点更多的节点来执行运行,则自动添加其他节点。缩放通常需要大约 5 分钟。

③ 运行:在这个阶段,必需的脚本和文件被发送到计算目标,挂载/复制数据存储,然后运行 entry_script。当作业正在运行时,STDUT 和 ./logs 目录流到运行历

史记录。您可以使用这些日志监视运行的进度。

④ 后处理：运行的./outputs 目录被复制到工作区中的运行历史中，以便您可以访问这些结果。

您可以以多种方式查看正在运行的作业的进度。本教程使用 Juyter 小部件以及 wait_for_completion 完成方法。

13. Jupyter 控件

用 Jupyter 小部件观看运行的进度。与运行提交一样，WIDGET 是异步的，并且每隔 10～15 秒提供实时更新，直到作业完成。

```
from azureml.train.widgets import RunDetails
RunDetails(run).show()
```

下面是在训练结束时显示的小部件的静态快照，如图 14-24 所示。

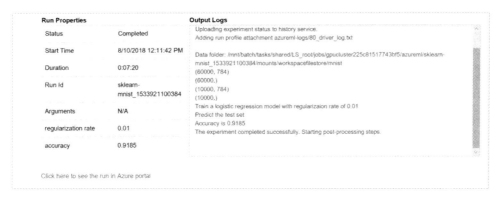

图 14-24 小部件的静态快照

14. 完成后获得日志结果

模型培训和监控发生在后台。等待模型在完成更多代码之前已经完成了训练。使用 wait_for_completion 完成模型训练时的显示：

```
run.wait_for_completion(show_output = False) # specify True for a verbose log
```

15. 显示运行结果

现在有一个在远程集群上训练的模型。检索模型的准确性：

```
print(run.get_metrics())
```

结果表明，由于训练过程中规则化率的增加，远程模型的精度略高于局部模型。

{'regularization rate': 0.8, 'accuracy': 0.9204}

在部署教程中，可以更详细地探究这个模型。

16. 寄存器模型

培训脚本中的最后一步是将文件 outputs/sklearn_mnist_model.pkl 写入执行作业集群 VM 中的名为 outputs 的目录中。输出是一个特殊目录,目录中的所有内容都自动上传到您的工作区。此内容显示在您的工作区下的实验运行记录中。因此,模型文件现在也可以在您的工作区中使用。

可以看到与该运行相关联的文件。

print(run.get_file_names())

在工作区中注册模型,以便您(或其他合作者)可以稍后查询、检查和部署该模型。

```
# register model
model = run.register_model(model_name = 'sklearn_mnist', model_path = 'outputs/sklearn_mnist_model.pkl')
print(model.name, model.id, model.version, sep = '\t')
```

14.12　在 Azure 容器实例(ACI)中训练图像分类模型

在本教程中,将学习如何使用自动机器学习(自动化 ML)生成机器学习模型。Azure 机器学习服务可以自动进行数据预处理、算法选择和超参数选择。然后,可以在部署模型教程的工作流中部署最终模型,如图 14 – 25 所示。

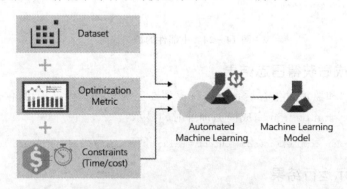

图 14 – 25　自动机器学习训练模型

本教程从 MNIST 数据集分类手写数字图像(0～9)。但是这次不需要指定一个算法或者调整超参数。自动 ML 技术能够运用迭代算法和超参数的许多组合,直到它根据您的标准找到最佳模型。

你将学会:

- 建立开发环境;
- 访问和检查数据;

- 在本地计算机上使用自动分类器进行训练；
- 探索结果；
- 复习培训结果；
- 注册最佳模型。

1. 拿笔记本

为了方便起见，本教程可作为 Jupyter 笔记本使用。运行 03.auto-train-models.ipynb 笔记本，要么在 Azure 笔记本上，要么在自己的 Jupyter 笔记本服务器上运行。

（1）Azure 笔记本——免费的木星在 Azure 笔记本

SDK 已经安装在 Azure 笔记本上，并为您配置。完成快速入门，创建工作区并启动 Azure 笔记本电脑。

（2）去 Azure 笔记本

在快速启动过程中创建的启动库中，转到教程文件夹。打开笔记本。

（3）你自己的 Jupyter 笔记本服务器

用 Python SDK 快速入门完成安装 SDK 并创建工作区。克隆 GITHUB 存储库。将快速启动期间创建的 aml_config 目录复制到克隆目录中。从克隆目录启动笔记本服务器。转到教程文件夹。打开笔记本。

2. 建立开发环境

您的开发工作的所有设置都可以在 Python 笔记本中完成。设置包括：

（1）导入 Python 软件包

配置工作区以实现本地计算机和远程资源之间的通信。创建一个存储训练脚本的目录。

（2）导入包

在本教程中导入您需要的 Python 程序包。

```
import azureml.core
import pandas as pd
from azureml.core.workspace import Workspace
from azureml.train.automl.run import AutoMLRun
import time
import logging
from sklearn import datasets
from matplotlib import pyplot as plt
from matplotlib.pyplot import imshow
import random
import numpy as np
```

3. 配置工作空间

从现有工作区创建工作区对象。Workspace.from_config()读取文件 aml_config/config.json 并将细节加载到一个名为 WS 的对象中。WS 在本教程的其余代码中使用。

一旦有了工作空间对象,就为实验指定一个名称,创建并注册一个带有工作区的本地目录。所有运行的历史记录都在指定的实验下。

```
ws = Workspace.from_config()
# project folder to save your local files
project_folder = './sample_projects/automl-local-classification'
# choose a name for the run history container in the workspace
experiment_name = 'automl-classifier'

import os

output = {}
output['SDK version'] = azureml.core.VERSION
output['Subscription ID'] = ws.subscription_id
output['Workspace'] = ws.name
output['Resource Group'] = ws.resource_group
output['Location'] = ws.location
output['Project Directory'] = project_folder
pd.set_option('display.max_colwidth', -1)
pd.DataFrame(data = output, index = ['']).T
```

4. 探索数据

最初的训练教程使用了 MNIST 数据集(28×28 像素)的高分辨率版本。由于自动 ML 训练需要多次迭代,因此本教程使用图像的较小分辨率版本(8×8 像素)来演示概念,同时加快每次迭代所需的时间。

```
from sklearn import datasets

digits = datasets.load_digits()

# Exclude the first 100 rows from training so that they can be used for test.
X_train = digits.data[100:,:]
y_train = digits.target[100:]
```

5. 显示一些样本图像

将数据加载到 numpy 数组中。然后使用 matplotlib 从数据集上绘制 30 个随机图像,加入上面的标签。

```
count = 0
sample_size = 30
plt.figure(figsize = (16, 6))
for i in np.random.permutation(X_train.shape[0])[:sample_size]:
    count = count + 1
    plt.subplot(1, sample_size, count)
    plt.axhline('')
    plt.axvline('')
    plt.text(x = 2, y = -2, s = y_train[i], fontsize = 18)
    plt.imshow(X_train[i].reshape(8, 8), cmap = plt.cm.Greys)
plt.show()
```

随机图像如图 14 - 26 所示。

6 2 8 0 9 7 1 2 3 0 6 4 3 5 1 4 1 0 3 6 8 2 3 1 8 7 1 2 8 4

图 14 - 26 另外一些随机图像

现在,必要的包和数据都准备好了,以便为您的模型自动训练。

6. 训练模型

为了自动训练模型,首先定义实验的配置设置,然后运行实验。

(1) 定义设置

定义实验设置和模型设置。训练用的参数如表 14 - 5 所列。

表 14 - 5 训练用的参数

性 能	本教程中的值	描 述
primary_metric	AUC Weighted	要优化的度量
max_time_se	12 000	每次迭代以秒为单位的时限
iterations	20	迭代次数。在每次迭代中,模型与特定管道的数据一起训练
n_cross_validations	3	交叉验证数
preprocess	False	true/false 使实验能够对输入执行预处理。预处理丢失的数据,并执行一些公共特征提取
exit_score	0.998 5	表示初等指标的双精度值。一旦目标超过则运行终止
blacklist_algos	['kNN','LinearSVM']	指示算法忽略的字符串数组

```
from azureml.train.automl import AutoMLConfig

## Local compute
```

```
Automl_config = AutoMLConfig(task = 'classification',
                             primary_metric = 'AUC_weighted',
                             max_time_sec = 12000,
                             iterations = 20,
                             n_cross_validations = 3,
                             preprocess = False,
                             exit_score = 0.9985,
                             blacklist_algos = ['kNN','LinearSVM'],
                             X = X_train,
                             y = y_train,
                             path = project_folder)
```

(2) 运行实验

开始实验在本地运行。将计算目标定义为本地,并将输出设置为true,以查看实验的进度。

```
from azureml.core.experiment import Experiment
experiment = Experiment(ws, experiment_name)
local_run = experiment.submit(Automl_config, show_output = True)
```

在每次迭代过程中,如以下输出一次出现一行。每隔10~15秒就会看到一条新线。

```
Running locally
Parent Run ID: AutoML_ca0c807b-b7bf-4809-a963-61c6feb73ea1
****************************************************************
ITERATION: The iteration being evaluated.
PIPELINE: A summary description of the pipeline being evaluated.
DURATION: Time taken for the current iteration.
METRIC: The result of computing score on the fitted pipeline.
BEST: The best observed score thus far.
****************************************************************
```

ITERATION	PIPELINE	DURATION	METRIC	BEST
0	Normalizer extra trees	0:00:15.955367	0.988	0.988
1	Normalizer extra trees	0:00:14.203088	0.952	0.988
2	Normalizer lgbm_classifier	0:00:15.089057	0.994	0.994
3	Normalizer SGD classifier	0:00:14.866700	0.500	0.994
4	Normalizer SGD classifier	0:00:13.740577	0.983	0.994
5	Normalizer DT	0:00:13.879204	0.937	0.994
6	Normalizer SGD classifier	0:00:13.379975	0.980	0.994
7	Normalizer lgbm_classifier	0:00:15.953293	0.997	0.997

Stopping criteria reached. Ending experiment.

7. 探索结果

用 Jupyter 小部件或通过检查实验历史来探索实验结果。使用 Jupyter 笔记本小部件来查看所有结果的图表和表格。

```
from azureml.train.widgets import RunDetails
RunDetails(local_run).show()
```

这是一个小部件的静态图像。在笔记本中,您可以使用图表上方的下拉视图来查看每个迭代的可用度量的图表,如图 14 - 27 和图 14 - 28 所示。

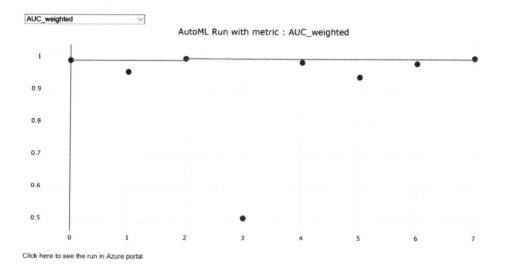

图 14 - 27 迭　代

图 14 - 28 基于权重的输出

8. 检索所有迭代

查看实验历史并查看每个迭代运行的单个度量。

```
children = list(local_run.get_children())
```

```
metricslist = {}
for run in children:
    properties = run.get_properties()
    metrics = {k: v for k, v in run.get_metrics().items() if isinstance(v, float)}
    metricslist[int(properties['iteration'])] = metrics

import pandas as pd
rundata = pd.DataFrame(metricslist).sort_index(1)
Rundata
```

表14-6显示了结果。

表14-6 结果表

	0	1	2	3	4	5	6	7
AUC_macro	0.988 094	0.951 981	0.993 606	0.5	0.982 724	0.936 998	0.979 978	0.996 639
AUC_micro	0.988 104	0.948 402	0.994 13	0.463 035	0.976 078	0.945 169	0.968 913	0.997 027
AUC_weighted	0.987 943	0.952 255	0.993 513	0.5	0.982 801	0.937 292	0.979 973	0.996 56
AUC_weighted_max	0.987 943	0.987 943	0.993 513	0.993 513	0.993 513	0.993 513	0.993 513	0.996 56
accuracy	0.852 093	0.666 464	0.898 057	0.070 128 4	0.832 662	0.701 827	0.833 25	0.925 752
average_precision_score_macro	0.929 167	0.786 258	0.961 497	0.1	0.917 486	0.685 547	0.906 611	0.977 775
average_precision_score_micro	0.932 596	0.728 331	0.964 138	0.090 903 1	0.880 136	0.757 538	0.859 813	0.980 408
average_precision_score_weighted	0.930 681	0.788 964	0.962 007	0.102 123	0.918 785	0.692 041	0.908 293	0.977 699
balanced_accuracy	0.917 902	0.814 509	0.944 91	0.5	0.909 248	0.833 428	0.907 412	0.959 351
f1_score_macro	0.850 511	0.643 116	0.899 262	0.013 092	0.825 054	0.691 712	0.819 627	0.926 081
f1_score_micro	0.852 093	0.666 464	0.898 057	0.070 128 4	0.832 662	0.701 827	0.833 25	0.925 752
f1_score_weighted	0.852 134	0.646 049	0.898 705	0.009 336 91	0.830 731	0.696 538	0.824 547	0.925 778
log_loss	0.554 364	1.157 28	0.517 41	2.303 97	1.940 09	1.576 63	2.184 8	0.250 725
norm_macro_recall	0.835 815	0.629 003	0.890 167	0	0.818 755	0.666 629	0.814 739	0.918 851
precision_score_macro	0.855 892	0.707 715	0.901 95	0.007 012 84	0.848 82	0.729 611	0.855 384	0.927 881
precision_score_micro	0.852 093	0.666 464	0.898 057	0.070 128 4	0.832 662	0.701 827	0.833 25	0.925 752
precision_score_weighted	0.859 204	0.711 918	0.903 523	0.005 006 76	0.861 209	0.737 586	0.863 524	0.928 403
recall_score_macro	0.852 234	0.666 102	0.901 151	0.1	0.836 88	0.699 966	0.833 265	0.926 966
recall_score_micro	0.852 093	0.666 464	0.898 057	0.070 128 4	0.832 662	0.701 827	0.833 25	0.925 752
recall_score_weighted	0.852 093	0.666 464	0.898 057	0.070 128 4	0.832 662	0.701 827	0.833 25	0.925 752
weighted_accuracy	0.851 054	0.666 39	0.895 428	0.049 121	0.829 247	0.702 754	0.833 464	0.924 723

9. 注册最佳模型

使用 local_run 对象获取最佳模型并将其注册到工作区中。

```python
# find the run with the highest accuracy value.
best_run, fitted_model = local_run.get_output()

# register model in workspace
description = 'Automated Machine Learning Model'
tags = None
local_run.register_model(description = description, tags = tags)
local_run.model_id # Use this id to deploy the model as a web service in Azure
```

10. 测试最佳模型

使用该模型来预测一些随机数字。显示预测值和图像。红色字体和逆图像(白色变黑色)是用来突出错误分类的样本。

由于模型精度高,所以可能需要运行以下代码数次才能看到错误分类的示例。

```python
# find 30 random samples from test set
n = 30
X_test = digits.data[:100, :]
y_test = digits.target[:100]
sample_indices = np.random.permutation(X_test.shape[0])[0:n]
test_samples = X_test[sample_indices]

# predict using the model
result = fitted_model.predict(test_samples)

# compare actual value vs. the predicted values.
i = 0
plt.figure(figsize = (20, 1))

for s in sample_indices:
    plt.subplot(1, n, i + 1)
    plt.axhline('')
    plt.axvline('')

    # use different color for misclassified sample
    font_color = 'red' if y_test[s] != result[i] else 'black'
    clr_map = plt.cm.gray if y_test[s] != result[i] else plt.cm.Greys

    plt.text(x = 2, y = -2, s = result[i], fontsize = 18, color = font_color)
    plt.imshow(X_test[s].reshape(8, 8), cmap = clr_map)
```

```
i = i + 1
plt.show()
```

如图14-29所示为出现错误的速记图像。

图14-29　出现错误的速记图像

14.13　在Azure容器实例(ACI)中部署图像分类模型

现在,您已经准备好将模型作为Azure容器实例(ACI)中的Web服务部署。Web服务是一种图像,在这种情况下是一个Docker图像,它封装了评分逻辑和模型本身。

ACI对于生产部署是不理想的,但它对于测试和理解工作流是很好的。

1. 拿笔记本

为了方便起见,本教程可作为Jupyter笔记本使用。在Azure笔记本或在自己的Jupyter笔记本服务器上运行02.deploy-models.ipynb。

(1) 蓝天笔记本——免费的木星在蓝天的笔记本

SDK已经安装在Azure笔记本上,并为您配置。完成快速入门,创建工作区并启动Azure笔记本电脑。

(2) 去Azure笔记本

在快速启动过程中创建的启动库中,转到教程文件夹。打开笔记本。

(3) 你自己的Jupyter笔记本服务器

用Python SDK快速入门完成安装SDK并创建工作区。克隆GITHUB存储库。将快速启动期间创建的aml_config目录复制到克隆目录中。从克隆目录启动笔记本服务器。转到教程文件夹。打开笔记本。

2. 先决条件

完成教程1中的模型训练:用Azure机器学习服务笔记本训练图像分类模型。

3. 创设环境

首先建立一个测试环境。

(1) 导入包

导入本教程所需的Python程序包。

```
% matplotlib inline
import numpy as np
```

```
import matplotlib
import matplotlib.pyplot as plt

import azureml
from azureml.core import Workspace, Run

# display the core SDK version number
print("Azure ML SDK Version: ", azureml.core.VERSION)
```

(2) 检索模型

在前一个教程中,您在工作区中注册了一个模型。现在,加载此工作区并将模型下载到本地目录。

```
from azureml.core import Workspace
from azureml.core.model import Model


ws = Workspace.from_config()
model = Model(ws, 'sklearn_mnist')
model.download(target_dir = '.')
import os
# verify the downloaded model file
os.stat('./sklearn_mnist_model.pkl')
```

4. 在本地测试模型

在部署之前,请确保您的模型在本地工作:
- 加载试验数据;
- 预测试验数据;
- 检验混淆矩阵。

(1) 加载试验数据

从培训教程中创建的./data/目录中加载测试数据。

```
from utils import load_data

# note we also shrink the intensity values (X) from 0-255 to 0-1. This helps the neural network converge faster

X_test = load_data('./data/test-images.gz', False) / 255.0
y_test = load_data('./data/test-labels.gz', True).reshape(-1)
```

(2) 预测试验数据

将测试数据集馈送到模型以获得预测。

```
import pickle
```

```
from sklearn.externals import joblib

clf = joblib.load('./sklearn_mnist_model.pkl')
y_hat = clf.predict(X_test)
```

(3) 检验混淆矩阵

生成混淆矩阵以查看测试集有多少个样本被正确分类。注意错误预测的错误分类值。

```
from sklearn.metrics import confusion_matrix

conf_mx = confusion_matrix(y_test, y_hat)
print(conf_mx)
print('Overall accuracy:', np.average(y_hat == y_test))
```

输出显示混淆矩阵：

```
[[ 960    0    1    2    1    5    6    3    1    1]
 [   0 1112    3    1    0    1    5    1   12    0]
 [   9    8  920   20   10    4   10   11   37    3]
 [   4    0   17  921    2   21    4   12   20    9]
 [   1    2    5    3  915    0   10    2    6   38]
 [  10    2    0   41   10  770   17    7   28    7]
 [   9    3    7    2    6   20  907    1    3    0]
 [   2    7   22    5    8    1    1  950    5   27]
 [  10   15    5   21   15   27    7   11  851   12]
 [   7    8    2   13   32   13    0   24   12  898]]
Overall accuracy: 0.9204
```

使用 matplotlib 将混淆矩阵作为图形显示。在该图中，x 轴表示实际值，y 轴表示预测值。每个网格中的颜色代表错误率。颜色越淡，误码率越高。例如，许多 5 的被误分类为 3。因此，可以看到一个明亮的网格 (5, 3)。

```
# normalize the diagnal cells so that they don't overpower the rest of the cells when visualized
row_sums = conf_mx.sum(axis = 1, keepdims = True)
norm_conf_mx = conf_mx / row_sums
np.fill_diagonal(norm_conf_mx, 0)

fig = plt.figure(figsize = (8,5))
ax = fig.add_subplot(111)
cax = ax.matshow(norm_conf_mx, cmap = plt.cm.bone)
ticks = np.arange(0, 10, 1)
ax.set_xticks(ticks)
```

```
ax.set_yticks(ticks)
ax.set_xticklabels(ticks)
ax.set_yticklabels(ticks)
fig.colorbar(cax)
plt.ylabel('true labels', fontsize = 14)
plt.xlabel('predicted values', fontsize = 14)
plt.savefig('conf.png')
plt.show()
```

以下是输出图像,如图 14-30 所示。

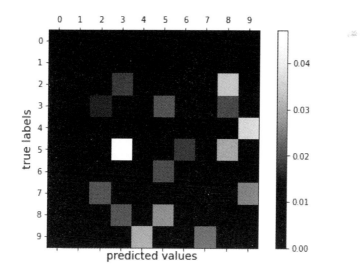

图 14-30　输出图像

5. 部署为 Web 服务

一旦测试了模型并对结果满意,就将该模型部署为在 ACI 中托管的 Web 服务。为 ACI 建立正确的环境,提供以下内容:
- 演示如何使用模型的记分脚本;
- 一个环境文件来显示需要安装什么包;
- 用于构建 ACI 的配置文件;
- 您以前训练过的模型。

6. 创建记分脚本

创建由 Web 服务调用使用的计分脚本,称为 SCOR.PY,以显示如何使用该模型。必须在记分脚本中包含两个所需的功能:

① Init()函数,它通常将模型加载到全局对象中。当启动 DOCKER 容器时,此函数只运行一次。

② 运行(输入数据)函数使用模型来预测基于输入数据的值。运行的输入和输出通常使用 JSON 进行序列化和去序列化，但支持其他格式。

```python
%%writefile score.py
import json
import numpy as np
import os
import pickle
from sklearn.externals import joblib
from sklearn.linear_model import LogisticRegression

from azureml.core.model import Model

def init():
    global model
    # retreive the path to the model file using the model name
    model_path = Model.get_model_path('sklearn_mnist')
    model = joblib.load(model_path)


def run(raw_data):
    data = np.array(json.loads(raw_data)['data'])
    # make prediction
    y_hat = model.predict(data)
    return json.dumps(y_hat.tolist())
```

7. 创建环境文件

接下来，创建一个名为 myenv.yml 的环境文件，该文件指定所有脚本的包依赖关系。此文件用于确保所有依赖项都安装在 DOCKER 映像中。该模型需要 scikit-learn 和 azureml-sdk。

```python
from azureml.core.conda_dependencies import CondaDependencies

myenv = CondaDependencies()
myenv.add_conda_package("scikit-learn")

with open("myenv.yml","w") as f:
    f.write(myenv.serialize_to_string())
```

查看 myenv.yml 文件的内容。

```python
with open("myenv.yml","r") as f:
    print(f.read())
```

8. 创建配置文件

创建部署配置文件并指定您的 ACI 容器所需的 CPU 数量和千兆字节 RAM。当它依赖于模型时，1 个内核和 1 千兆字节 RAM 的默认值通常对于许多模型来说是足够的。如果您觉得以后需要更多，则必须重新创建图像并重新部署服务。

```
from azureml.core.webservice import AciWebservice

aciconfig = AciWebservice.deploy_configuration(cpu_cores = 1,
                    memory_gb = 1,
                    tags = {"data": "MNIST", "method" : "sklearn"},
                    description = 'Predict MNIST with sklearn')
```

9. 在 ACI 中部署

预计完成时间：7~8 分钟。

配置图像并展开。使用以下文件构建图像：得分文件（得分）、环境文件（My-Env.YML）、模型文件。

下面的代码通过这些步骤：

① 在工作空间下注册该图像。
② 将图像发送到 ACI 容器。
③ 使用图像启动 ACI 中的容器。
④ 获取 Web 服务 HTTP 端点。

```
%%time
from azureml.core.webservice import Webservice
from azureml.core.image import ContainerImage

# configure the image
image_config = ContainerImage.image_configuration(execution_script = "score.py",
                                runtime = "python",
                                conda_file = "myenv.yml")

service = Webservice.deploy_from_model(workspace = ws,
                                name = 'sklearn-mnist-svc',
                                deployment_config = aciconfig,
                                models = [model],
                                image_config = image_config)

service.wait_for_deployment(show_output = True)
```

获取得分 Web 服务的 HTTP 端点，它接受 REST 客户端调用。此端点可以与想要测试 Web 服务或将其集成到应用程序中的任何人共享。

```
print(service.scoring_uri)
```

10. 测试部署服务

早些时候,您用模型的本地版本对所有测试数据进行评分。现在,您可以通过从测试数据中随机抽取 30 个图像来测试部署的模型。

下面的代码通过这些步骤:

① 将数据作为 JSON 数组发送到 ACI 中承载的 Web 服务。

② 使用 SDK 的运行 API 调用服务。您也可以使用任何 HTTP 工具如 CURL 进行原始调用。

③ 打印返回的预测并将它们与输入图像一起绘制。红色字体和逆图像(白色变黑色)是用来突出错误分类的样本。

由于模型精度高,所以您可能需要运行以下代码数次才能看到错误分类的示例。

```
import json

# find 30 random samples from test set
n = 30
sample_indices = np.random.permutation(X_test.shape[0])[0:n]

test_samples = json.dumps({"data": X_test[sample_indices].tolist()})
test_samples = bytes(test_samples, encoding = 'utf8')

# predict using the deployed model
result = json.loads(service.run(input_data = test_samples))

# compare actual value vs. the predicted values:
i = 0
plt.figure(figsize = (20, 1))

for s in sample_indices:
    plt.subplot(1, n, i + 1)
    plt.axhline('')
    plt.axvline('')

    # use different color for misclassified sample
    font_color = 'red' if y_test[s] != result[i] else 'black'
    clr_map = plt.cm.gray if y_test[s] !s = result[i] else plt.cm.Greys

    plt.text(x = 10, y = -10, s = result[i], fontsize = 18, color = font_color)
    plt.imshow(X_test[s].reshape(28, 28), cmap = clr_map)
```

```
        i = i + 1
plt.show()
```

这里是一个随机样本的测试图像的结果,如图 14 – 31 所示。

```
4 8 1 1 3 6 1 3 4 7 5 2 3 6 7 1 2 1 5 3 3 9 5 3 6 6 7 4 6 7
```

图 14 – 31 用来测试的随机样本图像

还可以发送原始 HTTP 请求来测试 Web 服务,如:

```
import requests
import json

# send a random row from the test set to score
random_index = np.random.randint(0, len(X_test) - 1)
input_data = "{\"data\": [" + str(list(X_test[random_index])) + "]}"

headers = {'Content-Type':'application/json'}

# for AKS deployment you'd need to the service key in the header as well
# api_key = service.get_key()
# headers = {'Content-Type':'application/json', 'Authorization':('Bearer ' + api_key)}

resp = requests.post(service.scoring_uri, input_data, headers = headers)

print("POST to url", service.scoring_uri)
# print("input data:", input_data)
print("label:", y_test[random_index])
print("prediction:", resp.text)
```